D1237839

MODERN BIOTECHNOLOGY

MODERN BIOTECHNOLOGY
Connecting Innovations in Microbiology and Biochemistry to Engineering Fundamentals

Nathan S. Mosier
Michael R. Ladisch

WILEY

A JOHN WILEY & SONS, INC., PUBLICATION

Published by John Wiley & Sons, Inc., Hoboken, New Jersey
Published simultaneously in Canada

For general information on our other products and services or for technical support, please contact
our Customer Care Department within the United States at (800) 762-2974, outside the United States
at (317) 572-3993 or fax (317) 572-4002.

Wiley also publishes its books in a variety of electronic formats. Some content that appears in print
may not be available in electronic formats. For more information about Wiley products, visit our web
site at www.wiley.com.

Library of Congress Cataloging-in-Publication Data:

Mosier, Nathan S., 1974-
   Modern biotechnology : connecting innovations in microbiology and biochemistry to engineering
fundamentals / Nathan S. Mosier, Michael R. Ladisch.
      p. cm.
   Includes index.
   ISBN 978-0-470-11485-8 (cloth)
   1. Biotechnology. I. Ladisch, Michael R., 1950- II. Title.
   TP248.2.M675 2009
   660.6–dc22
                                                              2009001779

Printed in the United States of America

10 9 8 7 6 5 4 3 2 1

CONTENTS

PREFACE

Biotechnology has enabled the development of lifesaving biopharmaceuticals, deciphering of the human genome, and production of bioproducts using environmentally friendly methods based on microbial fermentations. The science on which modern biotechnology is based began to emerge in the late 1970s, when recombinant microorganisms began to be used for making high-value proteins and peptides for biopharmaceutical applications. This effort evolved into the production of some key lifesaving proteins and the development of monoclonal antibodies that subsequently have provedn to be effective molecules in the fight against cancer. In the late 1980s and early 1990s biotechnology found further application in sequencing of the human genome, and with it, sequencing of genomes of many organisms important for agriculture, industrial manufacture, and medicine.

The human genome was sequenced by 2003. At about the same time the realization developed that our dependence on petroleum and other fossil fuels was beginning to have economic consequences that would affect every sector of our economy as well as the global climate. Modern biotechnology began to be applied in developing advanced enzymes for converting cellulosic materials to fermentable sugars. The process engineering to improve grain-to-ethanol plants and the rapid buildout of an expanded ethanol industry began. This provided the renewable liquid fuels in small but significant quantities.

Thus biology has become an integral part of the engineering toolbox through biotechnology that enables the production of biomolecules and bioproducts using methods that were previously not feasible or at scales previously thought impossible. We decided to develop this textbook that addresses modern biotechnology in engineering. We started with the many excellent concepts described by our colleagues by addressing bioprocess engineering and biochemical engineering from a fundamental perspective. We felt that a text was needed to address applications while at the same time introduce engineering and agriculture students to new concepts in biotechnology and its application for making useful products. As we developed the textbook and the course in which this textbook has been used, the integration of fundamental biology, molecular biology, and some aspects of genetics started to become more common in many undergraduate curricula. This further expanded the utility of an application-based approach for introducing students to biotechnology. This book presents case studies of applications of modern biotechnology in the innovation process that has led to more efficient enzymes and better understanding of microbial metabolism to redirect it to maximize production of useful products. Scaling up biotechnology so that large quantities of fermentation products could be produced in an economic manner is the bridge between the laboratory and broader society use.

Our textbook takes the approach of giving examples or case studies of how biotechnology is applied on a large scale, followed by discussion of fundamentals in biology, biochemistry, and enzyme or microbial reaction engineering. Innovations in these areas have occurred at an astounding rate since the mid-1990s. The current text attempts to connect the innovations that have occurred in molecular biology, microbiology, and biochemistry to the engineering fundamentals that are employed to scale up the production of bioproducts and biofuels using microorganisms and biochemical catalysts with enhanced properties.

The approach that we take treats microorganisms as living biocatalysts, and examines how the principles that affect the activity of microorganisms and enzymes are used in determining the appropriate scaleup correlations and for analyzing performance of living and nonliving biocatalysts on a large scale. Our textbook will hopefully provide the basis on which new processes might be developed, and sufficient background for students who wish to transition to the field and continue to grow with the developments of modern biotechnology industry. While we cannot hope to teach all the fundamentals that are required to cover the broad range of products that are derived using biotechnology, we do attempt to address the key factors that relate engineering characteristics to the basic understanding of biotechnology applied on a large scale.

October 7, 2008   NATHAN MOSIER AND MICHAEL LADISCH

ACKNOWLEDGMENTS

We wish to thank our family, colleagues, and Purdue University for giving us the time to focus on developing an organized approach to teaching the broad set of topics that define biotechnology. This enabled us to transform our teaching into a format that others may use to lecture and to gain from our experience. Special thanks go to Carla Carie, who worked diligently on preparing drafts of manuscripts, and assisted with the many processes involved in finalizing the manuscript for publication. We thank Dr. Ajoy Velayudhan for his development of the Runge-Kutta explanation and our many students, especially Amy Penner and Elizabeth Casey, for inputs and suggestions as well as assisting with making improvements in the various sections of the book. We also thank Craig Keim and Professor Henry Bungay (from RPI) for contributions to the Runge-Kutta code. We also thank the Colleges of Agriculture and Engineering, and specifically Dr. Bernie Engel, head of the Agricultural and Biological Engineering Department, and Professor George Wodicka, head of the Weldon School of Biomedical Engineering, for granting us the flexibility to complete this textbook and for providing encouragement and resources to assist us in this process.

One of the authors (Michael Ladisch) wishes to convey his appreciation to the heads of the Agricultural and Biological Engineering Department and Weldon School of Biomedical Engineering at Purdue University for facilitating a partial leave of absence that is enabling him to work as Chief Technology Officer at Mascoma Corporation. As CTO, he is a member of the team building the first cellulose ethanol plant. It is here that some of the lessons learned during the teaching of this material are being put into practice.

Most of all, we would like to thank the students in our mezzanine-level course ABE 580 (Process Engineering of Renewable Resources) with whom we developed the course materials. Their enthusiasm and success makes teaching fun, and keeps us feeling forever young. We also wish to thank John Houghton from the U.S. Department of Energy Office of Biological and Environmental Research for his review of a draft of this textbook and his helpful comments and suggestions.

LIST OF ILLUSTRATIONS

Figures

Tables

BIOTECHNOLOGY*

INTRODUCTION

By 2008, biotechnology touched the major sectors that define human activity: food, fuel, and health. The history of biotechnology starts with breadmaking, utilizing yeast, about 8000 years ago. Fermentation of grains and fruits to alcoholic beverages was carried out in Egypt and other parts of the ancient world in about 2500 BC. Other types of food fermentation practiced for thousands of years include the transformation of milk into cheeses and fermentation of soybeans. However, it was not until 1857 that Pasteur proved that alcoholic fermentation was caused by living cells, namely, yeasts. In the ensuing 100 years, the intentional manipulation of microbial fermentations to obtain food products, solvents, and beverages, and later, substances having therapeutic value as antibiotics gave rise to a large fermentation industry (Hacking 1986; Aiba et al. 1973; Evans 1965). Biotechnology emerged as an enabling technology defined as "any technique that uses living organisms (or parts of organisms) to make or modify products, to improve plants or animals, or to develop microorganisms for specific uses" (Office of Technology Assessment 1991).

A sea of change in biotechnology occurred in the midtwentieth century with discovery of the molecular basis of biology—DNA—and again in the twenty-first century, when it began to be used for obtaining renewable biofuels and enhanced production agriculture (Houghton et al. 2006). Biotechnology has helped to catalyze the growth of the pharmaceutical, food, agricultural processing and specialty product sectors of the global economy (National Research Council 1992, 2001). The scope of biotechnology is broad and deep. Biotechnology encompasses the use of chemicals to modify the behavior of biological systems, the genetic modification of organisms to confer new traits, and the science by which foreign DNA may be inserted into people to compensate for genes whose absence cause life-threatening conditions. Twenty-five years later the science of genetic engineering is finding

---

*Portions of this chapter and Chapter 2 are taken from a previously published section on bioprocess engineering in *Van Nostrand's Scientific Encyclopedia* (Ladisch 2002).

Modern Biotechnology, by Nathan S. Mosier and Michael R. Ladisch
Copyright © 2009 John Wiley & Sons, Inc.

applications in enhancing microbial and plant technologies to directly or indirectly fix CO_2 into renewable fuels (Kim and Dale 2008).

The engineering fundamentals required to translate the discoveries of biotechnology into tangible commercial products, thereby putting biotechnology to work, define the discipline of bioprocess engineering (National Research Council 1992). Bioprocess engineering translates biotechnology into unit operations, biochemical processes, equipment, and facilities for manufacturing bioproducts. The biotechnology addressed in this book provides a foundation for the engineering of bioprocesses for production of human and animal healthcare products, food products, biologically active proteins, chemicals, and biofuels. Industrial bioprocessing entails the design and scaleup of bioreactors that generate large quantities of transformed microbes or cells and their products, as well as technologies for recovery, separation, and purification of these products. This book presents the principles of the life sciences and engineering for the practice of key biotechnology manufacturing techniques and economic characteristics of the industries and manufacturing processes that encompass biotechnology, agriculture, and biofuels (Ladisch 2002; Houghton et al. 2006; NABC Report 19 2007; Lynd et al. 2008).

The principles and practice of bioprocess engineering are based in the biological sciences. The key technologies based on the biosciences are the (1) identification of genes, and the products that result from them, for purposes of disease prevention, remediation, and development of new medicines and vaccines; (2) application of molecular biology to obtain transformed microorganisms, cells, or animals having new and/or enhanced capabilities to generate bioproducts; and (3) development of biological sensors coupled to microprocessors or computers for process control and monitoring of biological systems (including humans).

The Directed Manipulation of Genes Distinguishes the New Biotechnology from Prior Biotechnology

New biotechnology represents a technology for manipulating genetic information and manufacturing products that are of biological origin or impact biological activity. It is based on methods introduced since 1970, applied in the laboratory since 1973, and has been used on an industrial scale since 1979. The combined use of restriction enzymes (that cut DNA in a directed manner) and ligases (i.e., enzymes that join foreign genes with the DNA of the host cell) was demonstrated during this time period. Stanley Cohen and Herbert Boyer showed that DNA could be cut and rejoined in new arrangements in a directed manner (Cohen et al. 1973). Their work gave birth to the field and the industry, based on new forms of organisms obtained through the sequencing, removal, insertion, and amplification of genes across different species of organisms. This gave rise to a new sector in the biotechnology industry based on genetically modified organisms.

New biotechnology enables directed manipulation of the cell's genetic machinery through recombinant DNA techniques. Recombinant DNA is defined by a "DNA molecule formed *in vitro* by ligating DNA molecules that are not normally joined" (Walker and Cox 1988). A recombinant technique is a method "that helps to generate new combinations of genes that were not originally present" in different organisms. By 1998, the impact of the new biotechnology on the pharmaceutical industry was becoming profound, while its potential effect on food processing and production agriculture resulted in multi-billion-dollar investments by some of the world's largest chemical companies (Fritsch and Kilman 1996; Kilman 1996, 1998). In 1997 Monsanto sold off its chemical business with sales of about $4 billion to form an entity known as Solutia, in order to focus on its more profitable agricultural, food ingredients, and pharmaceutical businesses with sales over $5 billion (Fritsch 1996a,b). Dupont, Dow Chemical, Novartis, and Monsanto were investing heavily or acquiring food technology, plant biotechnology, and seed companies by early 1998. Bioengineered food is one target of these acquisitions. Fat-free pork, vegetarian meat, bread enriched with cancer-fighting compounds, and corn products that fight osteoporosis were part of these companies' vision (Kilman 1998a,b). Industrial biotechnology for producing chemicals from plant sugars was perceived as the next emerging area (Ritter 2004). By 2007, Monsanto's strategy paid handsomely, with $8.5 billion in sales attained amid an increasing demand for grains due to an increasing global population and demand for grain-fed protein. Similarly, Dow AgroScience, with a total sales in 2007 of $3.4 billion, has been steadily increasing their share of sales in crop biotechnology to complement their core business in agrochemicals and home insect control chemicals.

The new biotechnology enables production of mammalian or plant proteins and other biomolecules having therapeutic value in quantities required for practical use. Recombinant methodologies also have potential for dramatically improving production efficiency of products already derived by fermentation through the directed modification of cellular metabolism: metabolic engineering and when applied to plants, enhanced agricultural production of food and feedstocks for renewable fuels. As summarized in the preface of the 1992 National Research Council Report, "scientists and engineers can now change the genetic make-up of microbial, plant, and animal cells to confer new characteristics. Biological molecules, for which there is no other means of industrial production, can now be generated. Existing industrial organisms can be systematically altered (i.e., engineered) to enhance their function and to produce useful products in new ways." The results of biotechnology, and the search for sustainable solutions for producing and fuels, has catalyzed debate on how this technology, and the agriculture that will provide the feedstock, might best be employed.

The focus of this book is microbiological engineering and the application of biotechnology to three major sectors of the economy: pharmaceuticals, food, and fuel. However, as the demand for the products of biotechnology move from bio-pharmaceuticals to biofuels, that is, from very high-value, very small-volume molecules to very high-volume lower-value molecules, debates on resource constraints, land use, alternative liquid fuels, and greenhouse gas generation have entered the technical considerations in rolling out new approaches to renewable liquid fuels such as cellulose ethanol. As important as this debate may be, the current text addresses the technical means of transforming renewable resources to fuels or chemicals via biological means. The larger and very important issues of sustainability and agriculture are the subject of other monographs (NABC Report 19 2007).

Growth of the New Biotechnology Industry Depends on Venture Capital

The pioneers of the biotechnology industry had a grasp not only of the science but also an understanding of the financial aspects of taking a new technology from test tube to market in less than 7 years. In the late 1970s and early 1980s, the technical success of the insulin project, and the apparent availability of venture capital for risky enterprises, converged to promote the industry. Venture capital was available in part because of government policy that promoted limited partnerships and relieved taxes on capital gains, although the laws have since changed. The financial climate of 2008/2009, in which major funds and banks approached insolvency, temporarily limited new investments.

Some of the pioneers had a sense that the large-scale production, clinical testing, obtaining government regulatory approval, and marketing required the infrastructure of a large pharmaceutical company. The first recombinant product approved by the FDA, human insulin, resulted from the cooperation of the biotechnology company Genentech and the pharmaceutical company Eli Lilly. Genentech ultimately merged with another major international pharmaceutical company, Hoffmann-LaRoche, in 1990 (see timeline in Table 1.1). Other mergers over the next 12 years resulted in consolidation of the industry, with acquisitions or licensing arrangements between biotechnology and pharmaceutical companies resulting in transfer of technology from small companies to larger ones.

A profound transition occurred around 2005, when the realization that demand for liquid transportation fuels, derived from petroleum could outpace demand. Rapid growth in global demand for petroleum-derived fuels and growth of large economies (China and India) caused demand for liquid fuels to increase rapidly. Coupled with the fuel ethanol mandate of 2005 that required the United States to use 7.5 billion gallons of fuel ethanol by 2012, this resulted in rapid expansion of the grain ethanol industry. The expansion was so rapid that the mandated requirements were met by 2008, 4 years ahead of schedule. Then corn prices increased and corn became too expensive for broader use in ethanol production. This was accompanied by the realization that ethanol derived from cellulose, a nonfood feedstock, would be needed to enable sustainable expansion to 22 billion gallons per year by 2020. Significant efforts are now underway to discover, develop, and implement biological and microbial catalysts that convert cellulose to ethanol.

Agricultural biotechnology differs from microbiological technology in the organisms that are modified. For example, the incorporation of a bacterial gene from *Bacillus thuringiensis* (Bt) gene, into cotton (Bollguard®) enabled the cotton to produce proteins that are toxic to cotton bollworm and budworm. The first trials of the transgenic cotton were a qualified success (Thayer 1997). Approximately 40% of the Bollguard-planted fields[1] had to be sprayed, although only one to two pesticide applications were required compared with a more typical four to five. Cotton growers reported an average 7% improvement in yield, as well as reducing or avoiding the use of pesticides. The farmers reported a $33/acre cost advantage of Bollguard cotton compared with insecticide-treated non-Bollguard cotton, after accounting for the technology licensing fee of $32/acre and supplemental pesticide

---

[1] Approximately 3% of Texas' 3 million acres of cotton and 60–70% of Alabama's 500,000 acres of cotton were planted with *B. thuringiensis* cotton in 1996. The potential economic and environmental benefits are evident when the size of the annual US cotton harvest (9 billion lbs, worth about $7.2 billion) and volume of insecticides ($400–$500 million/year) are considered. Bollguard plantings for 1997 were estimated at 2.5 million acres out of 14 million acres of cotton (Thayer 1997). By 2004, worldwide acreage of all genetically modified crops was 200 million (BIO 2005).

Table 1.1.

Timeline of Major Developments in the
Biotechnology Industries through 1998

Date	Event
6000 BC	Breadmaking (involving yeast fermentation)
3000 BC	Moldy soybean curd used to treat skin infections in China
2500 BC	Malting of barley, fermentation of beer in Egypt
1790 AD	Patent act provides *no* protection for plants and animals since they are considered to be products of nature
1857	Pasteur proves that yeasts are living cells that cause alcohol fermentation (Aiba, Humphrey, Millis, 1973)
	Birth of microbiology
1877	Pasteur and Joubert discover that some bacteria can kill anthrax bacilli
1896	Gosio discovers mycophenolic acid, an antibacterial substance produced by microbes; too toxic for use as antibiotic
1902	*Bacillus thuringiensis* first isolated from silkworm culture by Ishiwata
1908	Ikeda identifies monosodium glutamate (MSG) as flavor enhancer in Konbu
	Invertase adsorbed onto charcoal, i.e., first example of immobilized enzyme
1909	Ajinimoto (Japan) initiates commercial production of sodium glutamate from wheat gluten and soybean hydrolysates
1900–1920	Ethanol, glycerol, acetone, and butanol produced commercially by large-scale fermentation
1922	Banting and Best treat human diabetic patient with insulin extracted from dog pancreas
1923	Citric acid fermentation plant using *Aspergillus niger* by Charles Pfizer
1928	Alexander Fleming discovers penicillin from *Penicillium notatum*
1930	Plant Patent Act, which allows for patenting of asexually produced plants (except tubers)—i.e., plants reproduced by, tissue culture or propagation of cuttings
1943	Submerged culture of *Penicillium chrysogenum* opens way for large-scale production of penicillin
1945	Production through fermentation process scaled up to make enough penicillin to treat 100,000 patients per year
	Beginning of rapid development of antibiotic industry; during World War II, research driven by 85% tax on "excess" profits, encouraged investment in research and development for antibiotics—this led to their postwar growth
1953	DNA structure and function elucidated
	Xylose isomerase discovered

Table 1.1. *(Continued)*

Date	Event
1957	Commercial production of natural (L) amino acids via fermentation facilitated the discovery of *Micrococcus glutamicus* (later renamed *Corynebacterium glutamicum*)
	Glucose-isomerizing capability of xylose isomerase reported
1960	Lysine produced on a technical scale
1961	First commercial production of MSG via fermentation
1965	Corn bran and hull replaces xylose as inducer of glucose (xylose) isomerase in *Streptomyces phaeochromogenus*
	Phenyl methyl ester of aspartic acid and phenylalanine (aspartame) synthesized at G. D. Searle Co.
1967	Clinton Corn Processing ships first enzymatically produced fructose syrup
	A. E. Staley sublicenses technology
1970	Smith et al. report restriction endonuclease from *Haemophilis influenzae* that recognizes specific DNA target sequences
1971	Cetus founded
1973	Cohen and Boyer report genetic engineering technique (EcoRI enzyme) (Cohen et al. 1973)
	Aspartase in immobilized *E. coli* cells catalyzes L-aspartic acid production from fumarate and ammonia
1973–1974	Oil price increase (Yom Kippur war)
	High-fructose corn syrup (HFCS) market at ~500–600 million lbs/year in USA
	Sugar price peaks at 30¢/lb
1975	Kohler and Milstein report monoclonal antibodies
	Basic patent coverage for xylose (glucose) isomerase lost in lawsuit
	Opens up development of new HFCS processes
1976	Genentech founded
1978	Biogen formed; develops interferons
	Eli Lilly licenses recombinant insulin technology from Genentech
	3.5 billion lb HFCS produced in USA
1979–1980	Another major oil price increase (OPEC); sugar price at 12¢/lb
	Energy-saving method for drying ethanol using corn (starch) and cellulose-based adsorbents reported (Ladisch and Dyck 1979)
1977–1982	Fermentation ethanol processes adapted by wet millers for fuel-grade ethanol
1980	Amgen founded
	US Supreme Court rules that life forms are patentable

Table 1.1. *(Continued)*

Date	Event
1981	HIV/AIDS cases identified and reported in San Francisco
	Aspartame approved for food use by FDA
	Chiron founded
	Gene-synthesizing machines developed
	Sugar price at 30¢/lb
1982	FDA approves Humalin (human insulin) made by Lilly
	First transgenic mouse; rat gene transferred to mouse
1983	Aspartame comes on market sold by G. D. Searle as Nutrasweet®
	Worldwide antibiotic sales at about $8 billion
	First product sales of recombinant insulin
	HIV virus identified as cause of AIDS
1984	Oil at ~$40/barrel
	Process for industrial drying of fuel alcohol using a corn-based adsorbent in place of azeotropic distillation demonstrated on an industrial scale
	Transgenic pig, rabbit, and sheep by microinjection of foreign DNA into egg nuclei
	Polymerase chain reaction (PCR) developed at Cetus
	HIV genome sequenced by Chiron
1986	Phaseout of lead as octane booster in gasoline in USA creates demand for ethanol as a nonleaded octane booster for liquid fuels
	Ethanol production at 500 million gal/year for use as octane booster
1987	Merck licenses Chiron's recombinant hepatitis B vaccine
	Human growth hormone (Protropin®) introduced by Genentech
	Interleukin (IL-2), a protein used to treat cancer, by Cetus undergoes clinical trials
	Aspartame sales at $500 million
1988	Oil at $15–$20/barrel
	Ethanol production at 800 million gal/year
	Cetus requested approval for IL-2 to treat advanced kidney cancer
	Tissue plasminogen activator (TPA) introduced by Genentech
1989	Human Genome Project started
	American Home Products purchases A. H. Robbins
	Amgen introduces erythropoietin (EPO), a protein that stimulates red blood cell formation (produced in 2-L roller bottles)
	Merrill Dow combines with Marion Laboratories for $7.7 billion

Table 1.1. *(Continued)*

Date	Event
1990	Bristol Myers and Squibb merge ($12 billion)
	Beecham and SmithKline Beckman merge ($7.8 billion)
	FDA rejects IL-2 application of Cetus
	Hoffmann-LaRoche acquires 60% of Genentech for $2.1 billion
	Genentech's tissue plasminogen activator (TPA), for dissolving blockages that cause heart attacks, earns $210 million
	Protropin (human growth hormone) earns $157 million
	Amgen's EPO sales at $300 million/year in USA; licensed by Kilag
	(Johnson & Johnson in Europe) and Kirin (Japan)
1991	First attempt at human gene therapy
	HFCS world sales estimated about $3 billion (17 billion lbs)
	Amgen EPO sales exceed $293 million by August 1991
	Genetics Institute suit against Amgen for American rights to EPO
	American Home Products buys $666 million (60%) share in Genetics Institute; Chiron purchases Cetus for $650 million
1992	IL-2 (now owned by Chiron) approved for further testing by FDA
	TPA (from Genentech) under competitive pressure from less expensive product by Swedish Kabi
	Policy guidelines for the agricultural biotechnology established
1993	Chiron introduces Betaseron (a β-interferon) for treatment of multiple sclerosis; drug is marketed by Berlex (owned by Schering AG)
	Healthcare reform proposals create uncertainty in biotechnology industry
	Merck acquires Medco containment services for $6.6 billion
1994	American Home Products Buys American Cyanimid
	Roche Holding, parent of Swiss Drug company Hoffmann-LaRoche buys Syntex for $5.3 billion
	SmithKline Beecham merges with Diversified Pharmaceutical Services ($2.3 billion) and Sterling Winthrop ($2.9 billion)
	Eli Lilly acquires PCS Health Systems for $4.1 billion
	Bayer purchases Sterling Winthrop NA for $1.0 billion
	Process based on corn adsorbent dries half of fermentation ethanol in USA (750 million gallons/year)
1995	UpJohn and Pharmacia merge to form Pharmacia-UpJohn in a $6 billion stock swap
	Hoescht/Marion Merrell Dow merge for $7.1 billion
	Rhône-Poulenc Rorer/Fisons merge in a deal valued at $2.6 billion
	Glaxo/Wellcome merge in a $15 billion deal

Table 1.1. (*Continued*)

Date	Event
1996	Monsanto purchases Ecogen for $25 million, Dekalb Genetics for $160 million; Agracetus for $150 million; and 49.9% of Calgene
	American Home Products purchases the remaining 40% stake in Genetics Institute, Inc. for $1.25 billion
	Biogen introduces Avonex, to compete with Berlex's Betaseron for MS sufferers
	$27 billion merger of Ciba-Geigy AG and Sandoz AG to form Norvatis approved; estimated annual sales of $27.3 billion; US Federal Trade Commission (FTC) prevents monopoly that doesn't yet exist by requiring Norvatis to provide access to key genetic research discoveries; goal of FTC is to prevent company from dominating gene therapy research
1997	Monsanto completes purchase of Calgene for $320 million; agrees to buy Asgrow Agronomics Soybean business for $240 million, and Holden's Foundation Seeds (corn, sales of $50 million/year) for $1.02 billion
	Proctor & Gamble pays Regeneron $135 million to carry out research on small molecule drugs
	SmithKline Beecham forms joint venture with Incyte to enter genetic–diagnostics business
	Schering-Plough Corporation Acquires Mallinckrodt Animal-Health Unit for $405 million
	Merck, Rhône-Poulenc form animal healthcare 50/50 joint venture (Merial Animal Health); estimated annual sales are $1.7 billion)
	Novartis purchases Merck & Co.'s insecticide–fungicide business (sales of $200 million/year) for $910 million
	Roche Holding, parent of Swiss Drug Company Hoffmann-LaRoche buys Boehringer Mannheim for ~$11 billion
	Proctor & Gamble signs $25 million agreement with Gene Logic to identify genes associated with onset and progression of heart failure
	Dupont purchases Protein Technologies (division of Ralston Purina) for $1.5 billion as part of business plan to develop soy protein foods
	American Home Products discusses $60 billion merger with SmithKline Beecham PLC
	Monsanto spins off chemicals unit and becomes Monsanto Life Sciences
1998	SmithKline Beecham breaks off talks with American Home Products
	Glaxo enters merger discussions with SmithKline Beecham in a deal valued at $65–70 billion; merger discussions driven by successful hunt for human genes and opportunities for exploiting these findings for development of new pharmaceuticals; leads to formation of Glaxo SmithKline

Note: Developments subsequent to 1998 may be found in Table 1 of *Van Nostrand's Encyclopedia* (Ladisch 2002).

applications. By 2004, a total of 100 million acres transgenic crops were planted in the United States (BIO 2005).

This industry has the potential to surpass the computer industry in size and importance because of the pervasive role of biologically produced substances in everyday life (Office of Technology Assessment 1991; National Research Council 1992). The transition from basic discovery to production through scaleup of bioprocesses is a key element in the growth of the industry. Bioprocess engineering plays an important role in designing efficient and cost-effective production systems.

Although the industry has grown rapidly, profits were slow in coming. This is not surprising, given the 7–16-year time required to bring a new product to market. Nonetheless, the size of the US pharmaceutical and biopharmaceutical industry is large. It employed 413,000 people in 2004. Estimates suggest the industry will add 122,000 jobs and $60 billion in output to the US economy by 2014. Growth is likely to continue since every dollar spent on a pharmaceuticals is estimated to save $6.00 in hospital costs (Class 2004). The potential for growth is large, particularly when combined with the rapid emergence of a biofuels–chemicals production industry. Chapter 2, on new biotechnology, gives an overview of how advances in the biosciences have impacted the practice of biotechnology and the growth of the industry.

Submerged Fermentations Are the Industry's Bioprocessing Cornerstone

Submerged fermentations represent a technology in which microorganisms are grown in large agitated tanks filled with liquid fermentation media consisting of sugar, vitamins, minerals, and other nutrients. Many of the fermentations are aerated, with vigorous bubbling of air providing the oxygen needed for microbial growth. Since the microorganisms are in a liquid slurry, they are considered to be submerged, compared to microbial growth that occurs on a surface (such as on a moldy fruit or piece of bread).

The fermentation industry produced food and beverage products and some types of oxygenated chemicals by submerged fermentations prior to the first submerged antibiotic fermentations in 1943. Products in the early twentieth century included ethanol, glycerol, acetone, and butanol, with ethanol attracting renewed interest due to increasing, competitive uses for finite oil resources and the need to develop renewable energy sources. However, the growth of an efficient petrochemical industry in the years following World War II rendered some of these fermentation products economically unattractive. Petroleum sources supplanted many of the large-volume fermentation products with the exception of ethanol produced with government subsidies; yeasts for baking, vinegar, carboxylic acids, and amino acids for use as food additives; and in the formulation of animal feeds (Aiba et al. 1973). These large-volume fermentations depended on the availability of molasses or glucose from corn (starch) as the fermentation substrates (Hacking 1986). This sector of the fermentation industry has begun to reemerge, driven in part by high oil prices and concerns about energy security. This new growth is motivating agriculture to develop responses to the need for renewable feedstocks by growing more corn and carrying out research on cellulose energy crops.

Oil Prices Affect Parts of the Fermentation Industry

The manufacture of many oxygenated chemicals by fermentation was made uneconomical by the low prices of oil prior to 1973, and conversely economical by the oil price spike of 2006–2008. In the late 1980s large, aerated fermentations utilized methane or methanol (from petrochemical processing) as the main substrate in order to grow the cells. The goal was to propagate microorganisms whose protein content would make them attractive as a food source, using a relatively inexpensive substrate derived from petroleum. Since these fermentations were based on the growth of unicellular microorganisms, the source of the protein was appropriately named *single-cell protein*. However, concerns about carryover of harmful substances from the petroleum source into the fermentation where it would be incorporated into the edible cellular biomass, coupled with a sharp rise in oil prices in 1973 and subsequent decreases in soybean prices (a vegetable source of protein), made single-cell protein processes unattractive by the end of the twentieth century.

The prices of crude oil quadrupled in late 1973 and 1974, triggered in part by a war in the Middle East, decreased in the close of the twentieth century, and then increased very rapidly as the potential of demand exceeding supply became real. Reaction to these price spikes generated interest in fuel ethanol each time. After this initial shock in 1973, oil prices stabilized and interest in fermentation-derived fuels moderated until the twenty-first century. While the high-volume fermentation industry decreased in importance during this interlude, a new high-value, lower-volume biopharmaceutical fermentation industry evolved in the 1980s (Hacking 1986; Olsen 1986). However, by 2003 doubts began to surface on the ability of OPEC to keep a lid on oil prices (Bahree and Herrick 2003) since the discovery of new petroleum was slowing (Cummins 2004). Since the late 1990s, recombinant technology resulted in genetically engineered yeast that increased ethanol yields from cellulosic biomass by 50% and enhanced enzymatic activity and pretreatments to improve efficiency of transforming cellulose to fermentable sugars (Mosier et al. 2005; Ho 2004). Dramatic and growing oil price increases between 2004 and 2008 reinvigorated interest in alternative fuels, including ethanol from cellulose and diesel from soybean oil, and plant-derived sugars to make chemicals (Ritter 2004; Houghton et al. 2006; Lynd et al. 2008). The four fold rise in oil prices in 2004/07 was followed by an equally steep decrease in 2008/09. Concurrent contraction in financial markets made financing of renewable energy projects challenging. The story was still developing in 2009, but need for renewable energy for the sake of economic stability was evident.

GROWTH OF THE ANTIBIOTIC/ PHARMACEUTICAL INDUSTRY

The Existence of Antibiotics Was Recognized in 1877

In 1877 Pasteur and Joubert discovered that anthrax bacilli were killed by other bacteria. In 1896 Gosio isolated mycophenolic acid from *Penicillium brevi–compactum*. Mycophenolic acid inhibited *Bacillus anthracis* but was too toxic for use as a therapeutic agent in humans. By 1917 Grieg-Smith showed *actinomycetes* produce substances with antibacterial activity. In 1928 Alexander Fleming showed that *Staphylococcus* cultures were inhibited by growing colonies of *Penicillium notatum*. Unlike many of the other antibiotics at that time, penicillin was later found to be effective and suitable for systemic (internal) use. By 1937, actinomycetin, an antibacterial agent, had been isolated from a *streptomycete* culture but was too toxic for use as a therapeutic agent. In 1939 Dubos obtained gramacidin and tyro-cidin from the spore-bearing soil bacillus, *Bacillus brevis*. These agents proved to be effective for treatment of skin infections (topical use) but again were too toxic for internal systemic use (Evans 1965).

Penicillin Was the First Antibiotic Suitable for Human Systemic Use

Florey and Chain catalyzed the rediscovery of Fleming's penicillin, starting in 1939, during studies of compounds that would lyse bacteria. Their work was carried out as part of a search for new therapeutic agents for use in World War II. Fleming's strain of *Penicillium notatum* did not produce large amounts of penicillin, and penicillin itself was relatively unstable. Thus, significant efforts were carried out to develop isolation and purification procedures that minimized product loss. In a pattern that continues to be repeated to this day, the researchers developed methods for isolating small quantities of the therapeutic drug (penicillin, in this case) for trials with mice. The penicillin was shown to protect the mice against *Streptomyces haemolyticus* while being relatively nontoxic. Clinical trials in 1941 were most suc-cessful, and scaleup to obtain larger amounts of the antibiotic were then initiated by joint Anglo-American efforts that included the involvement of Merck, Pfizer, and Squibb with the help of government laboratories (Aiba et al. 1973). Experience with penicillin during World War II showed it to be one of the few antibiotics available at that time that was suitable for systemic use in fighting infections and saving thousands of lives (Evans 1965).

Genesis of the Antibiotic Industry

The genesis of the antibiotic industry that preceded the current era of engineered organisms was catalyzed by government incentives and support during World War II. The motivation was the discovery of the beneficial effects of penicillin and the difficulty of producing penicillin through chemical synthesis. Once the fermentation[2] route was chosen over chemical synthesis, events progressed rapidly. The rapid scaleup of production required development of a cost-effective medium on which the penicillin-producing microorganism, *Penicillin chrysogenum*, could be grown, design of large-scale aerated tanks that could be operated under sterile conditions, and recovery and purification of the final product (Aiba et al. 1973).

The first submerged fermentations were carried out on a corn steep liquor–lactose-based medium. Recovery of the labile penicillin molecule was achieved through a combination of pH shifts and rapid liquid–liquid extraction (Shuler and Kargi 1992). According to Hacking (1986), "the discovery of penicillin, which has now virtually passed into folklore, has been one of the major milestones of the twentieth century, and arguably one of the most beneficial discoveries of all time. Its impact on biotechnology, and the public perception of biotechnology should not

---

[2] Production was tried at first by growing the *Penicillium* on the surface of moist bran. This did not work well because it was difficult to control temperature and sterility. The second approach was to grow the microorganism on the surface of quiescent liquid growth medium in milk bottles and other types of containers. Yields were high, but potential for mass pro-duction was limited. This led to fermentations in large, agitated vessels that held a liquid volume of 7000 gallons or more (Shuler and Kargi 1992).

be underestimated." Hacking (1986, p. 6) also points out that the development of cost-effective processes for manufacture of penicillin provides an example of the impact of government support in starting up a new technology with the potential of widespread benefits for society:

> During the Second World War, the US government imposed an excess profits tax of 85% as one measure to pay for the war effort. As many pharmaceutical companies were liable for this tax, additional research cost them effectively only 15% of its total cost, "the 15¢ dollar." Several of the early large antibiotic screens and process development of penicillin were funded on this basis. This led to the growth of a fermentation-based antibiotic industry (see 1943 in timeline of Table 1.1).

Penicillin was at first produced by surface culture, in which the mold grew on the surface of a nutrient medium in 1–2-L batches in vessels resembling milk bottles. Antibiotic yield was improved by discovery of higher-yielding strains of *P. notatum* and *P. chrysogenum* and mutants obtained by ultraviolet and X-ray irradiation of several *Penicillium* strains (Coghill 1998).

Scaleup from 1 to 2 L to 100,000 L capacity was made possible by the discovery of a strain of *P. chrysogenum* (on a moldy cantaloupe in Peoria, Illinois) that could be grown in submerged culture. The development of submerged fermentation technology for this microbe in large agitated vessels by the US Department of Agriculture's Northern Regional Laboratories in Peoria followed. Mass production was now possible since the microorganism could be grown in large liquid volumes of nutrient media to which sterile air was provided to satisfy oxygen requirements for growth (Aiba et al. 1973; Evans 1965). Prices of penicillin quickly decreased because of productivity gains made possible by submerged fermentation (Fig. 1.1) (Hacking 1986; Mateles 1998) so that by the 1970s prices (and profits) had fallen precipitously and leveled off. Discovery, selection, and development of microorganisms suitable for submerged fermentations made penicillin's mass production possible.

Other Antibiotics Were Quickly Discovered after the Introduction of Penicillin

A different type of antibiotic, streptomycin, which is active against a wider range of pathogens than is penicillin, including *Mycobacterium tuberculosis*, was isolated from a strain of actinomycete from the throat of a chicken by Waksman at Rutgers in 1944 (Aiba et al. 1973; Evans 1965). Actinomycetes are commonly found in soil and are intermediate between fungi and bacteria. Numerous other metabolic products of actinomycetes were subsequently isolated and have made a transition from the bench scale to wide therapeutic use.

Between 1945 and 1965, about 30 new antibiotics[3] gained the status of established therapeutic agents. By 1981 about 5500 antibiotics were identified, but only

---

[3] Stinson (1996) gives a clear description of the difference between antibiotics and antibacterial drugs. Antibiotics are drugs that kill bacteria and are derived from natural sources such as molds. Chemically synthesized antibacterial drugs, whose structures are the same as antibacterials from natural sources, are also antibiotics. Hence, penicillin G from *Penicillium chrysogenum*, or organically synthesized ceftriaxone with the same 7-aminocephasosporanic acid nucleus as compounds derived from *Cephalosporium* molds, are antibiotics. "Antibiotics that are toxic to human cells, and to malignant human cells more so than normal ones," are

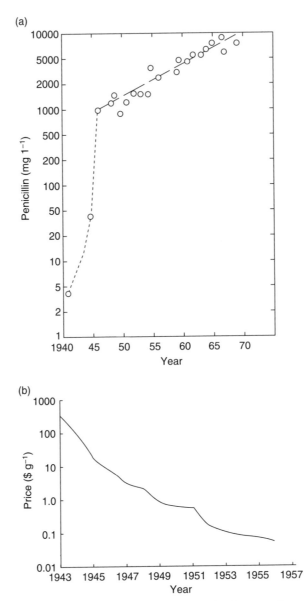

Figure 1.1. Example of process improvements of biotechnology product, and impact on cost: (a) changes in penicillin broth potency (i.e., product concentration) due to alteration of microorganisms regulation; (b) history of cost of penicillin from 1943 to 1956. Note that timescales on the plots do not coincide. [Figures reproduced from A. J. Hacking (1986), Figs. 5.1 and 5.2, respectively, p. 13. Data from A. L. Demain and P. P. King.]

---

anticancer antibiotics. Examples of antibacterial drugs are quinolones, oxazalidinones, and sulfa drugs. Some antibacterial drugs are also active against fungal, viral, or parasitic microorganisms. For example, *Helicobacter pylori* infections are now treated with antibacterial drugs; *H. pylori* infections of the stomach wall are found in 95% of patients with duodenal ulcers and 80% of patients with stomach ulcers (excluding patients who take nonsteroidal, antiinflammatory drugs).

100 made it to market. This number grew to about 160 variations of 16 basic compounds by 1996, but was accompanied by an alarming rise in antibiotic- and antibacterial-resistant bacterial infections[3] (Stinson 1996). The discovery of thousands of new antibiotics resulted in only a few hundred being characterized during the same time period (Aiba et al. 1973; Hacking 1986). This illustrates how successful development of therapeutic agents derived from microbial sources requires tremendous screening efforts. There is a continuing need for antibiotic discovery and development as drug-resistant infections have begun to appear, with drug-resistant bacteria viewed as a threat that is just as serious as the AIDS (HIV) epidemic (Stinson 1996; Tanouye 1996). There may also be lessons to be applied to the current status of pharmaceuticals where less than 30% of therapeutics in Phase II clinical trials make it to market (Whalen 2005).

Discovery and Scaleup Are Synergistic in the Development of Pharmaceutical Products

This synopsis of penicillin and antibiotic development illustrates the synergy between discovery and scaleup. Following discovery, enough of the new substance must be obtained for testing with animals, and later, if warranted, for treating humans. The ability to rapidly scale up production of a promising therapeutic agent complements the discovery process. Scale-up provides sufficient material so that it can be characterized and its activity defined. This helps to provide input on where future discovery efforts might be directed. Another role of scaleup is to provide larger quantities of pure drugs for clinical trials. Last, but not least, if the trials are successful, scaleup is needed to provide manufacturing capability to make the new drug available on a wide basis.

Success of the Pharmaceutical Industry in Research, Development, and Engineering Contributed to Rapid Growth but Also Resulted in Challenges

The pharmaceutical industry invests heavily in research to invent, screen, develop, and test new products. The industry has historically invested between 16% and 20% of its earnings in research, and has been continually concerned about developing new products and keeping its pipeline full of new pharmaceuticals. Costs of bringing a single drug to market are now between $800 million and $1.7 trillion (Landers 2003). Although the production of pharmaceuticals is sometimes perceived to be principally a fermentation–biotechnology based industry, many drugs are obtained through chemical synthesis or chemical modification of fermentation-derived molecules. Custom-drug discovery, synthesis, and marketing is carried out on a timescale similar to biopharmaceuticals, with 10 years often required to move from discovery to routine production, where routine production is classified as ≥500 kg/year (Stinson 1992).

A measure of the size of the industry is given by the estimated sales by large pharmaceutical companies of $259 billion in 2005 (Barrett et al. 2005). The combination of fermentation-derived and chemically synthesized drugs shows that there has been a significant increase in the overall sales of the top 10 companies. While the US sales of drugs by US pharmaceutical companies exceeded $32 billion in 1991, the US sales of protein biopharmaceuticals were only $1.2 billion. By 1996 the sales value of the biopharmaceuticals had grown to $7.5 billion

Thayer 1995), and by 2004, sales for the pharmaceutical sector and revenues for the biopharmaceutical sector were $259 billion and $51 billion, respectively (Barrett et al. 2005; Tsao 2005). Despite the large increase in sales, the cost of prescription drugs as a fraction of healthcare costs remained constant at 4–5% of health care expenditures during this period. Healthcare costs had risen rapidly in the period from 1982 to 1992—from about $300 billion/year to $820 billion/year, respectively, and were at $1.9 trillion in 2005.

The need and markets for new pharmaceuticals is large, and will continue to grow as the population grows and ages. The fit of new biotechnology and drug products into this growing market and the pending loss of patent protection of existing drugs help to explain the increasing pace of consolidation in the pharmaceutical industry since about 1995 (Thayer 1995). The pending loss of patent protection will affect, or has affected, major products from Pfizer, Merck & Co., Schering-Plough (plans to merge with Pfizer, 2009), Eli Lilly, Glaxo, Bristol-Myers Squibb, Syntex (now a part of Roche), UpJohn (now part of Pfizer), and Astra AB. An example of the effect of loss of patent protection is the heart drug Capoten (Bristol-Myers Squibb) whose US sales fell from $146 million to $25 million within a year after coming off patent. The price dropped from 57¢/pill to 3¢/pill when generic versions of the drug entered the market.

Drugs coming off patent in 2000 to 2002 had US sales of $6.8 billion for four US companies. By 2006, pharmaceuticals that generated $14 billion in revenue for Pfizer began to face generic competition. The magnitude of change is large, as is the challenge of managing it. Pfizer had selling, general, and administrative expenses of $16.9 billion, and research and development (R&D) expenses of $7.68 billion, and had to cut $2 billion in expenses in 2005 in the face of losing patent protection on some of its products (Hensley 2005). Declines have continued through 2009.

GROWTH OF THE AMINO ACID/ACIDULANT FERMENTATION INDUSTRY

Unlike biopharmaceuticals, whose emergence as a major biotechnology business sector has been relatively recent, the production of α-amino acids has a long history starting with the isolation of asparagine from asparagus juice in 1806. Other α-amino acids were isolated from a variety of natural substances. Threonine from fibrin was the last of these acids to be isolated and identified in 1935 (Anonymous 1985). Glutamic acid, in the form of monosodium glutamate (MSG), was identified by Ikeda in 1908 as the agent responsible for the flavor-enhancing properties of konbu, a traditional seasoning (food addition) in Japan. This discovery helped to initiate the development of an amino acid production industry starting in 1909, when amino acid hydrolysates of wheat gluten (protein) or soybean protein were used by Ajinomoto (Japan) to commercially produce MSG (Hirose et al. 1978).

The discovery of the glutamic acid–producing bacteria *Micrococcus glutamicus* (later renamed *Corynebacterium glutamicum*) in Japan in 1957 (Kinoshita 1985) marked the beginning of a 40-year period of growth in high-volume, fermentation-derived bioproducts. Initially the multiplicity of names given to glutamic acid bacteria discovered after 1957 was a source of confusion. On the basis of the conventional taxonomy, and the 73–100% DNA homology reported for several strains of *Corynebacterium, Brevibacterium*, and *Microbacterium* in 1980, the glutamic acid bacteria are now regarded as a single species in the genus *Corynebacterium* (Kinoshita 1985).

Production of Monosodium Glutamate (MSG) via Fermentation

The discovery of the L-glutamic acid producer, *M. glutamicus*, in 1957 (Hirose et al. 1978; Kinoshita 1985) and the realization that a critical concentration of biotin (~2 ppb), or biotin together with an antibiotic or nonionic detergent, was needed to maximize L-glutamic acid production (Aiba et al. 1973; Hirose et al. 1978), marks the transition of an amino acid industry based on processing of natural materials, to one dominated by fermentation-based technology. Production of food and feed additives, in addition to products used in therapeutic applications quickly developed. At about the same time, the first industrial production of another α-amino acid, L-lysine, was also initiated (Anonymous 1985).

By 1972, the world production of glutamate was 180,000 tons, 90% of which was made by fermentation (Aiba et al. 1973); and by 1976, Japan produced two-thirds of the world's amino acids with an estimated annual sales value of $3 billion and a volume of 200,000 tons/year. Half of this volume was MSG, which was used almost exclusively as a food seasoning. A small amount was employed in the production of a leather substitute (Hirose et al. 1978). While MSG remains the single largest fermentation-derived amino acid, a number of other valuable amino acids are obtained from mutants or related microorganisms of the genus *Corynebacterium*. Lysine (an animal feed additive) is the second largest amino acid obtained by fermentation. It had a worldwide production of 40,000 tons/year in 1983 compared to MSG at 220,000 tons/year. The worldwide lysine market was $600 million/year in 1996, with half of the sales attributed to Archer Daniels Midland. Some lysine producers also manufactured the carboxylic acid, citric acid, with annual sales of $1 billion in 1996. Citric acid is a widely used flavoring agent and acidulant in beverages (Kilman 1996a,b).

The Impact of Glutamic Acid Bacteria on
Monosodium Glutamate Cost Was Dramatic

In 1950 MSG was obtained from natural products (i.e., soybean) and cost $4/kg. When the first fermentation process started up in 1961, the price dropped to $2/kg. By 1970 the price settled to about $1/kg, and in 1983 approximately 220,000 tons of MSG/year was being produced by fermentation, worldwide, with a sales value of about $550 million. In 2001 Ajinomoto, a major amino acid producer, estimated total annual amino acid production was at 2 million tons, with MSG-based flavor enhancer to be 1 million tons/year (Kusumoto 2001).

Auxotrophic and Regulatory Mutants Enabled
Production of Other Amino Acids

Auxotropic and regulatory mutants are microorganisms whose metabolic pathways for a specific metabolic product have been altered to reduce or eliminate the production of an intermediate metabolite that is essential for cell growth. These organisms, also known as *nutritionally deficient mutants*, can be of commercial value since measured addition of the missing metabolite enables amplification of control and production of the desired product. The discovery of glutamic acid bacteria resulted in a fundamental change in the industry, not only for MSG production but also for other amino acids that could now be obtained via fermentation. Mutation of glutamic acid bacteria with ultraviolet light, X rays, γ rays, and chemicals resulted in amino acid producing auxotropic and regulatory mutants for the L forms of lysine, threonine, tyrosine, phenylalanine, ornithine, proline, leucine, citrulline, and homoserine, as well as glutamic acid. Many amino acids are small-volume (10–300 tons/year), high-value products obtained by fermentation, microbial conversion of intermediates, or enzyme synthesis (Hirose et al. 1978). As is the case for antibiotics, some amino acids are produced by chemical synthesis. In particular, DL-methionine, DL-alanine, and glycine are obtained via chemical synthesis.

The predominant suppliers of amino acids were Ajinomoto, Takeda, and Kyowa (in Japan) (Hacking 1986) and Archer Daniels Midland (abbreviated ADM) in the US (Kilman 1996b). The corn milling division of Cargill, a privately owned company, entered lysine production in the early 2000s. Degussa was "to build a 165 million lb/year synthetic[4] lysine production facility at Cargill's Blair, Nebraska complex," with production initiated in 2000 (Anonymous 1997b). The announcement stated that "Degussa says it is the world's leading supplier of methionine, and a leading supplier of threonine. Lysine, synthetically produced by cornfed microorganisms, is used as an additive for feeds for hogs and poultry." The syntax of this announcement is as interesting as its content. The term "synthetic," which has previously been associated with in vitro chemical methods, began to be used to describe an in vivo microbial method. This gives an indication of how chemistry and biotechnology is merging. By 2005, DeGussa was ready to buy out Cargill's share, subject to regulatory approval.

Some amino acids are derived by chemical routes, or through enzyme synthesis and microbial production starting with an intermediate. The highest-volume, chemically synthesized amino acid is DL-methionine (also an animal feed additive), which is obtained from propylene, hydrogen sulfide, methane, and ammonia (Anonymous 1985; Tanner and Schmidtborn 1981). Another major chemically synthesized amino acid is a glycine chemical building block that is an intermediate in the manufacture of Monsanto's herbicide, Round-up®. Round-up was estimated to generate about $2.2 billion in sales in 1995, with sales that grew with the successful introduction of "Round-up ready" soybean seeds in 1996.[5] These examples illustrate some of

---

[4] This is an example where a fermentation process, which may even be based on a recombinant microorganism, is used for production, is now labeled as a synthetic source. Previously, the term "synthetic" would have denoted a process based on chemical synthesis. This is obviously not the case here since Cargill will provide raw materials as well as large-scale *fermentation* and separation technology (Anonymous 1997b).

[5] Commercial quantities of these herbicide-resistant seeds were planted in 1996 and generated sales of about $45 million. A survey of 1058 farmers (~10% of the farmers who planted the seed) showed that 90% of those surveyed felt the product met or exceeded expectations (Fritsch 1996b). These seeds were genetically engineered by inserting a gene from petunia that confers herbicide resistance to soybean plants (Fritsch and Kilman 1996). This allows

the crossover and cross-disciplinary characteristics of biotechnology when applied to a specialty chemical business.[6]

Prices and Volumes Are Inversely Related

The price of biotechnology products decreases as the total sales volume increases; the relationship between price and volume falls on a characteristic line when the data are plotted on a double log scale (Fig. 1.2). This type of plot, previously used for correlating the cost of thermoplastics to their volume, is relevant for biotechnology products according to Hacking (1986). Although the price and volume data are from 1986, and there is scatter in the data, the plot illustrates the necessity of reducing production costs to achieve bulk sales. This is both a function of the technology that must be available if production costs are to be reduced, and market demand that would drive the production of larger volumes of the product. A variation of

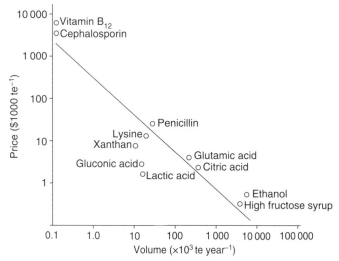

Figure 1.2.   Hierarchy of values represented as a log–log plot of price as a function of volume for biotechnology products [from Hacking (1986), Fig. 2.1, p. 18].

---

the fields to be sprayed with the herbicide, thereby killing all other vegetation, followed by planting of soybean seeds. The soybeans are able to grow since they are not affected by the herbicides.

[6] By late 1996, Monsanto had planned to divide itself into two entities: Monsanto Life Sciences and a chemical entity (later given the name Solutia) (Reisch 1996). This occurred within a year during which Monsanto purchased a plant biotechnology company (Calgene) that held patents for improved fresh produce, cotton seeds, specialty industrial and edible oils (derived from seeds), and plant varieties (Anonymous 1997). Analysis of 1995 sales revealed that the life sciences products accounted for annual sales of $5.3 billion based on the major products of Round-up herbicide, Round-up-resistant soybeans, Bollguard insect-protected cotton, Nutrasweet sweetener, and prescription drugs for arthritis and insomnia. This strategy resulted in Monsanto developing into a $8.6 billion company in 2007.

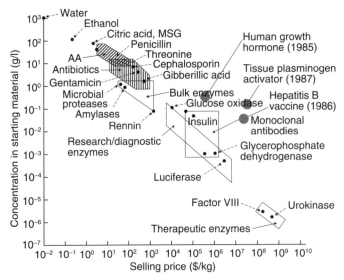

Figure 1.3. Log–log plot of concentration as a function of selling price for small and large molecules; and products used in a range of applications from food to therapeutic (reproduced with permission from Rachel Kerestes, "Global Biotech Crop Area Tops 200 Million Acres in 2004," *B10 News*, April/May 2005).

this plot is given for a range of products—therapeutic, biocatalytic, and food—in Fig. 1.3. Some of the single-use high-value biopharmaceutical products fall above this curve. Nonetheless, the take-home message is clear. Selling price determines minimum concentration of the product in the initial reaction mixture, extract, or fermentation broth.

Several examples are instructive. In the case of monosodium glutamate, technology enabled a reduction in cost, which in turn increased the volume used. The essential amino acid, tryptophan, would find uses in animal feed and therefore offer high-volume markets if its price were to decrease from $100/kg to $10/kg [based on 1986 prices, per Hacking (1986)]. The dramatic decrease in the cost of penicillin within 12 years after its introduction illustrates that the volume–price relationship will also affect a high-value drug. Penicillin was a wonder drug when first introduced in 1941–1945, but then decreased in price from $500/g in 1943 to $0.10/g by 1956 (Hacking 1986). Advances in manufacturing technology, improvements in productivity and product concentration (Fig. 1.1a), rampup of its volume, and economies of scale drove the price down.

A more recent and developing story is that of fuel ethanol derived from cellulosic biomass. When pretreated celluloses are mixed with cellulase enzymes and fermented, fuel-grade ethanol results. Feedstocks are wood, switchgrass, corn stover, and other agricultural residues. In this case the reductions in the cost of enzyme and development of microorganisms that can ferment both hexoses and pentoses, as well as generate enzymes that hydrolyze cellulose to fermentable sugars, are key developments (Mosier et al. 2005; Wyman 2007).

While the production costs for a high-value therapeutic contribute as little as 10% to the overall price of the product, production economics are secondary only to the timeline of drug development, competitive pressures, and marketing issues normally associated with introduction of a new product. The loss of patent protection and the exclusivity that it provides will challenge the business models of the biopharmaceutical industry as generic drugs manufactured in a global arena enter the market. For drugs, function can be mapped according to chemical structure and generics can be introduced with respect to identical chemical structures. Unbranded versions of protein pharmaceuticals—that is, biogenerics or biosimilars—face a more complex situation. The manufacturing process can affect the finished product, and hence full clinical testing may be needed before approval. Because of the cost of such testing, biogenerics may not offer as much of a discount as generic versions of drugs (Class 2004).

Biochemical Engineers Have a Key Function in All Aspects of the Development Process for Microbial Fermentation

Biochemical engineering was defined by Aiba, Humphrey and Millis in 1973 as "conducting biological processes on an industrial scale, providing the link between biology and chemical engineering," with its heart being "the scale-up and management of cellular processes." It was viewed as the interaction between two disciplines. The definition was extended by Bailey and Ollis (1977) to encompass "the domain of microbial and enzyme processes, natural or artificial," including "wastewater treatment as well as industrial fermentation and enzyme utilization." Shuler and Kargi (1992) described biochemical engineering as "the extension of chemical engineering principles to systems using a biological catalyst to bring about desired chemical transformations," and as being "subdivided into bioreaction engineering and bioseparations." Blanch and Clark (1996) similarly present an interdisciplinary viewpoint.

Challenges that a biochemical engineer faces are the prevention of contamination (sterility), control and promotion of microbial growth and productivity through media composition, control of growth conditions (pH, temperature, product precursors), oxygen transfer (for aerobic fermentation), and directing metabolic pathways of organisms to enhance rates and extents of product accumulation (i.e., metabolic engineering). The engineer must devise efficient and robust methods for recovering and purifying an often unstable or labile product. These disciplines and skills apply to manufacture of products from mammalian cells as well as microorganisms that form the basis of the biopharmaceutical and fermentation industries and that are enabling components of biotechnology. These fall under the umbrella of bioprocess engineering, which is the "sub-discipline within biotechnology that is responsible for translating life-science discoveries into practical products capable of serving the needs of society. It is critical in moving newly discovered bioproducts into the hands of the consuming public," (National Research Council 1992). This applies to food, fuels, and biopharmaceuticals.

REFERENCES

Aiba, S., A. E. Humphrey, and N. F. Millis, *Biochemical Engineering*, Academic Press, 1973, pp. 3–6.

Aiba, S., A. E. Humphrey, and N. F. Millis, *Biochemical Engineering* (2nd edition), Academic Press, New York, 1–11 (1973).

Anonymous, "Agriculture: Biotech Crops Put Down Global Roots," *Business Week* **3913**, 12 (Dec. 20, 2004).

Anonymous, *Amino Acids for Animal Nutrition*, A. G. DeGussa, 1985.

Anonymous, "Cargill, Degussa to Form Lysine Venture," *Chem. Eng. News* 75(28), 17 (1997b).

Bahree, B. and T. Herrick, "Officials Say OPEC Can't Keep Lid on Rising Oil Prices," *Wall Street J.* A1 (Feb. 28, 2003).

Bailey, J. E. and D. F. Ollis, *Biochemical Engineering Fundamentals*, McGraw-Hill, New York, 1977, pp. xiii, 3–23, 45–57, 225.

Barrett, A., J. Carey, and M. Arndt, "Drugs: More Bitter Pills for Big Pharma," *Business Week* **3915**, 113–114 (2005).

BIO (Biotechnology Industry Organization), http://www.bio.org/news, News Item, Washington, DC (Jan. 12, 2005).

Blanch, H. W. and D. S. Clark, *Biochemical Engineering*, Marcel Dekker, New York, 1996.

Class, S., "Health Care in Focus," *Chem. Eng. News* 81(49), 18–24 (2004).

Coghill, R. D., in *Penicillin: A Paradigm for Biotechnology*, Mateles, R. I., ed., Candida Corp., 1998, pp. 19–25.

Cohen, S. N., A. C. Y. Chang, H. W. Boyer, and R. B. Helling, "Construction of Biologically Functional Bacterial Plasmids *In Vitro*," *Proc. Natl. Acad. Sci. USA* 70(11), 3240–3244 (1973).

Council on Competitiveness, Office of the Vice President, *Report on National Biotechnology Policy*, Washington, DC, 1991.

Cummins, C., "Data Cast Doubt on Oil Reserves," *Wall Street J.* A2 (Jan. 23, 2004).

Evans, R. M., *The Chemistry of Antibiotics Used in Medicine*, Pergamon Press, Oxford, 1965.

Fritsch, P. and S. Kilman, "Seed Money: Huge Biotech Harvest Is a Boon for Farmers—and for Monsanto," *Wall Street J.* A1, A8 (Oct. 24, 1996).

Fritsch, P., "Monsanto May Shed Its Chemical Unit," *Wall Street J.* A3 (Oct. 11, 1996a).

Fritsch, P., "Biotech Boosts Monsanto Stock to Record Height," *Wall Street J.* A3 (Oct. 4, 1996b).

Hacking, A. J., *Economic Aspects of Biotechnology*, Cambridge Univ. Press, 1986, pp. 6–8, 18–20, 39–72.

Hensley, S., "Pfizer Plans $2 Billion in Cost Cuts," *Wall Street J.* A3 (Feb. 11, 2005).

Hirose, Y. K. Sano, and H. Shibai, "Amino Acids," in *Annual Reports on Fermentation Processes*, Academic Press, New York, 1978, Vol. 2, pp. 155–189.

Houghton, J., S. Weatherwax, and J. Ferrell, *Breaking The Biological Barriers To Cellulosic Ethanol: A Joint Research Agenda*, Office of Science and Office of Energy Efficiency and Renewable Energy, US Department of Energy, June 2006.

Kilman, S., "Green Genes: If Fat-Free Pork Is Your Idea of Savory, It's a Bright Future," *Wall Street J.* A1, A10 (Jan. 29, 1998).

Kilman, S., "Ajinomoto Pleads Guilty to Conspiring with ADM, Others to Fix Lysine Prices," *Wall Street J.* B5 (Nov. 15, 1996a).

Kilman, S., "ADM Settles Two Suits for $65 Million, Raising Hopes for Plea in Federal Probe," *Wall Street J.* A1, A6 (Sept. 30, 1996b).

Kim, S. and B. E. Dale, "Life Cycle Assessment of Fuel Ethanol Derived from Corn Grain via Dry Milling," *Bioresource Technology* 99, 5250–5260 (2008).

Kinoshita, S., "Glutamic Acid Bacteria," in *Biology of Industrial Microorganisms*, A. L. Demain and N. A. Solomon, eds., Benjamin/Cummings, Menlo Park, CA, 1985, pp. 115–142.

Kusumoto, I., "Industrial Production of L-Glutamine," *J. Nutr.* 131, 25525–25555 (2001) (see Website: http://www.nutrition.org).

Ladisch, M. R., Bioprocess Engineering (biotechnology), *Van Nostrand's Scientific Encylopedia*, 9th ed., Vol. 1, G. D. Considine, ed. (editor-in-chief), 2002, pp. 434–459.

Ladisch, M. R. and K. Dyck, "Dehydration of Ethanol: New Approach Gives Positive Energy Balance," *Science* 205(4409), 898–900 (1979).

Landers, P., "Drug Firms See Costs Increase to Bring a Product to Market," *Wall Street J.* B4 (Dec. 8, 2003).

Lynd, L. R., M. S., Laser, D. Brandsby, B. E. Dale, B. Davison, R. Hamilton, M. Himmel, et al., "How Biotech Can Transform Biofuels," *Nature Biotechnol.* 26(2), 169–172 (2008).

Mateles, R. J., ed., *Penicillin: A Paradigm for Biotechnology*, Candida Corp., 1998.

Mosier, N. S., C. Wyman, B. Dale, R. Elander, Y. Y. Lee, M. Holtzapple, and M. Ladisch, "Features of Promising Technologies for Pretreatment of Lignocellulosic Biomass," *Bioresource Technol.* 96(6), 673–686 (2005).

NABC Report 19, *Agricultural Biofuels: Technology, Sustainability, and Profitability*, A. Eaglesham and R. W. F. Hardy, eds., 2007, pp. 3–11.

National Research Council (NRC) Committee on Bioprocess Engineering, *Putting Biotechnology to Work: Bioprocess Engineering*, National Academy of Sciences, Washington, DC, 1992.

National Research Council (NRC) Committee on Opportunities in Biotechnology for Future Army Applications, *Opportunities in Biotechnology for Future Army Applications*, National Academy of Sciences, Washington, DC, 2001.

Office of Technology Assessment, US Congress, *Biotechnology in A Global Economy*, B. Brown, ed., OTA-BA-494, US Government Printing Office, Washington, DC, 1991.

Olsen, S., *Biotechnology: An Industry Comes of Age*, National Academy Press, Washington, DC, 1986.

Reisch, M., "Monsanto Plans to Split in Two: Chemical and Life Sciences Firms Will Be Formed, up to 2500 Jobs May Be Cut," *Chem. Eng. News* 74(51), 9 (1996).

Ritter, S. K., "Biomass or Bust," *Chem. Eng. News* 82(22), 31–32 (2004).

Sedlak, M. and N. W. Y. Ho, "Characterization of the Effectiveness of Hexose Transporters for Transporting Xylose During Glucose and Xylose Co-Fermentation by a Recombinant Saccharomyces Yeast." *Yeast* 21, 671–684 (2004).

Shuler, M. L. and F. Kargi, *Bioprocess Engineering. Basic Concepts*, Prentice-Hall PTR, Englewood Cliffs, NJ, 1992, Vol. 2, pp. 4–8.

Stinson, S. C., "Custom Chemical Producers Set to Meet Industry Standards," *Chem. Eng. News* 70(5), 25–50 (1992).

Stinson, S. C., "Drug Firms Restock Antibacterial Arsenal, Growing Bacterial Resistance, New Disease Threats Spur Improvements to Existing Drugs and Creation of New Classes," *Chem. Eng. News* 74(39), 75–100 (1996).

Tanner, H. and H. Schmidtborn, "DL-*Methionine—the Amino Acid for Animal Nutrition*, A. G. Degussa, 1981.

Tanouye, E., "Drug Makers Go All Out to Squash 'Superbugs,'" *Wall Street J.* B1, B6 (June 25, 1996).

Thayer, A. M., "Betting the Transgenic Farm, First Plantings of Engineered Crops from Monsanto and Others Are Serving as Agbiotech's Proving Ground, *Chem. Eng. News* **75**(17), 15–19 (1997).

Thayer, A., "Consolidation Reshapes Drug Industry, Shifts Employment and R&D Outlook: In Response to Global Pressures on the Industry, Companies Are Merging to Create Entities with the Critical Mass to Compete," *Chem. Eng. News* **73**(42), 10 (1995).

Tsao, A., "Biotech: One Sector with Strong Vital Signs," *Business Week* **3915**, 114 (2005).

Walker, J. W. and M. Cox, with A. Whitaker as a contributor, *The Language of Biotechnology, A Dictionary of Terms*, ACS Professional Reference Book, American Chemical Society, Washington, DC, 1988, p. 204.

Whalen, J., "Glaxo to Report on Many Drug Trials," *Wall Street J.* B2 (Feb. 10, 2005).

Wyman, C., "What Is (and Is Not) Vital to Advancing Cellulosic Ethanol," *Trends Biotechnol.* **25**(4), 153–157 (2007).

CHAPTER ONE

HOMEWORK PROBLEMS

1.1. Plot the growth of total annual revenues of the biotechnology industry from 1900 to 2004. Compare this to the total annual revenues of the new biotechnology industry from 1991 to 2004.

    (a) Briefly discuss the relative growth rates of the two industries.

    (b) How do the two sectors differ with respect to products? Technologies?

1.2. When did agriculture begin to feel a major impact from developments in the "new" biotechnology? On the basis of the data in this chapter, and the BIO Website, plot the growth in the acreage of genetically engineered crops. What has been the growth rate? What might limit the rate of growth of these crops? What are the benefits of genetically modified crops? What are potential concerns?

1.3. What were the major technological and engineering hurdles that had to be overcome to enable production of penicillin on a large scale? Categorize your answer into biological-versus-engineering issues. What happened to the cost of penicillin over a 10 year time span (starting in 1943)?

1.4. The flavor enhancer MSG has a current usage estimated to be on the order of 1 million metric tons/year. If the price of MSG had stayed constant, since 1950, what would be the value of MSG, sold? What actually happened (plot the cost of MSG vs. time, as best possible). What does this suggest for the value and magnitude of sales of more recently developed biotechnology products?

1.5. Current concerns in the health sector address cost of healthcare and cost of biopharmaceuticals (some of which cost $10,000/year or more per patient). If prior history is any indication, are these costs likely to decrease? Briefly discuss. Be sure to indicate differences between the manufacture and use of biopharmaceuticals as compared to antibiotics.

1.6. Determine an algebraic correlation for concentration of a bioproduct or biotherapeutic compound as a function of selling price. How does this correlate with price as a function of volume of product (demand)?

NEW BIOTECHNOLOGY

INTRODUCTION

New biotechnology **is defined by the** science that "enabled directed manipulation of the cell's machinery through recombinant DNA techniques and cell fusion. Its application on an industrial scale since 1979 has fundamentally expanded the utility of biological systems." Scientists and engineers change the genetic makeup of microbial, plant, and animal cells to confer new characteristics. Biological molecules, for which there is no other means of industrial production, are generated. Existing industrial microorganisms are systematically altered (i.e., engineered) to enhance their function and to produce useful products in new ways. The new biotechnology, combined with the existing industrial, government, and university infrastructure in biotechnology and the pervasive influence of biological substances in everyday life, has set the stage for unprecedented growth in products, markets, and expectations. This description, taken from the National Research Council report on Bioprocess Engineering (1992), correctly projected the major impact that genetic engineering would have on biologically based (biobased) methods for producing tangible commercial products.

Bioprocess engineers and scientists have put the new biotechnology to work, first in the biopharmaceutical industry, and now in agriculture and industrial manufacture and/or use of biobased products. The laboratory methods developed as part of the quest for new biopharmaceutical therapeutics starting in 1976, and the goal of sequencing the human genome starting in 1991 (Table 1.1), provided tools for obtaining new bioproducts and directing the evolution of plants, animals, and microorganisms to generate bioproducts in amounts not previously achievable. New biotechnology has displaced, to some extent, traditional plant and animal breeding methods with tools that enable cloning and the directed insertion of genes from one species or genetic donor into another. One example is the cloning of Dolly, the sheep, by removing the genetic contents of an egg cell, replacing it with the genetic contents of the cell of another sheep, and then nurturing the cell to birth after the fertilized egg is implanted in a surrogate mother. Another is the integration of a gene derived from a microorganism into a corn or cotton plant, so that these plants produce a protein with pesticide properties [i.e., Bt (*Bacillus thuringiensis*) corn or Bt cotton), and resulted in a new agricultural biotechnology industry as described in Chapter 1.

This chapter describes the basis and impact of the new biotechnology on the industry. Chapter 3 presents development of processes and bioproducts that occurred both before and after the introduction of genetic engineering techniques.

GROWTH OF THE BIOPHARMACEUTICAL INDUSTRY

Traditional biotechnology is based on indirect methods of selecting, improving, and propagating organisms. Its transition to a new biotechnology, based on genetic engineering, could be interpreted as an abrupt change in technology when viewed from a historical (8000-year) perspective. Actually, the technological change in the industry followed an evolutionary path in both methodology and perceptions that occurred over a period of about 40 years. Microorganisms, cells, plants, and animals (collectively referred to as *organisms*) with enhanced characteristics for the production of bioproducts and antibiotics have been selected, manipulated, and improved since the beginning of the industry. Genetic structures have been purposely altered using mutagens, radiation, selective pressures, breeding, and strain improvement. The discovery and development of the structure and function of DNA over a period of a century allowed genes to be identified, and traits to be correlated to an organism's genetic characteristics. These findings advanced both the fundamental understanding and practical applications of organisms for generating bioproducts (food, beverages, medicines, and chemicals). The developments of recombinant technology, which now benefits the pharmaceutical industry, was catalyzed by, or was "a direct result of generous governmental funding for basic biomedical research since World War II" (Olsen 1986).

The discovery of enzymes that cut DNA at specific locations, and the development of practical methods for achieving alterations in a cell's genetic makeup, changed the field by allowing genetic manipulations—and the determination of the effects of these alterations—to be achieved in a matter of days to weeks, rather than years or decades. These methods, carried out in a directed and rational manner, allowed introduction of completely new traits by design, rather than by chance, into the host organisms. This gave rise to the expression "genetic engineering," meaning that genes could now be engineered and altered in a predetermined manner to confer a specified end result.

This translated to the birth of a new industry at a time when US government policy (in the 1980s) regarding limited partnerships in risky ventures and availability of venture capital encouraged the formation of new startup companies. An estimated 1000 biotechnology companies were formed during 1980–1990. The major product groups addressed by these companies were (1) diagnostics (human and/or animal health), (2) therapeutics (human and/or animal health), (3) Ag-bio (agrobiotechnology; plant genetics and/or microbial crop protectants), and (4) suppliers to the industry (Burrill 1988).

A survey of the use of biotechnology in the US industry, published in 2003, and biotechnology industry facts provided by BIO (Biotechnology Industry Organization) in 2005 gives an updated snapshot. Websites of these organizations give current statistics and trends in this rapidly evolving industry. Revenues of publicly traded companies in 2003 were $39.2 billion (BIO 2005), with the industry employing 198,300 as of December 2003 (BIO 2005). According to the BIO Report (2005), there were 1473 biotechnology companies in the United States, 314 of which were publicly held in 2003. The US Department of Commerce (2003) survey addressed 3000 firms engaged in biotechnology-related industries (but not all are biotech companies, i.e., companies whose sole products are biotechnology-related). According to the Department of Commerce survey, 66,000 employees in the industry (in 2002) could be classified as biotech-related technical workers where scientists accounted for 55% of the total; technicians, 30%; engineers, 8%; and "R&D focused computer" specialists (6%) the rest. Employees with other biotech-related responsibilities include administration, production, and legal workers (US Department of Commerce 2003), and may form two-thirds of the workforce if the difference between the BIO and US Department of Commerce statistics are an indication. These statistics are changing rapidly, hence presenting the authors with a challenge: referring to employee and revenue estimates that change before the "ink is dry." The reader is referred to the BIO Website for the latest estimates.

About 72% of the biotechnology companies are in human health, while the major part of the remaining companies are in activities related to animals husbandry, agriculture, or industrial or agriculturally derived processing (US Department of Commerce 2003). The overall industry was capitalized at $311 billion, and its aggregate revenues of $33.5 billion in 2001 translated to 0.33% of the $10 trillion US gross domestic product (US Department of Commerce 2003). By 2005, aggregate revenues were $50 billion. There is significant room for growth, in both human health and biofuels. There are 370 biotech drug products and vaccines in clinical trials targeting more than 200 diseases (BIO 2005). Growth in sectors of the industry, other than human health, will become more important as biotechnology is applied to ending hunger, meeting energy and environmental needs, defending our homeland, and catalyzing new innovations in biomaterials.

The analysis of these sales in the context of the growth potential of biopharmaceuticals is complicated since the effect of major pharmaceutical companies buying smaller biotechnology companies, and the merging of sales figures is difficult to place into categories.[1] The adaptation of biotechnology as a means of production by some of the large and long-established chemical, agricultural, and food companies further obscures the definition of a biotechnology company. Furthermore, a significant volume of pharmaceuticals is manufactured by chemical synthesis, and these are not considered to be biotechnology products. Examples are antiinflammatory agents and cholesterol-lowering drugs.

---

[1] The projections and financial performance of the biotechnology industry did not uniformly coincide with companies that manufacture and sell biotechnology products. For example, Eli Lilly and Company and Novo Nordisk are major suppliers of biosynthetic human insulin derived from recombinant microorganisms, yet these companies are identified as drug or industrial enzyme companies—not biotechnology companies. Consequently, a complete accounting of their sales derived from recombinant technology may not be possible. Revenues will vary significantly and depend on a particular survey's definition of the biotechnology industry. A thorough survey by the US Department of Commerce of the industry gives an excellent accounting.

The analysis of a new industry, and the terminology used to describe excellent performance, may be at times be confusing to the student. For example, only *10* out of the 1300 biotechnology companies reported profits in 1997. All the others had negative earnings as of the first half of 1997 (Thayer 1997a). Nonetheless, the professional services firm of Ernst and Young, which annually analyzes and reports on the biotechnology industry, characterized the period between July 1, 1995 and June 30, 1996 as showing "stunning results" for the industry. This was based on sales of biotechnology-related products[2] that rose by 16% to $10.8 billion/year. The top 10 recombinant therapeutics accounted for nearly $5 billion in sales, and the top 16 had sales of about $6.8 billion (Thayer 1996). By 2004, the top 10 therapeutics had a value of $15 billion. The market capitalization and cash flows of the top companies tripled between 1996 and 2004. Consequently, an understanding and anticipation of trends taken in the context of a snapshot of industry revenue is needed for a sector dominated by startup companies. Similar comments apply to the emergence of the cellulosic biofuels industry.

Product sales, revenues, R&D expenses, and market value increased significantly for the industry between 1991 and 2004. The industry as a whole continues to show negative earnings and a deferred profit margin. This is consistent with the characteristics of the "present phase" of the biotechnology industry's life cycle (Fig. 2.1). For example, in 1997, six leading biotechnology companies dominated the earnings figures. The single major contributing company was Amgen. In 1992, Amgen's total revenue was $1.09 billion with $358 million net income. By 1997 revenue was $2.4 billion and net income $644 million. (Amgen's revenue for calendar year 2008 was $15 billion with net income of $4.2 billion (Amgen, 1997, 1998, 2009)). The combined 1997 earnings of Biochem Pharma, Biogen, Chiron, Genentech, and Genzyme contributed another $98 million. The remaining 24 firms among the top 30 combined for a net loss of $120 million. This accounting helps to communicate the stage at which the industry found itself in the late 1990s. An understanding of the technology and insights into potential applications and markets is needed if an industry's tremendous potential for growth is to be understood and windows of opportunity identified. The growth of the biopharmaceutical industry from 1997 to 2008 provides examples ranging from vaccines to chemotherapeutic

---

[2] Biotechnology products, as defined here, are products sold by the biotechnology companies and exclude many bioproducts sold by established pharmaceutical, food, or chemical industries, even though these products are biotechnology-related. While the financial status of pharmaceutical companies is reported in the context of sales, biotechnology companies report revenues, which include sales as well as income received for services and contract research. This method of accounting helps the financial analysis of the industry to clearly track the progress of companies started up within the last 30 years with 80% of these started up since 1986. The distinction is clearly described by Burrill and Lee (1991):

> Biotech companies will transform traditional industries. Pharmaceutical companies will become biopharmaceutical companies, being themselves transformed as they partner with biotech companies, acquire biotech companies and their technologies, and marry traditional pharmaceutical R&D methodologies with biotechnology's understanding of cellular behavior. ... we will see emergence of bioagricultural, biochemical, and bioenergy companies. Biotech companies will continue to exist with traditional companies transformed through their partnership.

Biotechnology industry life cycle and value drivers

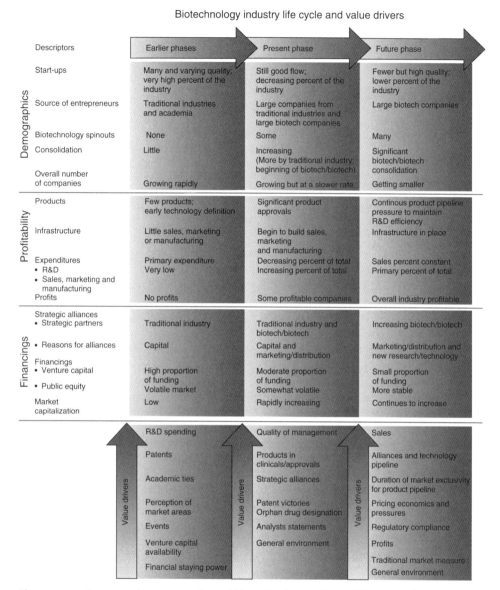

Figure 2.1. Conceptual representation of biotechnology industry life cycle. The early phase coincides with period of approximately 1975–1990. The present phase started in about 1990 and is likely to continue through the beginning of the twenty-first century. [Figure from Burrill and Lee (1991), Ref. 98, p. 3, with permission.]

proteins (monoclonal antibodies), as well as the maturing of many of its early products.

The Biopharmaceutical Industry Is in the Early Part of Its Life Cycle

The assessment of an industry doing well when it is losing money may seem unusual to a student in engineering or biological sciences. However, this is consistent with

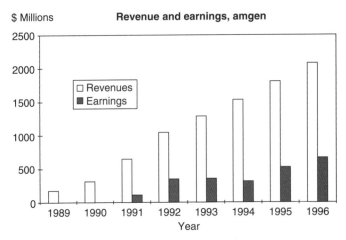

Figure 2.2. Cash flows for Amgen during its early growth.

the startup of an entire new sector of the economy, where negative cash flow during the beginning years of the individual companies cannot be offset until approved products reach the market and sales are generated. It is only in the later phase of the industry's lifecycle that the overall industry becomes profitable.

The key phases of the biotechnology industry's life cycle was schematically represented by Burrill and Lee (1991) in Fig. 2.1. This differs from an established industry where most companies have positive earnings, with only a few startups that might be experiencing a negative cash flow. The rapid changes between negative earnings and profitability can be illustrated by the early earnings history of Amgen, one of the largest biotechnology companies at that time (Fig. 2.2). Kirin Brewery partnered with Amgen in the mid-1980s to manufacture a human protein that stimulates production of red blood cells in the body. This protein, Epogen, was approved by the FDA in 1989. This molecule and related products grew to annual sales of $8 billion (Weintraub 2005). The biotechnology industry, while risky, has an enormous earnings potential. Research investments exceed $100,000 per employee (about 10× higher than in other industries), and its patent award rate is on the order of 7000–8000 biotech patents/year (BIO 2005; US Department of Commerce 2003). Some investors were willing to risk capital to get the industry started. The industry is now in the "future phase" depicted in Fig. 2.1.

An important transition from wild-type to genetically modified industrial microorganisms occurred approximately 45 years after the discovery of penicillin, and 30 years after its first large-scale manufacture through submerged fermentation. The first therapeutics were antibiotics—microbial substances made by microbes for activity against other microorganisms. Biologically active proteins derived from humans, referred to as *biopharmaceuticals*, were made possible by advances in molecular biology that enabled microorganisms to express mammalian proteins having therapeutic utility.

Molecular biology and genetic techniques have made it practical to discover how biological systems function, and how various molecules, not just proteins, moderate the biological chemistry of living cells. New types of therapeutic approaches are made practical by this expanded knowledge of biological systems.

The means by which molecules for therapeutic use are manufactured has proved to be a key technological shift that differentiated fermentation and pharmaceutical industry of the twentieth century from the developing biotechnology industry. Ultimately, the application of the scientific methods that enable manufacture of high-molecular-weight biotherapeutic molecules to small-molecule production will change the methods for production of liquid transportation fuels and commodity chemicals from renewable resources.

Discovery of Type II Restriction Endonucleases Opened a New Era in Biotechnology

In 1970 Kelly, Smith, and Wilcox discovered enzymes, known as *Type II restriction endonucleases*, which recognize a particular target sequence in duplex DNA. These enzymes enabled researchers to obtain discrete DNA fragments of defined length and sequence, and therefore facilitated a key step in carrying out genetic engineering (Old and Primrose 1981). Unlike mutation and screening procedures that identify microorganisms that may have randomly attained better properties, genetic engineering entails insertion of a foreign gene into the genetic material of a microbial cell, so that the cell can produce proteins that would otherwise not be found in the cell. The impact and commercial possibilities of this technique are credited to Cohen, Chang, and Boyer, who in 1974 established practical utility of constructing biologically functional plasmids in vitro and transforming *Escherichia coli* to express a foreign protein (Ryu and Lee 1988; MacQuitty 1988). The basis of the technique is illustrated in Fig. 2.3.

The developments associated with restriction enzymes that facilitate genetic engineering and hybridomas that produce monoclonal antibodies opened up vistas for a new range of potential products. Isolated from humans directly, these proteins can now be produced in microorganisms or by cell culture, and be used in large concentrations to enhance the body's ability to fight heart disease, cancer, immune disorders, and viral infections.

The Polymerase Chain Reaction (PCR) Is an Enzyme-Mediated, In Vitro Amplification of DNA

Another significant development in enzyme technology was the polymerase chain reaction (PCR). Since about 1985, this method has enabled the sequencing of DNA in vitro. PCR was developed by Cetus Corporation (now part of Chiron, which in turn was bought by Novartis in 2006) scientists in 1984 and 1985. It is an enzyme-catalyzed reaction that facilitates gene isolation and avoids the complex process of cloning which requires the in vivo replication of a target DNA sequence by integrating it into a cloning vector in a host organism.

Polymerase chain reaction is initiated by DNA denaturation at $90\,°C$, followed by primer annealing at $70\,°C$, and then addition of a DNA polymerase and deoxynucleoside triphosphates to form a new DNA strand across the target sequence. This cycle, when repeated n times, produces 2^n times as much target sequence as was initially present. Thus 20 cycles of the PCR yields about a millionfold increase or amplification of the DNA. Applications include comparison of altered, uncloned genes to cloned genes, diagnosis of genetic diseases, detection of pathogens, and retrospective analysis of human tissue (Arnheim and Levenson 1990).

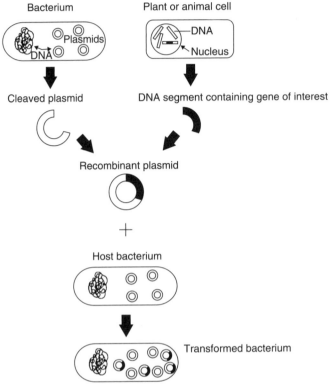

The genetic engineering of bacteria

Figure 2.3. "One common way to genetically engineer bacteria involves the use of small, independently replicating loops of DNA known as *plasmids*. Certain enzymes can cleave these plasmids at specific sequences in their genetic codes. DNA from other organisms that has been treated with the same enzymes can then be spliced into the plasmids with enzymes that join the cut ends of DNA. These recombinant plasmids are reinserted into bacteria, where they can reproduce themselves many times over. At the same time, the bacteria can divide, creating millions of copies of the introduced DNA. This DNA can then be studied through analytical techniques, or, if a gene within the introduced DNA can be made to produce the same protein it did in its original location, the genetically engineered bacteria can be used as microbial factories to make large quantities of the protein." [Reprinted with permission from Olsen (1986). Copyright, p. 17, National Academy Press.]

IMPACTS OF THE NEW BIOTECHNOLOGY ON BIOPHARMACEUTICALS, GENOMICS, PLANT BIOTECHNOLOGY, AND BIOPRODUCTS

There are a number of disciplines that impact biotechnology. The fields of biology, molecular biology, biochemistry, chemistry, and biochemical engineering have made individual contributions to the rapid advance of biotechnology in a commercial sense. The demonstration that a gene from human pancreas cells could be introduced into a β-galactosidase operon, which in turn was introduced into *Escherichia coli*, enabling *E. coli* to produce human insulin, was a key event. The first steps in initiating this chain of events include the development of techniques for transgenosis, the artificial transfer of genetic information from bacterial cells to eukaryotic cells by transducing phages; isolation of a restriction enzyme that specifically cleaves certain parts of a small circular DNA known as a *plasmid*; the first cloning experiments in 1972 to 1973; electrophoresis and blotting techniques; and in vitro translation systems for converting mRNA (messenger RNA) into proteins (Old and Primrose 1981; Alberts et al. 1989). Other important enzymes were those that could ligate pieces of DNA (i.e., ligases), and would otherwise allow manipulation of DNA strands to obtain predetermined DNA sequences, namely, genes, to be introduced into a bacterial plasmid.

The plasmid pBR322 was one of the most versatile artificial plasmid vectors since it contained ampicillin (Ap®) and tetracycline (Tc®)-resistant genes, as well as desirable replication elements. These genes, when introduced into the cell, enable the bacteria to make enzymes that hydrolyze the two antibiotics. Hence, when a foreign gene is purposely engineered into the plasmid, and the plasmid is placed back into the bacterial host cell to obtain a transformed cell, the transformed cell will also be able to resist the two antibiotics as well as express a foreign protein corresponding to the gene. Transformed cells will grow in the presence of the antibiotic, while all other cells will be killed. Since transformations are not 100% efficient, the presence of ampicillin and tetracycline resistant genes in a recombined plasmid enable the researcher to select for the transformed cells by growing them in a medium that contains the antibiotics. Once a transformation is successful, the removal of the markers and stably integrating the cloned gene(s) into the microbial chromosome is a second challenge, since it is undesirable to utilize antibiotic resistant bacteria on a large scale, where there is a chance these may be released.

The pBR322 plasmid is 4362 base pairs long, and was completely sequenced during the early days of the new biotechnology industry (Old and Primrose 1981). Plasmid pBR322 was the starting point for constructing the plasmid pSomII-3 containing a chemically synthesized gene for the small hormone, somatostatin (Itakura et al. 1977). These developments resulted in a new language, as well as a new technology. Bioprocess engineers involved in applying recombinant technology to commercial practice must know the vocabulary to communicate with bioscientists.

At a meeting convened by Eli Lilly and Co. in Indianapolis in 1976, these tools together with a method for producing copious quantities of mRNA for insulin by a special tumor cell known as an *insulinoma* apparently helped encourage a race for the cloning of the human insulin gene into *E. coli*. The ensuing 3 years of intense activity (ca. 1976–1979) make up the fascinating story told by Hall (1987).

Biotechnology Developments Have Accelerated Biological Research

Genetic engineering techniques have facilitated studies of proteins in plant and animal cells, which otherwise would not have been possible since these proteins (such as DNA polymerases, reverse transcriptase, and ribozymes) are normally present in very small amounts. Their study led, in turn, to their use as tools in further understanding cellular function and genetics. Genetic engineering techniques have been used to amplify mammalian proteins in bacterial systems, so that sufficient protein becomes available for classical study using established enzymology techniques. Some examples are neural peptides that impact hypertension, and co-reductases implicated in heart disease.

Drug Discovery Has Benefited from Biotechnology Research Tools

Drug discovery was thought to be the area that would benefit enormously from techniques developed through genetic engineering. Proteins having possible therapeutic value are often difficult to study because their in vivo concentration in mammalian systems is very low, or because they are difficult to isolate because of limited specimen availability. Once a gene can be identified, molecular biology techniques can be used to produce larger quantities of the protein, thus facilitating its study. The interleukins are an example of this approach, where it is estimated that more interleukin was produced during several years of its discovery phase than was generated since the beginning of mankind. Cytokines examined or used in treatment of wound healing, cancer, and immune deficiencies since 1990 (Thayer 1991) have evolved to include proteins against non-Hodgkins lymphoma (Rituximas) and rheumatoid arthritis [Humira (adalimumab) and Remicade (infliximab)] (Mullin 2004; Abboud 2004).

Site-specific mutations can also be achieved using recombinant techniques. This facilitates study of a protein's activity if given amino acids are replaced in its structure. Thus, the protein's binding properties with respect to inhibitors, where the inhibitors represent an experimental drug, can be studied.

The Fusing of Mouse Spleen Cells with T Cells Facilitated Production of Antibodies

Kohler and Milstein are responsible for the third significant development during this time. In 1975 they demonstrated that cells derived from mouse B lymphocytes, which secrete antibodies, were fused with myeloma tumor cells from a mouse, which can grow indefinitely in culture (Fig. 2.4). The resulting fused cell, is a *hybrid myeloma* or hybridoma cell (illustrated in Fig. 2.5) that can grow in cell culture and produce large quantities of chemically identical antibodies, specifically, monoclonal antibodies, which are excellent probes for diagnostic purposes (Olsen 1986), and also are being developed for therapeutic uses (Johannes 1996; Mullin 2004). The history of the early development of monoclonal antibodies is unique in that they were never patented (Springer 2002).

Regulatory Issues Add to the Time Required to Bring a New Product to Market

While most scientists and engineers may not be directly involved in regulatory affairs, students of biotechnology should be aware that these issues have a major impact on the development of new biotechnology-derived products. The biotechnology industry is regulated by the US Food and Drug Administration (FDA), the Environmental Protection Agency (EPA), and the Department of Agriculture (USDA). This will continue to be the case for stem cell research, and with respect to the introduction of biotechnology for biomedical implants and wound repair. The

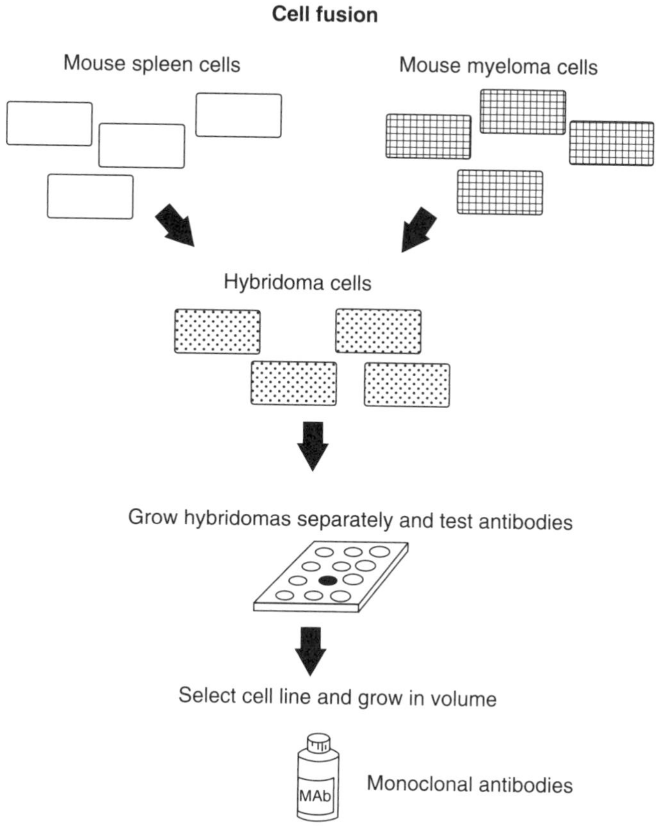

Cell fusion

Mouse spleen cells   Mouse myeloma cells

Hybridoma cells

Grow hybridomas separately and test antibodies

Select cell line and grow in volume

MAb   Monoclonal antibodies

Figure 2.4. "To produce monoclonal antibodies, antibody-producing spleen cells from a mouse that has been immunized against an antigen are mixed with mouse myeloma cells. Under the proper conditions, pairs of the cells fuse to form antibody-producing hybrid myeloma ("hybridoma") cells, which can live indefinitely in culture. Individual hybridomas are grown in separate wells, and the antibodies that they produce are tested against the antigen. When an effective cell line is identified, it is grown either in culture or in the body cavities of mice to produce large quantities of chemically identical, monoclonal antibodies." [Reprinted with permission from Olsen (1986), p. 26. Copyright, National Academy Press.]

student should be cognizant of the impact of regulation, since it touches on many parts of the discovery and development processes in biotechnology, as is illustrated by case studies presented in a book edited by Chiu and Gueriguian (1991). While orthopedic implants and tissue engineering products are regulated as devices, biopharmaceuticals and drugs are regulated as drugs (chemically identifiable entities) or biologics (where the manufacturing process is part of the product's characteristic).

A new drug must move through three stages of testing and review. These stages follow an investigational new drug (IND) application (Thayer 1991):

Phase I Clinical Trials. Safety and pharmacological activity of a drug is established using 100 or fewer healthy human volunteers. These require about 1 year.

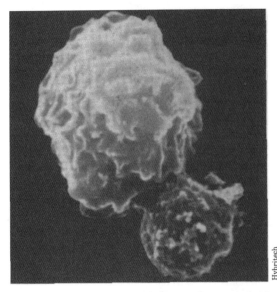

Figure 2.5. A mouse spleen cell and tumor cell fuse to form a hybridoma. As the hybridoma divides, it gives rise to a "clone" of identical cells, giving the name "monoclonal" to the antibodies that those cells produce. [Reprinted with permission from Olsen (1986), p. 27. Copyright, National Academy Press.]

> *Phase II Clinical Trials.* Controlled studies are carried out on 200–300 volunteer patients to test drug efficacy together with animal and human studies for safety. This stage requires about 2 years.
>
> *Phase III Clinical Trials.* Efficacy results are confirmed using about 10 times more patients than used in Phase II trials. Low-incidence adverse effects are also identified. This phase requires about 3 years.

A product license application is then filed with the FDA. This type of license application requires between 15 months and 2.5 years. Nonetheless, a new product may require up to 7–10 years to pass from preclinical trials to commercial introduction (BIO 2005). For serious or life-threatening diseases, promising Phase I results can lead to combined Phase II and Phase III trials, thus reducing the time by 2 or 3 years.

The approval process is thorough, lengthy, and expensive. Bioprocess engineering plays a key role in quickly developing means of production for supplying the needed amounts of pure material required for clinical trials. In this case, the scaleup of production must follow a timescale that is less than that required for each trial. Otherwise, completion of the trials can be delayed by a shortage of the medicine being tested.

The challenge for bioprocess engineers transcends scaleup issues and also encompasses the drive for a company to be the first to market. As articulated by Bailey and Ollis (1977), "modifications in the organism or in the process require some degree of reiteration of earlier tests, or re-examination of the system by regulatory authorities. Therefore, a premium exists for engineering strategies that provide effective *a priori* guidance … in organism, product, and process development." Because of the associated regulatory costs, incremental process improvements may not be commercially significant until relatively late in the product life cycle, when there are several suppliers and competition for a given drug develops. A similar analysis was also presented by Wheelwright (1991) in the context of purifying proteins in a commercial setting.

Barriers to business competitiveness, identified by the US Department of Commerce (2003) survey, were the regulatory approval process and costs, research costs, and access to startup capital. In addition, companies in agricultural, animal health, marine, and environmental biotechnology are also concerned about foreign and US laws as well as public acceptance and ethical considerations. This was an issue identified by the National Research Council biotechnology report in 1992, and continues to be an issue in 2009. The introduction of a cellulosic biofuels industry faces the same barriers identified for biopharma in 1992 and re-iterated in 2003: regulatory challenges, research costs and capital. The impact of success is even larger given the pervasive use of liquid transportation fuels an annual cost that may exceed a trillion dollars in the global economy.

New Biotechnology Methods Enable Rapid Identification of Genes and Their Protein Products

The discovery of the biochemical nature of diseases and identification of ways by which drugs can interfere with the disease process have been greatly accelerated by the new biotechnology and the identification of genes that represent underlying mechanisms of disease. The new genes are patentable, hence giving a 20-year monopoly to the companies that are the first to identify and patent them (Tanouye and Langreth 1998). This capability also speeds up drug discovery. The resulting candidates hold the promise of new and more potent medicines, but often at high cost. One report places the cost of a 10-month treatment of colon cancer using Genentech's Avastin (bevacizumab) at $49,000, and Imclone's Erbitux (cetuximab) at $38,400 (Abboud 2005). This has fostered attempts to develop less expensive biogenerics or biosimilars.

While genomics (i.e., gene-identifying technology) provides information on the targets or sites on which drugs can act, combinatorial chemistry and automation of chemical synthesis makes it possible to rapidly generate thousands of candidates that could be tested against targets identified through genomic methods. This, in turn, has driven drug companies to commit resources and billion-dollar research budgets to screen, develop, and test the new drugs resulting from the combination of computerized gene discovery and automated synthesis of molecules with potential as drugs.

Genomics Is the Scientific Discipline of Mapping, Sequencing, and Analyzing Genomes

The definition of all of the human genes was the goal of the Human Genome Project that was first proposed in 1980 and initiated in 1991 after the National Institutes of Health (NIH) and the US Department of Energy (DOE) developed a joint research plan (Watson et al. 1992; Collins and Galas 1993). The goal was to sequence the human genome by 2005, but this task was completed about 5 years ahead of schedule. This resulted in significant excitement on how this information could be used to identify targets for drugs as well as design new drugs through knowledge of the proteins that correspond to the genes. At first, scientists hypothesized that there were 80,000–100,000 genes in the 3% of the human genome that represents genes. The actual number turned out to be 34,000. The impact of the sequencing achievement will require years of research to become clearly defined. Not only must the protein(s) derived from a gene be identified; its contribution in disease processes must be understood. This is a "long and painstaking" process (Carey et al. 2000), and is one that is still in process as of the publication of this book in 2009.

The objectives of this project were to

1. Sequence the genomes of humans and selected model organisms
2. Identify all of the genes
3. Develop technologies required to achieve all of these objectives.

Physical maps of the genes were developed so that locations of the genes could be identified to within 100 kb (kilobases) of their actual position (1 kb = 1000 bases). In addition, the function of each of these genes must be identified. Finally, the sequence of the bases that make up the DNA in each gene is needed to give the details of the map. An important parameter in judging success is that finished DNA sequences have 99.99% accuracy, so that they contain no more than one error in every 10,000 bases. According to Pennisi (1998), this goal was initially difficult to meet. This became apparent when different laboratories exchanged pieces of sequenced DNA, and found that they could not exactly reproduce each others sequences (Pennisi 1998). This resulted in the need to resequence the DNA, and slowed progress.

The final step in DNA sequencing, known as the "finishing step," required considerable intuition as well as the assistance of computer programs to piece the sequence data back into contiguous pieces of DNA. While automated instruments (robots) carried out many repetitive operations involved in determining nucleotide sequences, some regions of the DNA could not give a clear sequence of nucleotides, or could not be determined because of the inability of the enzymes and biochemical methods to accurately and/or completely depolymerize the DNA into its individual nucleotides (Pennisi 1998). The most time-consuming step occurred near the completion of the process. This entailed piecing together the sequenced fragments into the proper order. When the sequences of DNA fragments did not overlap, the experiments had to be repeated (Pennisi 1998; Marshall and Pennisi 1998).

Although only about 2–3% of the human genome had been sequenced by early 1998, significant progress had been reported for genomes for microorganisms of industrial importance including yeast (Schuler et al. 1996), *E. coli* (Blattner et al. 1997), and *Bacillus subtilis*, and for microorganisms that cause some types of diseases, including ulcers (*Helicobacter pylori*) and lyme disease (*Borrelia burgdorferi*). Table 2.1 lists some of the microorganisms whose genomes had been mapped and sequenced at that time (Pennisi 1997). By the beginning of 2008 the list of completed genomes sequenced exceeded 710 species, with an additional 1300 genome projects in various stages of completion. The development of powerful database management software has improved the ability of researchers to analyze, explore, and compare this enormous amount of data. The basics of cloning are discussed in Chapter 12, and developments in genomes in Chapter 14.

The goal to sequence the entire human genome resulted in many technical advances in identifying genes, mapping locations of markers, amplifying 100–300-

Table 2.1.

Progress in Sequencing Genomes of Microorganisms

Organism	Genome Size in Millions of Base Pairs
Saccharomyces cerevisiae	12.1
Escherichia coli	4.6
Bacillis subtilis	4.2
Synechocystis sp.	3.6
Archaeoglobus fulgidus	2.2
Haemophilus influenzae	1.8
Methanobacterium thermoautotrophicum	1.8
Helicobacter pylori	1.7
Methanococcus jannaschii	1.7
Borrelia burgdorferi	1.3
Mycoplasma pneumoniae	0.8
Mycoplasma genitalium	0.6

Source: From E. Pennisi, "DNA Sequencers' Trial by Frie: With 97% of the Human Genome to Be Deciphered for a Place in World's Largest Genome Project; Many Have Stumbled in the Early Going,"*Science* **280**(5365), 814–817 (1998), 1433, and modified with Updates. Reprinted with permission from AAAS.

kb DNA fragments, sequencing the fragments, and reassembling the sequences of the fragments into contiguous segments of the larger pieces of DNA from which the fragments had originally been obtained. By mid-1998, two US agencies—the Department of Energy and the National Human Genome Research Institute (NHGRI; headed by Francis Collins)—and the Wellcome Trust (Britain) had spent over $1.5 billion. Costs of sequencing, that started at $3–$5 per base (pair) in 1988, decreased to an average of $1.35 per base by April 1998. A cost of $0.35 per base or less was the ultimate goal, as the technology developed (Pennisi 1998). If an overall average cost of $1.00/per base were assumed, sequencing of the entire human genome would have required $3 billion. Now that the genes were sequenced, the challenge remained to understand and assign functions to all of them. This was a formidable task when it is considered that the complete function of the genome of *E. coli*, with 4288 genes, was not yet resolved even though the entire genome of 4.6 million base pairs was sequenced in 1998 (Pennisi 1997; Blattner et al. 1997). However, by 2007 44% of the genes had their function unified experimentally. This represents the complete functions for known metabolic functions. On going efforts now focus on using genomics to drive biological discovery (Feist and Palsson 2008).

The development of computational technology helped catalyze the application of new methodologies to keep track of the information on the location and sequences of individual genes. The rapid and steady improvements in capabilities of personal computers have enabled the development of laboratory information management systems for tracking and recording reagents, protocols, and sequencing machine performance, as well as automating many of the operations needed to sequence genes. The need to handle all of this information also resulted in creation of a new discipline—bioinformatics. Bioinformatics addressed computer based methods for acquisition, storage, analysis, modeling, and distribution of the many types of information embedded in sequence data of DNA and protein biomolecules (Rowen et al. 1997).

Products from the New Plant Biotechnology Are Changing the Structure of Large Companies that Sell Agricultural Chemicals

The chemical, food, and agricultural industries also began to experience major changes by the mid-1990s, catalyzed by advances in plant biotechnology, and the ability to identify and manipulate traits of economically important crops at the cellular level. The first commercial success of plant biotechnology came in 1996, after about 20 years of research and development in the genetic engineering of crops. In this case, Monsanto had shown that soybean seeds, genetically engineered to resist the herbicide Roundup®, gave higher yields. The herbicide could be applied to kill all other plants in a field while having no effect on the soybean plants. The introduction of genetically engineered seeds fostered concerns about transgenic supercrops. Could these breed superweeds if genes from widely planted and cultivated, weed resistant plants were to migrate to wild relatives (Kling 1996)? The risk was perceived to be small. By 2004, when 75 million acres of soybeans were grown, 89% "were planted with seed that contained a Monsanto gene" (Kilman 2005).

Another genetically altered product by Monsanto, introduced in 1996, was cotton seed that contained a gene from bacterium, *Bacillus thuringiensis*. The gene enables the bacterium—and now the cotton plants—to make a protein that is toxic to bollworms and tobacco budworms, thereby reducing the need for pesticides that are otherwise used to control insect infestations (Fritsch and Kilman 1996). The *B. thuringiensis* gene has also been incorporated into potato so that it produces a protein toxic to the Colorado potato beetle. The potato was first approved for use in Canada (Yanchinski 1996). By 1996, DowElanco, through its investment in Mycogen (a San Diego–based biotechnology company that Dow AgroSciences purchased in 1998), had marketed transgenic corn with *B. thuringiensis*–based resistance to the European Corn Borer (Thayer 1996). France, a major corn producer, prohibited planting (but not consumption) of transgenic corn (Balter 1997). By 2004, "51% of all corn planted in the US contained a Monsanto gene" (Kilman 2005). The Dow–Elanco joint venture of Dow Chemicals and Lilly was bought by Dow Chemicals in 1997 and was renamed Dow AgroSciences, a wholly owned subsidiary of the Dow Chemical Company.

There were 30 crop species that, by 1996, had been engineered to express *B. thuringiensis* endotoxins. These proteins are highly toxic to specific insect pests. There are concerns of the evolution of insects resistant to these toxins may occur since millions of acres are now being planted with these crops (Ives 1996). If insect species develop that are resistant to *B. thuringiensis* endotoxin, strategies for overcoming this resistance will need to be developed, resulting in a cycle of further manipulation of the plants' genes to alter the toxins. Success of the transgenic soybean seeds, insect-resistant cotton seeds, and other genetically engineered seeds led to predictions in 1996 that annual sales would grow to $6.5 billion, by 2006 (Fritsch and Kilman 1996). In actuality, genetically engineered seed sales were $6.2 billion in 2006 (ISAAA 2006).

The current growth of the agricultural biotechnology sector may, in part, have benefited from a 1992 ruling in which policy guidelines were defined on how the Food and Drug Administration would oversee new plant varieties, including those developed through biotechnology. Their basis was "that products would be judged by their risks ... not by the process which produces them." Hence if the composition of a food product is not significantly altered, the FDA's premarket approval would not be needed, and special labeling would not be required. However, if "genetic manipulation were to lead to a product that has increased allergenicity, or that contains new proteins, fats, or carbohydrates, it will have to pass rigorous pre-market review. And if approved its new traits will have to be labeled" (Ember 1992a). The widespread tests and apparent initial success of several genetically engineered crops, about 4–6 years later, may have benefited from this policy, proposed under the first Bush administration. The timeline between 1992 and 1997 (see Table 1.1)[3] shows that a combination of technical, business, and regulatory factors coincided at about the same time, and culminated in the first significant commercialization of plant biotechnology products by 1997. While public opposition to genetically modified crops has been strong since 1997, especially in Europe, this technology has been the fastest adopted crop production technology in recent history. In 2006, genetically modified crops were planted in 22 countries for a total of 250 million acres, 60 times the number from 1997 (ISAAA 2006).

Bioproducts from Genetically Engineered Microorganisms Will Become Economically Important to the Fermentation Industry

The economic impact of bioproducts and value-added chemicals derived through fermentations was initially modest, but the potential was clear. In 1997, Dupont and Genencor developed an inexpensive and environmentally friendly fermentation for making polyester building blocks from glucose. This process was based on a proprietary, recombinant microorganism that combines bacterial and yeast pathways so that the fermentable sugar is converted to glycerol, and the glycerol to trimethylene glycol, which, in turn, is polymerized to polyester. The resulting polymer is reported to have properties that are superior to existing polyesters derived from petrochemical building blocks (Potera 1997).

Another example is a genetically engineered bacteria that converts naphthalene (priced at $1.00/kg) to indigo (priced at $8–$17/kg). Indigo is used to dye blue

---

[3] Hileman (1995) concisely summarized the status of regulation, in 1995, as follows. The statutes divide agricultural products into six categories: plants engineered for (1) herbicide resistance, (2) pesticide resistance, (3) delayed (tomato) ripening; (4) plants modified to make products otherwise obtained from other crops; (5) plants engineered so their crop can be processed more easily; and (6) bacteria enhanced to fix nitrogen or control insects. Genetically modified plants and animals for purposes of producing pharmaceuticals are not included in these categories. While few products from transgenic plants require agency approval, most companies have voluntarily consulted the FDA to obtain its stamp of approval. Transgenic plants to be grown on a large scale are regulated by the US Department of Agriculture (USDA). The US Environmental Protection Agency (USEPA), regulates gene-modified organisms that express a pesticide or function as a pesticide. Genetically engineered microorganisms, such as a bacterium engineered to produce ethanol, would also be regulated by the USEPA. *Bacillus thuringiensis* corn is regulated by all three agencies: insecticidal properties by the EPA, large-scale growing by the USDA, and corn as a food product by the FDA.

jeans, has an annual market of $250 million, and accounts for 3% of all of the dye colors currently used. This dye, given the name "bio-indigo," was proposed to replace dye currently made using cyanide and formaldehyde (Hamilton 1997). The enzyme that enables this conversion and the genetically engineered bacteria that gave the result were discovered at Amgen. The technology was sold to Genencor International Inc. in 1989, when one of Amgen's biopharmaceuticals began to take off and caused Amgen to drop its research on industrial enzymes (see timeline of Table 1.1).

The realization that there is an interface where biology, fermentation, and recovery technologies all impact each other (Swartz 1988), gave rise to new challenges. Questions that may arise as a process is designed, scaled up, and operated are: If recovery is difficult, can the protein be modified to make it easier to purify? If protein refolding gives low yields, can the molecular biologist clone the gene to obtain a proactive form? If test tube scale production gives low yields, can the bioprocess engineer optimize conditions to give a better yield? If the product is intracellular, how can the cell best be programmed to respond to a step input to the fermentation, to turn on protein production machinery? How can a separation system be modified to handle complex mixtures containing low titers of the final product? Another important question in a practical context is: How much will this cost, and how long will it take?

This philosophy applies to the development of a reinvigorated biofuels and bioproducts industry, which will be based on renewable and sustainable cellulosic feedstocks, as well as sugarcane and corn. The development of this industry will face similar challenges as those encountered in the biotechnology (biopharma) industry. In addition, to regulatory, research, and financial hurdles, the cellulose (non-food) based feedstocks such as wood grasses, straws, agricultural residues, and energy crops will require solids processing, where solids are processed or fermented in an aqueous solution. Significant challenges in scale-up may be encountered based on the analysis of Merrow (1985, 1988) and Bell (2005). In the past, first of a kind plants with refined or raw solid feeds initially operate at 50% or less of design capacity, with start-up times that may be two to four times longer than planned. Most problems are solved, and subsequent plants designed with the benefit of commercial experience. These are more readily started up and attain design capacity more quickly. Bioprocess engineering will play a key role in defining scale-up strategies, mitigating scale-up risk, and empirically linking solids processing to the cost-effectiveness of commercial production (Merrow, 1985, Bell, 2005). The relevance of solids handling and processing is illustrated by Hennessey (2009) in accessing impacts of solids handling in the scale-up of a biorefinery. Corn stover, wheat straw, bagasse and switchgrass are defined as a raw solid feed. Hence pilot plant operations are important if not critical to the scale-up process.

The response to these types of questions has fostered evolution of a team, cross-disciplinary approach to problem solving in the biotechnology industry. This book attempts to present such an approach by combining basics of biochemistry and biology, with bioprocess engineering, fundamentals of cell growth, enzyme reactions, and product purification. The impact of the basic science or engineering fundamentals on process configurations is illustrated by example. Where possible or appropriate, this approach is also presented in a manner that facilitates a first calculation of costs.

REFERENCES

Abboud, L., "One Start-up's Path to Making Biotech Clones," *Wall Street J.* B1 (Aug. 16, 2005).

Abboud, L., "Battle Royalties: Abbott Fights over Biotech Licensing," *Wall Street J.* C1 (Nov. 22, 2004).

Alberts, B., D. Bray, J. Lewis, M. Raff, K. Roberts, and J. D. Watson, *Molecular Biology of the Cell*, 2nd ed., Garland Publishing, New York, 1989, pp. 258–284.

Amgen Annual Reports for 1996, 1997, 2008 published in 1997, 1998, 2009.

Arnheim, N. and C. H. Levenson, "Polymerase Chain Reaction," *Chem. Eng. News* **68**(40), 38–47 (1990).

Bailey, J. E. and D. F. Ollis, *Biochemical Engineering Fundamentals*, McGraw-Hill, New York, 1977, pp. xiii, 3–23, 45–57, 225.

Balter, M., "Transgenic Corn Sparks a Furor," *Science* **275**(5303), 1063 (1997).

Bell, T. A., Challenges in the scale-up of particulate processes—an industrial perspective, *Powder Technology* **150**, 60–71 (2005).

BIO (Biotechnology Industry Organization), "Biotechnology Industry Facts," http://www.bio.org/speeches/pubs, (Feb. 6, 2005).

Blattner, F. R., G. Plunkett III, C. A. Block, N. T. Perna, V. Burland, M. Riley, J. Collado-Vides, et al., "The Complete Genome Sequence of Escherichia coli K-12," *Science* **277**(5331), 1453–1462 (1997).

Burrill, G. S. and K. B. Lee, Jr., *Biotech '92: Promise to Reality—an Industry Annual Report*, Ernst & Young, San Francisco, 1991.

Burrill, G. S. with the Arthur Young High Technology Group, *Biotech 89: Commercialization*, Mary Ann Liebert Publishers, New York, 1988.

Carey, J., A. Barrett, and E. Licking, "The Genome Gold Rush," *Business Week* **3685**, 147–162 (2000).

Chiu, Y.-Y. H. and J. L. Gueriguian, *Drug Biotechnology Regulation: Scientific Basis and Practices*, Marcel Dekker, New York, 1991.

Collins, F. and D. Galas, "A New Five Year Plan for the US Human Genome Project," *Science* **262**(5130), 43–49 (1993).

Ember, L., "No Special Rules Needed for Biotech Food Products," *Chem. Eng. News* **70**(22), 5–6 (1992a).

Ember, L., "FDA to Speed Approval of Biotechnology Drugs," *Chem. Eng. News* **70**(11), 6 (1992b).

Feist, A. M. and B. Ø. Palsson, "The growing scope of applications of genome-scale metabolic reconstructions using *Escherichia coli*," *Nature Biotech.* **26**, 659–667 (2008).

Fritsch, P. and S. Kilman, "Seed Money: Huge Biotech Harvest Is a Boon for Farmers—and For Monsanto," *Wall Street J.* A1, A8 (Oct. 24, 1996).

Hall, S. S., *Invisible Frontiers, the Race to Synthesize a Human Gene*, Atlantic Monthly Press, New York, 1987.

Hamilton, J., "A Gene to Make Greener Blue Jeans," *Business Week* **3520**, 82 (1997).

Hennessey, S. M., T. A. Bell, E. L. Chilels, F. G. Gallagher, W. D. Partem, R. G. Taylor, P. M. Vrana, Scaling-up: From Bench to Biorefinery at Penn State Cellulosic Biofuels Short Course, San Francisco, CA., March 24, 2009.

Hileman, B., "Views Differ Sharply over Benefits, Risks of Agricultural Biotechnology," *Chem. Eng. News* **73**(34), 8–17 (1995).

ISAAA (International Service for the Acquisition of Agri-Biotech Applications), *ISAAA Brief 35-2006*, 2006.

Itakura, K., T. Hirose, R. Crea, A. D. Riggs, H. L. Heyneker, F. Bolivar, and H. W. Boyer, "Expression in Escherichia coli of a Chemically Synthesized Gene for the Hormone Somatostatin," *Science* **198**, 1056–1063 (1977).

Ives, A. R., Technical Comments—"Evolution of Insect Resistance to Bacillus thuringiensis-Transformed Plants," *Science* **273**(5280), 1412 (1996).

Johannes, L., "Magic Bullet Drugs Try for a Comeback, Magic Bullet Theory, Tries Comeback in 1996 after Big Flops," *Wall Street J.* B31 (Jan. 1996).

Kelly, T. J. and H. O. Smith, "A restriction enzyme from *Hemophilms influenzae*: II. Base sequence of the recognition site," *Journal of Molecular Biology* **51**(2), 393–409 (1970).

Kilman, S., "Monsanto Receives Illinois Subpoena on Modified Seeds," *Wall Street J.* A7 (April 21, 2005).

Kling, J., "Agricultural Ecology: Could Transgenic Supercrops One Day Breed Superweeds?" *Science* **274**, 180–181 (1996).

Kohler, G. and C. Milstein, "Continuous cultures of fused cells secreting antibody of predefined specificity," *Nature* **256**(5517), 495–597 (1975).

MacQuitty, J. J., in *The Impact of Chemistry on Biotechnology*, ACS Symp. Ser. 362, M. Phillips, S. P. Shoemaker, R. D. Middlekauf, and R. M. Ottenbrits, eds., ACS (American Chemical Society, Washington, DC), 1988, pp. 11–29.

Marshall, E. and E. Pennisi, "Hubris and the Human Genome," *Science* **280**(3566), 994–995 (1998).

Merrow, E. W., Estimating Startup Times for Solids-Protessing Plants, *Chemical Engineering*, 89–92 (Oct. 24, 1988).

Merrow, E. W., Linking R&D to Problems Experienced in Solids Processing, *CEP*, 14–22 (May, 1985).

Mullin, R., "Biopharmaceuticals," *Chem. Eng. News* **82**(17), 19–24 (2004).

National Research Council (NRC), Committee on Bioprocess Engineering, *Putting Biotechnology to Work: Bioprocess Engineering*, National Academy of Sciences, Washington, DC, 1992.

Old, R. W. and S. B. Primrose, "Principles of Gene Manipulation—an Introduction to Genetic Engineering," *Studies in Microbiology*, Vol. 2, Univ. California Press, Berkeley, 1981, pp. 1–47.

Olsen, S., *Biotechnology: An Industry Comes of Age*, National Academy Press, Washington, DC, 1986.

Pennisi, E., "DNA Sequencers' Trial by Frie: With 97% of the Human Genome to Be Deciphered for a Place in World's Largest Genome Project; Many Have Stumbled in the Early Going," *Science* **280**(5365), 814–817 (1998).

Pennisi, E., "Laboratory Workhorse Decoded," *Science* **277**(5331), 1432–1434 (1997).

Potera, C., "Genencor & Dupont create 'green' polyester," *Genetic Engineering, & Biotechnology News* **17** 17, (1997).

Rowen, L., G. Mahairas, and L. Hood, "Sequencing the Human Genome," *Science* **278**(5338), 605–607 (1997).

Ryu, D. D. Y. and S-B. Lee, in *Horizons of Biochemical Engineering*, S. Aiba, ed., Oxford Univ. Press, 1988, pp. 97–124.

Schuler, G. D., M. S. Boguski, E. A. Stewart, L. D. Stein, G. Gyapay, K. Rice, R. E. White, et al., "A Gene Map of the Human Genome," *Science* **274**(5287), 540–546 (1996).

Smith, H. O. and K. W. Wilcox, "A restriction enzyme from *Hemophilus influenzae*: I. Purification and general properties," *Journal of Molecular Biology* **51**(2), 379–391 (1970).

Springer, T. A., "Retrospective: Immunology, César Milstein (1997–2002)," *Science* **296**(5571), 1253 (2002).

Swartz, J. at the 1988 Engineering Foundation Conference coined the phrase "fermentation/recovery interface" (April 1988).

Tanouye, E. and R. Langreth, "Genetic Giant: Cost of Drug Research Is Driving Merger Talks of Glaxo, SmithKline," *Wall Street J.* A1, A8 (Feb. 2, 1998).

Thayer, A., "Drug, Biotech Firms Strong in the First Half," *Chem. Eng. News* **75**(34), 18–19 (1997a).

Thayer, A. M., "Market Investor Attitudes Challenge Developers of Biopharmaceuticals," *Chem. Eng. News* **74**(33), 13–22 (1996).

Thayer, A. M., "Biopharmaceuticals Overcoming Market Hurdles," *Chem. Eng. News* **69**(8), 27–48 (1991).

US Department of Commerce, Technology Administration, Bureau of Industry and Security, *A Survey of the Use of Biotechnology in U.S. Industry*, Washington, DC, 2003, pp. xi–xiv, 1–3, www.technology.gov/reports, (Oct. 2003).

Watson, J. D., M. Gilman, J. Witkowski, and M. Zoller, *Recombinant DNA*, 2nd ed., Scientific American Books, W. H. Freeman, New York, 1992, pp. 1–11, 20–48, 50–54, 63–77, 135–138, 236–237, 273–285, 460–461, 521–523, 561–563, 590–598, 603–618.

Weintraub, A., "Voices of Innovation: Amgen's Giant," *Business Week* **3951**, 22 (2005).

Wheelwright, *Protein Purification*, Hansen Publishers, Munich, 1991.

Yanchinski, S., Canadian Health Agency OK'S Monsanto's Genetically Engineered Potatoes," *Genet. Eng. News* **17**(4), 24 (1996).

CHAPTER TWO
HOMEWORK PROBLEMS

2.1. Explain how Type II restriction enzymes enabled the development of the new biotechnology.

2.2. How does the new biotechnology differ from the biotechnology that came before it? Why is it important?

2.3. Plot the growth of revenues of the new biotechnology industry. What was the compounded rate of growth? Express your answer in percent (%) per year.

2.4. The plasmid pBR322, is used to construct a vector for growth hormone. The *E. coli* cells, after transformation with pBR322, are placed in a growth medium containing ampicillin. Despite best efforts on the researcher's part, the cells do not grow. Provide an explanation.

2.5. What are hybridomas? Why are they important to the biotechnology industry? How is the term "monoclonal" derived?

2.6. Phase III trials are carried out for a new protein biopharmaceutical. Several patients became ill and treatment is immediately stopped, even though Phase I and Phase II trials were successful. Explain why adverse effects may be observed during Phase III trials, even though Phase I and II trials were successful.

2.7. Describe some of the regulatory steps that a new biopharmaceutical will face as it is brought to market. How do regulations differ for drugs as compared to biologics?

2.8. A new biotechnology product is to be introduced to the market. However, the company that is fielding the new product decides to save $100,000 by *not* hiring two more process scientists and an engineer, and rather letting the existing staff to set up for production of clinical lots of the pharmaceutical. This slows down product development by 6 months. Sales of the product are expected to generate revenues of $10,000 the first month, $50,000 the second, and $100,000 the third before doubling in sales every month thereafter. What is one possible impact of this decision on revenues? On the company? Is this a wise decision?

2.9. What are genomics? What are potential impacts of the sequencing of the human genome? Of the genomes of yeast or bacteria?

2.10. The experiment described in problem 2.4 is repeated. This time, the cells grow, and some product is detected by centrifuging down the cells, and analyzing the cell's intracellular contents. One analysis requires 50×10^{12} cells. If the doubling time of the cells is 40 min, and the number of transformed cells initially present is 100 cells/mL, how long will 50 mL of the broth need to be incubated to obtain enough cells to do an analysis?

BIOPRODUCTS AND BIOFUELS

INTRODUCTION

New biotechnology was initially driven by the quest to generate pharmaceutical products not obtainable by other means. This quest required development of "omics" including genomics, proteomics, and metabolomics. The products of the research were biologically inspired and biologically manufactured pharmaceuticals. Hence these were given the name-biopharmaceuticals. The new biotechnology has resulted in efficient and cost-effective techniques of gene sequencing, metabolic profiling and pathway engineering of microorganisms that produce other economically important products. Consequently, foods, biochemicals, and agriculture are benefiting from the tools of the new biotechnology.

Biofuels, industrial sugars, antibiotics, sweeteners, amino acids used as flavor enhancers, and enzymes for detergents are examples of bioproducts already produced prior to the discovery of restriction enzymes and genetic engineering. The ability to modify or direct pathways that lead to bioproducts enables biocatalytic routes that have a smaller environmental footprint and result in enhanced products and more efficient processes. This chapter presents examples of how old meets new biotechnology to create old products in new ways. These products will impact society through the food that people eat, renewable low carbon biofuels that they use, the environment in which they live, and production of drugs that fight infections and other diseases.

BIOCATALYSIS AND THE GROWTH OF INDUSTRIAL ENZYMES

Catalysis in biotechnology has traditionally been associated with enzymes. At first glance the enzyme market would appear to be modest but growing. The value of enzymes sold doubled from $650 million (worldwide) in 1996 to $1.4 billion in the United States alone in 2004. About 40% of the enzymes are used in detergents, 25% in starch conversion, and 19% in the dairy (cheese) industry. Most enzymes are employed as catalysts in hydrolysis of starches (amylases), protein (proteases), and milk (chymosin and related proteases). Examples of how the new biotechnology has affected the enzyme business is given by the production of the milk-clotting enzyme chymosin through a recombinant yeast and the improvement of cellulase enzyme productivity in fungal fermentations.

Recombinant chymosin is in commercial use in the dairy industry, where it supplants an enzyme previously derived from calf stomachs. Cellulase enzyme systems have been used in the clarification of citrus juices, and to impart a "stone-washed" appearance to cotton fabrics by limited hydrolysis of the cellulose-containing cotton fabric (Tyndall 1991). More recently, the development of cellulases, microorganisms capable of fermenting glucose and xylose to ethanol, and microbial systems that generate both hydrolytic enzymes and ethanol (i.e., consolidated bioprocessing) have gained renewed emphasis (Wyman et al. 2005a,b; Lynd et al. 2008; Mosier et al. 2005a). Schoemaker et al. (2003) point out that broad application of biocatalysis has yet to develop. Advances in "genomics, directed evolution, gene and genome shuffling, and the exploration of earth's biodiversity aided by bioinformatics and high throughput screening," facilitate development of enzymes for bioprocess and chemical catalysis.

Technology for cost-effectively converting cellulosic wastes to fermentable sugars will be enabled by application of cellulase enzymes and cellulolytic microorganisms. A large market for these enzymes and/or microorganisms that combine hydrolytic and fermentative functions will develop as biotechnology is able to reduce costs. In the interim, a number of smaller, specialty niches will make up the demand for enzymes, although the impact of such products in processing bulk agricultural commodities into value-added products should not be overlooked.

Enzymes for the hydrolysis of cellulosic plant biomass is a major and growing market that depends on successful development of other technologies needed for cellulose bioprocessing. These include cellulose pretreatments, fermentation of both pentoses and hexoses to ethanol, and energy-efficient processing and purification of the fermentation products (Mosier et al. 2005a; Sedlak and Ho 2004; Gulati et al. 1996; Lynd et al. 2008).

The growth of biofuels has resulted in a new area of growth for biotechnology for the development of advanced enzymes and microorganisms. The United States and Brazil are the world's largest producers of fuel ethanol, with outputs of close to 7 and 5 billion gallons by 2008. Brazil used about 5 billion gallons of ethanol in 2007/2008 and produced sufficient quantities of ethanol from sugarcane to satisfy its own demand, and export 0.43 billion gallons to the United States (2006). In Brazil, the price of ethanol is tied to the price of gasoline (EIA 2007; RFA 2007). Annual gross earnings from sugar and ethanol in Brazil was $20 billion in 2007/2008, with 54% derived from ethanol (Unica 2009).

Transitioning from sugar and grain crops to nonfood feedstocks (sawdust, wood, straw, etc.) for producing biofuels in an economical manner requires improvements in processing technology. Pretreatment and hydrolysis converts the cellulose and hemicellulose portions of inedible plant material into sugars (Figure 3.1). Pretreatment increases the accessibility of the cellulose to enzymes or microbes

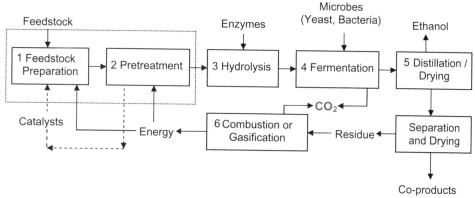

Figure 3.1.   Unit operations of a biorefinery. Biorefinery as represented here is viewed as being energy self-sufficient through combustion or gasification of residual lignocellulose. If only corn grain were processed, the remaining solids that are high in protein are recovered as coproducts and sold as animal feed, which has a higher value than use as a boiler fuel. If pretreatment uses added chemical catalysts (acid or base) these must be recycled. CO_2 is recycled into plant matter through production agriculture. [Adapted from Eggeman and Elander (2005).]

that hydrolyze it into sugars or ferment it to ethanol. Enzymes are required for the conversion of cellulose and hemicelluloses into monomeric sugars, and recombinant yeast or bacteria are needed for subsequent fermentation of both five-carbon and six-carbon sugars into ethanol. The growth of biofuels has resulted in a new growth of biotechnology tools for the development of advanced enzymes and microorganisms.

Figure 3.1 gives a schematic representation of the key steps in a sequence of pretreatment, hydrolysis, and fermentation that make up a biorefinery for converting both starch and lignocelluloses to ethanol.

Pretreatment of biomass renders cell wall structure accessible to enzymes, and is necessary to achieve maximal monosaccharide and fermentation product yields. As fundamental knowledge in systems biology enables modifications in cell wall structure to enable more effective bioprocessing, the need for pretreatment will decrease (Chapple et al. 2007). The graphic of Fig. 3.2 [from Houghton et al. (2006)] illustrates how pretreatment opens up the structure. This enables enzymes to access the structural carbohydrates so that rapid and extensive hydrolysis of plant cell wall components to sugars occurs. When fermentative microorganisms secrete the necessary hydrolytic enzymes and directly convert hydrolysis products to ethanol, hydrolysis and fermentation is said to be consolidated (Shaw et al. 2008; Lynd et al. 2005). When hydrolysis and fermentation occur through the same microorganism consolidated bioprocessing occurs (Mascoma 2009).

Glucose Isomerase Catalyzed the Birth of a New Process for Sugar Production from Corn

Glucose isomerase is an example of an enzyme that catalyzed the development of several new industries by enabling isomerization reactions for converting glucose

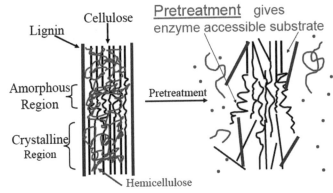

Figure 3.2. Schematic of pretreatment disrupting physical structure of biomass. [in Houghton et al. (2006); Fig. 1 of Mosier et al. (2005a).]

(from starch) into fructose and xylose from cellulosic biomass into xylulose. Its first impact was in the sweetner industry. After being introduced in 1967 (see timeline of Table 1.1 in Chapter 1), fructose-containing corn syrups grew from 17 billion lb/year in 1991, 21 billion lb in 1997 and 24 billion lb in 2003 and 2005 (Lastick and Spencer 1991; USDA Economics Research Service 2009).

Glucose isomerase was initially isolated and identified in 1953 as a xylose isomerase (Hochester and Watson 1953). Four years later the glucose-isomerizing capability of xylose isomerase from *Pseudomonas hydrophilia* was discovered although arsenate was required to enhance the reaction. This enzyme was thus impractical for food production (Marshall and Kooi 1957). Commercial prospects improved in 1961, when it was found that not all glucose isomerases required arsenate, and that at least one of the arsenate-requiring enzymes was in fact a glucose phosphate isomerase that converted glucose-6-phosphate to fructose-6-phosphate. While *Escherichia intermedia* had glucose phosphate isomerase (Natake 1966, 1968; Natake and Yoshimura 1964), several other organisms (*Lactobacillus brevis*, *Aerobacter cloacae*, and *Streptomyces phaeochromogenus*) yielded isomerases that did not require arsenate (Yamanaka 1963a; Tsumura and Sato 1965a,b, 1970).

Glucose and xylose isomerase activities were proposed to be due to the same enzyme (Yamanaka 1963b, 1968). Evaluation of the activities of a highly purified crystallized glucose isomerase from *Bacillus coagulans* was consistent with this hypothesis (Danno 1970b). It has been suggested that two enzymes, xylose isomerase (which requires xylose as an inducer, but does not need arsenate for activity), and glucose phosphate isomerase (which does not require xylose as an inducer, but needs arsenate for activity), were copurified in the early work, hence giving the observed properties (Tsumura et al. 1967).

The industry is now based on an enzyme known as *glucose isomerase* that catalyzes the conversion of glucose (dextrose) to its chemical isomer (fructose). The equilibrium mixture of glucose and fructose (58% glucose, 41% fructose) has a sweetness approaching that of sugar derived from sucrose (cane sugar). Development of large scale separation technology for partially separating glucose from fructose to give a product with 55% fructose, enabled this sugar to be used in consumer products ranging from soft drinks to donuts in place of sucrose, since 55% fructose has the same sweetness. Since the glucose was derived from corn, which at the time was a less expensive agricultural commodity than sucrose from sugarcane grown in the United States, high-fructose corn syrups displaced hydrolyzed sucrose (invert) in the majority of high-volume applications. In 2007 to 2009, volatility in corn prices have complicated the economics of HFCS, but the strong position of this corn derived sugar as an industrial sugar continues.

Identification of a Thermally Stable Glucose Isomerase and an Inexpensive Inducer Was Needed for an Industrial Process

The discovery of glucose isomerase from *Streptomyces phaeochromogenus*, which was thermally stable at 90 °C in the presence of substrate and Co^{2+} (Tsumura and Sato 1965a,b; Tsumura et al. 1967), was important in setting the stage for commercialization. A thermally stable enzyme reduces cost by increasing its useful lifetime, and therefore its overall productivity. The last major barrier to large-scale production of the enzyme was replacement of the expensive xylose media on which *S. phaeochromogenus* had to be grown to obtain glucose (xylose) isomerase enzyme, with a less expensive carbohydrate source. This turned out to be xylan containing wheat bran, corn bran, and corn hull (Takasaki 1966; Takasaki et al. 1969). Later it was found that D-xylose could induce glucose isomerase activity in *Bacillus coagulans* cells previously grown on glucose (Danno 1970a). The stage was set for a new biocatalyst to transform the corn wet-milling industry.

A number of years later the discovery that this enzyme could be used to transform xylose into xylulose, which, in turn, is fermentable to ethanol by yeast helped to set the stage for growth in technologies for producing cellulosic ethanol (Gong et al. 1981). This observation opened the way to research on the engineering of yeast and *E. coli* to coferment glucose and xylose to ethanol so that almost all of the structural carbohydrates in corn stover (Schwietzke et al. 2008) may be converted to ethanol (Sedlak and Ho 2004; Mohagheghi et al. 2004; Hahn-Hagerdal et al. 2007; Kim et al. 2007). In principle, pentose-fermenting microorganisms enable ethanol yields from biomass to be increased by 40% over the case where only glucose from cellulose is fermented to ethanol.

The Demand for High-Fructose Corn Syrup (HFCS) Resulted in Large-Scale Use of Immobilized Enzymes and Liquid Chromatography

In 1965–1966, Clinton Corn Processing Company of Iowa (at that time a division of Standard Brands, Inc.) entered into an agreement with the Japanese government to develop the glucose isomerase technology for commercial use in the United States (Tsumura et al. 1967; Lloyd and Horvath 1985). The glucose isomerase was really a xylose isomerase derived from *Streptomyces rubiginosus* and had the properties of thermal stability and high activity, which made it amenable for use at industrially relevant conditions in batch reactions. The first enzymatically produced fructose corn syrup was shipped by Clinton Corn in 1967. A. E. Staley licensed the technology from Clinton Corn and entered the market.

The basic patent coverage for use of xylose isomerase to convert glucose to fructose was lost in 1975 as the result of a civil action suit between CPC International and Standard Brands. This enabled development of alternate processes. By 1978, the estimated US production volume was 3.5 billion lb and consisted mostly of syrups containing 42% D-fructose sold at a 71% solids level (Antrim et al. 1979). The introduction of large-scale liquid chromatographic purification of the fructose through simulated moving-bed (SMB) technology enabled production of a 55% fructose that could be used in soft drinks in place of invert from sucrose.

The growth of the industry both promoted and followed development of large-scale technologies for carrying out conversion in reactors packed with immobilized glucose isomerase, and large-scale liquid chromatographic separations of the resulting sugars. Immobilized enzyme technology took advantage of the heat stability of the glucose isomerase by fixing the enzyme within, or on the surface of, solid particles. These particles, when packed in large vessels, enabled the enzyme, which was otherwise water-soluble, to be held immobile while the glucose solution (at concentrations of ≤40% solids) was passed through the reactor. This made it possible to continuously process an aqueous glucose solution by passing it through a fixed-bed, biocatalytic reactor to convert it to fructose (Fig. 3.1). However, equilibrium and reaction kinetics limited the conversion to 42% fructose.

Large-scale liquid chromatography enabled purification of 42% fructose to the 55% required for sweetness equivalent to invert (a hydrolyzed form of sucrose). Liquid chromatographic separation technology (Sorbex) for continuous sorptive separation of xylenes in a simulated moving bed was adapted to glucose/fructose separation by UOP to give the SAREX process for enriching the fructose content. Part of the effluent from the immobilized enzyme reactor is diverted to the chromatography column. The glucose-rich stream from the chromatography step can be recycled to the immobilized bed reactor to be converted back to 42% fructose. The fructose-rich is mixed with the remaining 42% fructose stream to give the 55% product.

The combination of an enzyme reaction with separation made the 55% HFCS product economically and technically feasible in about 1975. Subsequently, HFCS 42% and 55% syrups gradually displaced sucrose for commercially produced beverages and baked products.

Figure 3.1 gives a schematic representation of the process. The immobilized reactor is filled with pellets or particles in which the glucose isomerase is either entrapped or attached. Glucose solution contains ions required for enzymatic activity and about 3% residual soluble starch fragments (maltodextrins). The glucose feed is passed through the reactor, which is maintained at 50–60 °C. The immobilized enzyme converts the glucose to fructose. The product leaving the reactor contains about 42% fructose as well as the other ions and maltodextrins that do not react. The effluent from the reactor enters a large liquid chromatography column, packed with an adsorbent that selectively, but weakly, retains the fructose over glucose. The 42% fructose feed is introduced to different parts of the column in a time-varying manner, through a special valve, in a manner that simulates a moving bed, so that the use of adsorbent is maximized (Ladisch 2001). Sections of the bed that are rich in glucose and fructose are formed and maintained at steady-state operation. This system allows a feed of 42% fructose to be continuously separated and to upgrade the high-fructose corn syrups from 42% fructose to 55% fructose.

The glucose that is separated out is returned to the reactor so that it will be converted to fructose.

Rapid Growth of HFCS Market Share Was Enabled by Large-Scale Liquid Chromatography and Propelled by Record-High Sugar Prices

The "Pepsi taste challenge" showed consumers preferred soft drinks that were made from 55% HFCS compared to a competing product made from sucrose. By 1984, production of the 42% and 55% syrups exceeded 8 billion lb annually, and accounted for 30% of all nutritive sweeteners in the United States (Lloyd and Horvath 1985). The 55% fructose was quickly displacing industrial sugars previously derived from sucrose and accounted for over 50% of the HFCS shipped in the United States in 1984. By 1990 the world market for HFCS had grown to about 17 billion lb/year. The US production of high-fructose corn syrup 42% was 8.57 billion lb, while high-fructose corn syrup 55% was 12.5 billion lb for a total of 21 billion lb in 1997 and 24 billion pounds in 2002 through 2005 (Corn Refiners Association 1997, 2003, 2006). The value of the xylose (glucose) isomerase enzyme used in these processes is estimated at between $15 million and $70 million/year (Petsko 1988; Lastick and Spencer 1991). Most of the glucose isomerase was produced for captive or contract use.

The rapid adaption of the combined enzyme reactor/product separation technology (see Fig. 3.3) in the wet-milling industry coincided with dramatic, but short-lived, increases in world sugar (sucrose) prices. The demand for industrial sugars exceeded the supply. This shortfall not only raised prices, but also provided a window of opportunity for a competing product, HFCS, to enter the market. The properties of HFCS were the same as dissolved invert (from sucrose) in applications where sucrose would otherwise be used. The cost was less. By positioning the price of the HFCS slightly below that of sucrose (invert), HFCS producers were able to win and retain market share.

When the sucrose prices increased to extraordinary levels, the price of HFCS was about double its cost of production. Consequently, investments in building new plants were quickly paid off, making further rounds of expansion possible, and further entrenching the competing product in the market (Antrim 1979). This is not typical of other biotechnology products, where higher prices are based on a unique product, protected by a patent.

Although the annual per capita (US) consumption is 130 lb of sugars, the price spike behavior has moderated. The acceptance of starch-derived and low-calorie sweeteners being widely accepted as sugar substitutes was a factor. The introduction of aspartame, in 1981, and several other low-calorie, high-intensity sweeteners coincided with a significant increase in the US per capita consumption of low-calorie sweeteners, with low-calorie sweetener use being equivalent to 20 lb per capita in sugar sweetness equivalent (SSE). In 1980, this was about 6% of the total sugar consumption compared to 13% in 1988. Starch-based sweeteners, and to a less extent low-calorie sweeteners, are poised to take advantage of sugar shortfalls and high prices, should they recur (Barry et al. 1990).

The history of sugar prices and development of HFCS illustrate the complex interactions between technology, commodity prices, and market demand for a product derived from a renewable resource (see Fig. 3.4). It has been suggested that

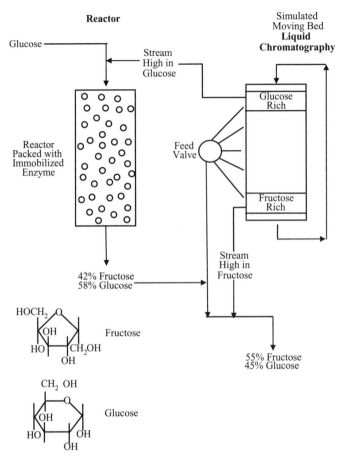

Figure 3.3. Schematic diagram of combined immobilized enzyme reactor and simulated moving-bed chromatography for producing 55% high-fructose corn syrup (HFCS). The composition shown in the figure is normalized for the total glucose and fructose content, and does not reflect other dissolved solids.

the impact of weather on commodity prices is now moderated by the ability of weather forecasters to predict El Niño weather patterns and their effects on rainfall in sugarcane-producing regions. The history of sugar production and prices may serve as a case study for the coming development of the fuel ethanol industry, with increasing prices and price volatility of oil motivating cyclical patterns of expansion in fuel ethanol production.

Biocatalysts Are Used in Fine-Chemical Manufacture

Although the total US enzyme market is about $650 million/year, only about 5% is for biocatalysts used in fine-chemical manufacture. The main biocatalysts are hydantoinase, β-lactamases, acylases/esterases, and steroid-transforming enzymes (Polastro et al. 1989). The discovery and development of catalytic antibodies may expand the role of biocatalysts, since this type of biocatalyst combines specificity

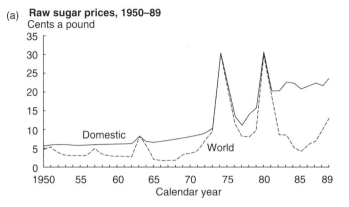

(a) **Raw sugar prices, 1950–89**
Cents a pound

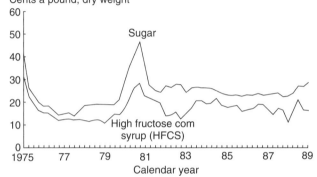

(b) **Wholesale HFCS and sugar prices, by quarter, 1975–89**
Cents a pound, dry weight

Source: U.S. Dept. Agr., Economic research service

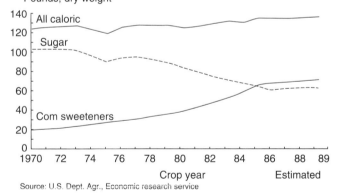

(c) **U.S. per capita consumption of sugar and sweeteners**
Pounds, dry weight

Source: U.S. Dept. Agr., Economic research service

Figure 3.4. Trends in sugar prices and consumption: (a) price spikes in sugar prices, 1950–1989 (Barry et al. 1990, Fig. 5, p. 29); (b) wholesale fructose prices are strategically positioned below world sugar prices (Barry et al. 1990, Fig. 4, p. 27); (c) market share of HFCS increased dramatically between 1972 and 1985, followed by a leveling off (Barry et al. 1990, Fig. 2, p. 20).

of binding of an antibody with the ability to chemically transform a target molecule. Hybridoma technology allows generation of antibodies or haptens to stabilized transition-state analogs that have catalytic side chains in an antibody-combining site. Hybridomas also facilitate synthesis of gram quantities of the catalysts with very high specificity. The antibodies are able to catalyze a large number of different reactions, including at least one type (Diels–Alder) for which an enzyme has yet to be isolated (Schultz et al. 1990).

Enzymes in low-water environments may carry out selective transesterification reactions. Examples in the literature include polymerization of phenols in an organic solvent, and the formation of the biodegradable polymer sucrose polyester (Ryu et al. 1989; Patil et al. 1991). This is currently a developing field, but holds promise for applications in organic synthesis, since some enzymes have been shown to be active in organic solvents. Fundamental studies have shown the enzyme subtilisin Carlsberg, which ordinarily hydrolyzes proteins, will carry out transesterification of N-acetyl-l-phenylalanine 2-chloroethyl ester (NAPCE) with 1-propanol, when the reactants and enzyme are mixed into tetrahydrofuran containing a small amount of water (Affleck et al. 1992). This research, and work that has preceded it, gives a first indication of the potential of proteins, i.e., enzymes, as catalysts in organic synthesis.

GROWTH OF RENEWABLE RESOURCES AS A SOURCE OF SPECIALTY PRODUCTS AND INDUSTRIAL CHEMICALS

There are already large industries associated with corn processing and food manufacture that utilize enzymes and microbial fermentation on an extremely large scale. In these cases, production and substrate costs are 70% of the total product costs, and cost-efficient engineering becomes paramount. The development of the industry that produces fuel-grade ethanol, used as a nonleaded octane booster and environmentally acceptable fuel oxygenate, had its roots in the development of high-fructose corn syrup and the oil price shocks of 1973–1980 (see timeline in Table 1.1, Chapter 1). Continuing growth of biofuels is attributable to environmental regulations relating to clean air and transportation fuels, continued government subsidy, and tightening of supplies of oil and refined petroleum products.

Success in submerged fermentation of microbes that produced the intracellular enzyme glucose isomerase was a key step in obtaining a cost-effective enzyme. When combined with development of processes for utilizing the enzyme at a commercial scale and an economic separation technology, a new product resulted. The enzyme glucose isomerase also set the stage for the US fuel alcohol industry, since the major HFCS-producing companies had the infrastructure and large-scale processing know how to establish it. The overall US fermentation ethanol industry (both wet and dry grind processes) grew from about 100 million gal in 1980 to 7.0 billion gal in 2008.

Future potential exists for production of consumer products in an environmentally acceptable manner from renewable and agricultural commodities. These products include plastics, paints, finishes, adsorbent, and biomaterials. As the impact of biologically based production grows, so will the need for engineering. Bioprocess engineering—the discipline that deals with development, design, and operation of biologically based processes—will have a major role in the development of bioprocesses for transforming renewable resources into industrial chemicals. Numerous possibilities exist as shown in Figs. 3.3 and 3.4 (USDA 1993).

A Wide Range of Technologies Are Needed to Reduce Costs for Converting Cellulosic Substrates to Value-Added Bioproducts and Biofuels

The future growth of renewable resources as a source of value-added biochemicals and oxygenated molecules will be driven by technologies rather than markets, as long as oil prices remain in the range of $15–$20/barrel. However, in 2005, the cost of oil jumped to $67/barrel. This enhanced interest in renewables. For most of 2006 and 2007, oil prices remained above $70/barrel, with short spikes above $135/barrel. In 2008, oil prices approached $140/barrel but then quickly retreated back to $40 to $50/barrel. The market was extremely volatile. Nonetheless, fermentation processes or enzyme transformations that achieved low carbon fuels at a reduced environmental impact become attractive. An example is the production of fumaric acid by fermentation for fumaric acid used as a food additive, compared to fumaric acid by chemical synthesis for other purposes (Cao et al. 1997). Some fermentation processes might offer attractive economics if there were a sustained increase in oil prices. Temporary spikes in oil prices could also offer windows of opportunity, although long-term cost advantages would be required to sustain growth of a renewable-resource-based industry.

The attractiveness of renewable resources lies in their ability to sequester and/or recycle CO_2 and their potential to be processed into useful products in an environmentally friendly manner. This is illustrated by corn, which is used for food, animal feed, and specialty starches, and as a source of sugars (principally glucose) that can be fermented or converted to numerous value-added products (See Figs. 3.5 and 3.6). Fats and oils derived from plants and animals provide examples of another type of renewable resource. These are the basis of industries that already produce a number of products that range from soaps to lubricants and biodiesel.

A thorough analysis of industrial uses of agricultural materials shows their potential (USDA 1993). Current products range from fuel ethanol to starch adhesives. Emerging technologies in ethanol production have decreased the cost of production. The total energy required to process one gallon of ethanol with an energy value of 76,000 Btu/gal (British thermal units per gallon) was as high as 120,000 Btu when the fuel ethanol industry started in 1977 (Hohmann and Rendleman 1986). Even when an energy credit of 32,000 Btu was assigned to the coproducts of ethanol production (corn oil, gluten feed, gluten meal, and CO_2), there was still a net energy loss of 12,000 Btu/gal. Research, the rapid implementation of the resulting process improvements on an enormous scale by some of the ethanol producers, and process integration decreased energy costs and improved energy efficiency (USDA 1993). By 1993, the total energy requirement for an average efficiency ethanol plant was 75,811 Btu/gal. With a coproduct energy credit of 24,950 Btu, a net energy *gain* of 25,139 Btu/gal was achieved for modern, large-scale corn-to-ethanol routes (USDA 1993). This was improved further in the subsequent 10 years. By 2001 the energy output:input ratio increased to 1.34 (Shapouri et al. 2002).

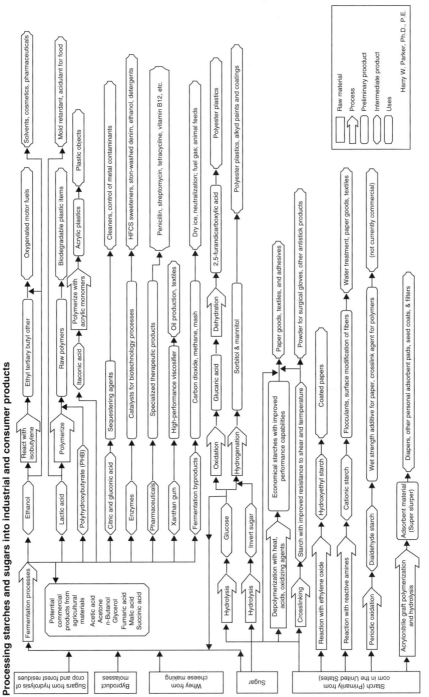

Figure 3-5. Chart showing industrial chemicals derived from starches and sugars [from USDA (1993), Fig. 4, p. 13].

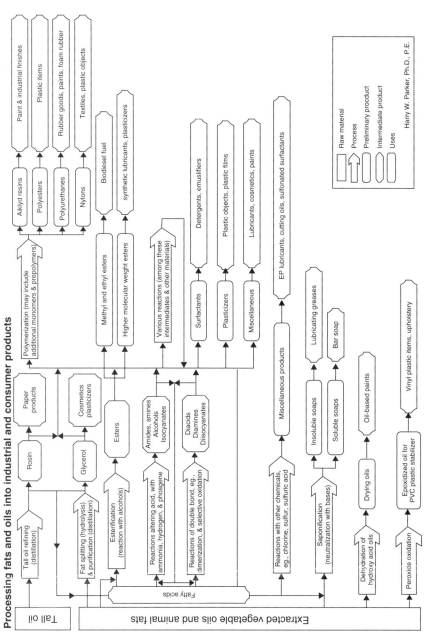

Figure 3.6. Chart showing products derived from renewable sources of fats and oils [from USDA (1993), Fig. 5, p. 16].

Further improvements and decreased costs were anticipated because of better enzymes, bioreactor designs, and membranes for enhanced recovery of high-value coproducts. Further significant reductions in cost, however, require lower feedstock prices. Thus, for example, the conversion of corn fiber, a cellulose containing coproduct of ethanol production, to fermentable sugars could reduce costs by 3.0–7.5¢/gal.[1] Further research and analysis has shown that a fiber conversion process, integrated into a corn-to-ethanol plant could produce ethanol at an incremental cost of $1/gal or less (Mosier et al. 2005d).

Cellulosic residues have the potential to be a less expensive substrate, but their composition and physical structure are different from those of starch. The cellulose structure is crystalline and resistant to hydrolysis. Consequently, a pretreatment is necessary to soften the cellulose prior to subjecting it to hydrolysis. Less expensive enzymes (that hydrolyze cellulose to glucose) are also needed to achieve a cost-effective hydrolysis step. Pentoses and hexoses occur in approximately 4:5 ratio, in cellulosic materials, compared to a 1:10–1:15 ratio in corn grain. Conversion of sugars from the pentosan and hexosan fractions of cellulosic materials will require microorganisms that ferment both pentoses and hexoses, to a single product (Hohman and Rendleman 1986). A number of effective glucose and xylose fermenting microorganisms were developed in the subsequent 20 years, with pentose fermenting yeast and bacteria ceasing to become a gating item by 2009 (Klembara 2009).

Progress has been made in pretreatment and enzyme hydrolysis (Mosier et al. 2005a,b; Weil 1994; Kohlmann et al. 1995; Ladisch et al. 1983), as well as obtaining genetically engineered microorganisms that convert both pentoses and hexoses to ethanol (Lynd et al. 1991; Sedlak and Ho, 2004). However, an economically viable process had not yet been achieved in 1996 (Gulati et al. 1996). Major improvements were still needed for all three technologies. By 2005, advances had begun to occur, including pentose-fermenting yeasts, cost-effective pretreatments, and less expensive enzymes (Ho et al. 1998; Mosier et al. 2005a). The integration of cellulose technologies into ethanol plants was anticipated in the mid-1980s (Hohnman and Rendleman 1986), and is now beginning to occur (Klembara 2009). The emergence of a cellulose-based industry will open the way for production of many other types of oxygenated chemicals and lead to the growth of a fermentation based, renewable resources industry. However, development of this industry continues to require government policies and mandates much as was the case for other biotechnologies and in particular, HFCS.

The type of processing plants that produce HFCS are also the major source of fermentation ethanol, large volumes of amino and carboxylic acids, and high-volume specialty products. These plants are well positioned to implement improvements in cellulose conversion as another source of fermentable sugars, and then utilizing these sugars in an optimal manner for production of a diverse product mix. The future of the industry is likely to evolve from being based only on hexoses, to an industry that utilizes and transforms both hexoses and pentoses to value-added products. Bioprocess engineering will play a major role in implementing technology

---

[1] These cost comparisons were made relative to the cost of producing fermentation ethanol from corn in 1992, which was at $1.25/gal, down from $1.35 to $1.45/gal in 1980.

for this transition. The current and massive restructuring and consolidation of the high-volume bioproducts sector [see timeline from 1992 to 1998 in Table 1.1, Chapter 1 (Ladisch 2002) and the BIO Website] and the financial industry (starting in 2007) will shape the business government alliances needed for the evolution of a biobased industry.

Renewable Resources Are a Source of Natural Plant Chemicals

Chemicals derived from plant products are referred to as *phytochemicals*. These chemicals are formed as either primary or secondary metabolites. Balandrin et al. (1985) proposed that chemicals from primary metabolism could be classified as high-volume low-value bulk chemicals with prices below $2/lb. Examples are vegetable oils, fatty acids (for making soaps or detergents), and carbohydrates (sucrose, starch, pectin, or cellulose). Secondary metabolites are classified as high-value, with prices ranging from $100/lb to $5000/g. Examples are pyrethins and rotenone (used as pecticides); steroids and alkaloids such as *Digitalis* glycosides, anticancer alkaloids, and scopolamine. Secondary metabolites in plants often have an ecological role as pollination attractants or for mounting chemical defenses against microorganisms and insects. Secondary natural products often consist of complex structures with many chiral centers that are important to the biological activity of these molecules. Plant-derived proteins can also be classified as high-value products. For example, papain is a proteolytic enzyme used as a meat tenderizer. Bromelain (from pineapple) has applications in milk clotting. Malt extract from barley has amylolytic enzymes used in beer brewing.

The summary of valuable products derived from plant materials by Balandrin et al. (1985) is as valid to day as it was in 1985. Their analysis shows that the recovery of bioactive compounds, classified as secondary metabolites, will continue to be an important source of medicines, pesticides, and specialty chemicals. The naturally occurring compounds could be numerous. These include pesticides (pyrethrum from flowers, rotenone from roots, and nicotine), growth inhibitors of soil microorganisms (allelochemicals) plant growth regulators (brassinolide promotes growth at 1 gram per 5 acres), insect growth regulators, and medicinal (anticancer and analgesic drugs).

An example is given by the insecticidal properties of the neem tree (Stone 1992). Neem seeds have chemicals that ward off more than 200 species of insects, while exhibiting less toxicity to humans than synthetic pesticides, and having little effect on predators of the insects. Neem seeds were valued at $300/ton in 1992. However, a lower price is needed if the extracts are to find widespread use as insecticides. The economic potential for naturally derived insecticides is large considering that about $2 billion of synthetic insecticides are sold annually in the United States.

Another class of plant-derived compounds is known as *saponins*. Saponins have an aglycone consisting of a titerpene, a steroid, or a steroidal alkaloid attached to a single- or multisugar chain containing ≤11 monosaccharide units. One saponin derivative (Abrusoside E) is 150 times sweeter than sugar. Another group of saponins from *Yucca schidigera*, a medicinal plant used by Native North Americans, have antifungal activity and could be used as preservatives for processed foods. The plant itself is reported to be recognized by the FDA as safe for human use. A saponin from a member of the lily family, *Ornithogalum saundersiae*, has been found to exhibit little toxicity with respect to human cells, but is "remarkably toxic to malignant tumor cells" (Rouhi 1995). Saponins from the South American tree *Q. saponaria*, make animal vaccines more effective. Extracts from two types of African trees have been used to control schistosomiasis, by eliminating host snails that harbor the disease.

Genetic engineering of bacteria or yeast to produce complex secondary metabolites remains a challenge, since these are formed via the concerted action of many enzymes (i.e., many gene products). The culturing of plant cells, or plant tissues, to produce secondary metabolites could be used to produce some of the more valuable and bioactive compounds, and has been proposed for obtaining the chemotherapy agent, Taxol (paclitaxel). It was estimated (in 1985) that a yield of 1 g of product per liter of culture for compounds exceeding $1/g in value is needed to obtain an economically viable process. Such an estimate may be optimistic. Even if successful, such cultures would still owe their start to the plants from which the compounds were identified, and the cells derived.

The advances in methods for genetic engineering of plants, and propagating the transformed species could someday make use of knowledge on the structure and function of plant metabolites. Numerous projects for mapping plant genomes have been completed or are working. The combination of these factors is leading to development of other types of transgenic, pest-resistant species, and perhaps the development of plant species grown to generate pharmaceuticals that are difficult or too costly to obtain by other means. These possibilities give strong rationale for preserving the biodiversity of Earth's plant life, estimated between 250,000 and 750,000 species (Balandrin et al. 1985). Plants are also being modified to integrate enzymes into plant tissue. Enzymes are packaged into the plant cell walls so that the structural carbohydrates may be converted to fermentable sugars after the plants are harvested and enzymes are activated in a bioprocessing facility (Pollack 2006; Agrivida 2009; Syngenta 2009). This represents another form of consolidated bioprocessing.

Bioseparations Are Important to the Extraction, Recovery, and Purification of Plant-Derived Products

The recovery and processing of plant-derived molecules and the processing of renewable resources can be classified by the type of product. Primary metabolites are generally biopolymers. Natural rubber, condensed tannins and lignin, and high-molecular-weight polysaccharides, pectin, starch, cellulose, hemicellulose, and gums, are included in this category. Their processing into high-volume, relatively low-value products is subject to economies of scale (bigger is better), cost constraints (large volumes = low price), a physical structure that is resistant to processing, and/or low solubility in water. The separation of the plant material into value-added components may require a combination of physical treatments combined with a change in the material's structure. For example, the processing of cellulose (in the cell walls of wood or agricultural residues) requires that the cellulose and hemicellulose be hydrolyzed to their monomers glucose and xylose and arabinose, respectively. This can be readily achieved only if the crystallinity of cellulose structure and its close association with lignin are disrupted and an enzyme or acid catalyst is used to depolymerize the cellulose and hemicellulose polymers into their constituent mono-saccharides. This processing, referred to as *pretreatment/hydrolysis*, must be carried out in a very inexpensive manner if it is to be practical. One approach that has proved to be successful is the pressure cooking of cellulose in water at temperatures between 160 and 210 °C and a pH between 4 and 7 (Ladisch 1989; Mosier et al. 2005b–d; Weil et al. 1994, 1997, 1998).

The recovery of secondary metabolites can be achieved by steam distillation, aqueous extraction, or extraction with an organic solvent. However, these bioproducts, whose molecular weights are below 2000 Da, may be chemically unstable or only sparingly soluble in the extractant, thereby making their processing difficult. An example is given by annonaeous acetogenins from the Indiana banana (pawpaw) (*Asimina triloba*) tree (McLaughlin 1995). These compounds have potential as chemotherapy agents against breast cancer, and are isolated from 100-g samples of natural materials. These are chromatographed over a 25 cm × 2 m-long silica gel column, in a gradient method that requires over a week to complete. The challenge is to increase the rate of preparative-scale chromatography so that samples can be isolated and tested more quickly.

BIOPROCESS ENGINEERING AND ECONOMICS

Evaluation of the engineering economics of bioprocesses is important as new products move from the research and development phase into commercial production and test marketing. Products of current interest include a wide range of biological molecules for human or animal health, monoclonal antibodies for diagnostic purposes, enzymes, biochemical intermediates, anbibiotics, amino acids, and biofuels. Engineers and scientists sometimes need to obtain a first estimate of relative costs to choose process alternatives and establish magnitude-of-order product costs. Since there is such a broad range of products with different economic criteria, a single type of cost calculation approach cannot be used for all these cases. Consequently, procedures for estimating the relative economics for each type of product on an internally consistent basis must be used.

Bioprocess economics require the engineer or scientist to be able to quickly recognize limiting technical parameters that are likely to be encountered on scaleup. Processing schemes that minimize operational difficulties and costs must be selected accordingly. Hence, process conceptualization is an important part of the evaluation process and must reflect regulatory constraints as well as transport and kinetic phenomena required to size key pieces of equipment. Examples are oxygen transfer, mixing, and heat removal in a fermentation (i.e., compressor and heat exchange costs), maintaining sterile conditions in a continuous bioreactor (productivity and regulatory issues), or predicting resolution in a chromatography column (separation costs). In some cases it is difficult to carry out calculations since physical properties (e.g., broth viscosities and heat capacities, or diffusion coefficients of biomolecules) are not well defined during the early stages of a R&D program. Hence, approaches and heuristic rules that might be used in this situation are also needed. This book addresses the basics of microbial growth, biocatalysis, and product properties so that first estimates of process boundaries may be made.

BIOSEPARATIONS AND BIOPROCESS ENGINEERING

The large-scale purification of proteins and other bioproducts is the final production step, prior to product packaging, in the manufacture of therapeutic proteins, specialty enzymes, diagnostic products, and value-added products from agriculture, including biofuels. These biochemical separation steps purify biological molecules or compounds obtained from biological sources and hence are referred to as *bioseparations* (Ladisch 2001). The essence of large-scale bioseparations is both art and science. Bioseparations often evolve from laboratory-scale techniques. These are adapted and scaled up to satisfy the need for larger amounts of extremely pure material, test quantities of the product for analysis, characterization, testing of efficacy, clinical or field trials, and finally, full-scale commercialization. The uncompromising standards for product quality, driven by commercial competition, end-use applications, and regulatory oversight, provide many challenges to the scaleup of protein purification and biochemical separations. The rigorous quality control of manufacturing practices and the complexity of the macromolecules being processed provide other practical issues that must be addressed.

The emphasis of research and development in bioseparations is currently on the purification of proteins derived from recombinant fermentations and cell culture. This is to be expected since the purification of proteins and other bioproducts is a critical and expensive part of most biotechnology based manufacturing processes, and may account for 50% or more of production costs (Ladisch 2001; National Research Council 1992). While overall production costs have been considered to be secondary to being the first to market, this perspective is changing as the price—and value—of new bioproducts is decreasing. When the volume of the products is small and the price is high, being the first to market together with attaining high product quality (in terms of purity, activity, dependability, or flexibility) are the major competitive advantages (National Research Council 1992; Wheelwright 1991; Bisbee 1993). Bioseparations are important in ensuring product quality, but manufacturing cost is likely to be secondary for these types of products. When the scale of production of new bioproducts increases from kilograms to tons, the need for cost-effective purification schemes is also increased in importance. High-volume products could range from serum (blood) proteins produced by recombinant organisms, to organic acids, enzymes, and food additives obtained from large-scale fermentations or from enzyme transformations. The need for low costs is even more pronounced for commodity bioproducts such as biofuels or industrial sugars.

A major technical challenge in the production of biopharmaceuticals is the "development of high-resolution protein purification technologies that are relatively inexpensive, are easily scaled-up and have minimal waste-disposal requirements" (National Research Council 1992). Separation processes for bioprocessing of renewable resources and agricultural products will benefit from development of "more efficient separations for recovering fermentation products, sugars, and dissolved materials from water," and in particular, lowering the cost of separating water from the product in the fermentation broth. The bioprocess engineer must therefore be familiar with principles of a wide range of bioseparations techniques, and the practice of the dominant methods that are likely to be used in the industry.

The most popular method for purifying protein biopharmaceuticals is some form of liquid chromatography, due to its resolving power. Large-scale liquid chromatography can also be used to purify high-volume bioproducts including sugars, antibiotics, vitamins, nucleosides, organic acids, and alcohols. However, chromatography is only one of the many steps needed to purify bioproducts. Recovery and purification steps preceding chromatography may include centrifugation or filtration in order to remove solids; membrane filtration or fractionation to separate large molecules from small ones; and precipitation, crystallization, extraction, and/or adsorption to concentrate the product.

Bioseparations engineering combines the disciplines of engineering, life sciences, chemistry, and medicine in order to match the unique properties of biomolecules with the most appropriate techniques for their large-scale purification. The rapid development of new methods, organisms, and molecules means that these products will be continually changing, as will the methods by which they are purified. Consequently, the design of bioseparation processes requires that the engineering principles of individual separation steps be understood, so that they can be combined into sequences of steps that result in the needed product purity at a cost that is clearly defined and calculated. This is the subject of another book (Ladisch 2001).

REFERENCES

Affleck, R., Z.-F. Zu, V. Suzaura, K. Focht, D. S. Clark, and J. S. Dordick, "Enzymatic Catalysis and Dynamics in Low-Water Environments," *Proc. Natl. Acad. Sci. USA* **89**, 1100–1104 (1992).

Agrivida, http://www.agrivida.com/, March, 2009.

Antrim, R. L., W. Colilla, and B. J. Schnyder, "Glucose Isomerase Production of High Fructose Syrups," *Appl. Biochem. Bioeng.*, Vol. **2**, L. B. Wingard, ed., Academic Press, New York, 1979, pp. 98–154.

Balandrin, M. F., J. A. Kloche, E. S. Wurtele, and W. H. Bollinger, "Natural Plant Chemicals: Sources of Industrial and Medicinal Materials," *Science* **228**, 1154–1160 (1985).

Barry, R. D., L. Angela, P. J. Buzzanell, and F. Gray, *Sugar, Background for 1990 Farm Legislation*, AGES 9006, USDA Economic Research Service, Feb. 1990, pp. 19–35.

BIO (Biotechnology Industry Organization), www.bio.org, 2005.

Bisbee, C. A., "Current Perspectives on Manufacturing and Scale-up of Biopharmaceuticals," *Genet. Eng. News*, **13**(14), 8–9 (1993).

Cao, N., J. Du, C. Chien, C. S. Gong, and G. T. Tsao, "Production of Fumaric Acid by Immobilized *Rhizopus* Using Rotary Biofilm Contactor," *Appl. Biochem. Biotechnol.* **63–65**, 387–394 (1997).

Chapple, C., M. R. Ladisch, and R. Meilan, "Loosening lignin's grip on biofuel production," *Nature Biotechnology* **25**(7), 746–748 (2007).

Committee on Bioprocess Engineering, National Research Council (NRC), *Putting Biotechnology to Work: Bioprocess Engineering*, National Academy of Sciences, Washington, DC, 1992.

Corn Refiners Association, *Shipments of Products of the Corn Refining Industry—2002* (www.corn.org) (May 15, 2003).

Corn Refiners Association, *Corn Annual*, Washington, DC, 1997.

Danno, G., *Agric. Biol. Chem.* **34**(11), 1658–1667 (1970a).

Danno, G., *Agric. Biol. Chem.* **34**(12), 1805–1814 (1970b).

Eggeman, T. and R. T. Elander, "Process and Economic Analysis of Pretreatment Technologies," *Bioresource Technol.* **96**(18), 2019–2025 (2005).

EIA (Energy Information Agency), *US Fuel Ethanol Oxygenate Production*, US Department of Energy, Washington, DC, 2007.

Gong, C.-S., L.-F. Chen, M. C. Flickinger, L.-C. Chiang, and G. T. Tsao, "Production of Ethanol from D-xylose by Using D-xylose Isomerase and Yeasts," *Appl. Environ. Microbiol.* **41**(2), 430–436 (1981).

Gulati, M., K. L. Kohlmann, M. R. Ladisch, R. Hespell, and R. J. Bothast, "Assessment of Ethanol Production Options for Corn Products," *Bioresource Technol.* **58**, 253–264 (1996).

Hahn-Hagerdal, B., K. Karhumaa, M. Jeppsson, and M. F. Gorwa-Grauslund, "Metabolic Engineering for Pentose Utilization in Saccharomyces cerevisiae," *Adv. Biochem. Eng./ Biotechnol.* **108**, 147–177 (2007).

Ho, N. W. Y., Z. Chen and A. P. Brainard, "Genetically engineered *Saechoromyes* yeast capable of effective cofermentation of glncose and xylose," *Applied and Environmental miciobilogy*, **64**(5), 1852–1859 (1998).

Hochester, R. M. and R. W. Watson, *J. Am. Chem. Soc.* **75**, 3284–3285 (1953).

Hohmann, N. and C. M. Rendleman, *Emerging Technologies in Ethanol Production*, Agriculture Information Bulletin 663, 1986.

Houghton, J., S. Weatherwax, and J. Ferrell, *Breaking the Biological Barriers to Cellulosic Ethanol: A Joint Research Agenda*, DOE SC-0095, Department of Energy, Washington, DC, 2006.

Kim, Y., L. O. Ingram, and K. T. Shanmugam, "Construction of an Escherichia coli K-12 Mutant for Homoethanologenic Fermentation of Glucose or Xylose without Foreign Genes," *Appl. Environ. Microbiol.* **73**(6), 1766–1771 (2007).

Klembara, M, DOE Peer Review Public Session: Integrated Biorefinery Projects, Westin Hotel, Outer Harbor, MD, March 19, 2009.

Kohlmann, K. L., A. Sarikaya, P. J. Westgate, J. Weil, A. Velayudhan, R. Hendrickson, and M. R. Ladisch, "Enhanced Enzyme Activities on Hydrated Lignocellulosic Substrates," in *Enzymatic Degradation of Insoluble Carbohydrates*, ACS Symp. Ser. 618, J. N. Saddler and M. H. Penner, eds., 1995, pp. 237–255.

Ladisch, M. R., "Bioprocess Engineering (Biotechnology)," *Van Nostrand Scientific Encyclopedia*, 9th ed., Vol. **1**, Glenn D. Considine, ed. (editor-in-chief), 2002, pp. 434–459.

Ladisch, M. R., *Bioseparations Engineering, Principles, Practice, and Economics*, Wiley, New York, 2001, pp. 423–429.

Ladisch, M. R., "Hydrolysis," in *Biomass Handbook*, C. W. Hall and O. Kitani, eds., Gordon & Breach, London, 1989, pp. 434–451.

Ladisch, M. R., K. W. Lin, M. Voloch, and G. T. Tsao, "Process Considerations in Enzymatic Hydrolysis of Biomass," *Enzym. Microbial. Technol.* **6**, 82–100 (1983).

Lastick, S. M. and G. T. Spencer, "Xylose-Glucose Isomerases, Structure, Homology, Function," in *Enzymes in Biomass Conversion*, G. F. Leatham and M. E. Himmel, eds., ACS Symp. Ser. 460, 1991, pp. 487–500.

Lloyd, N. E. and R. O. Horvath, "Biotechnology and Development of Enzymes for the HFCS Industry," in *Bio-Expo 85*, O. Zaborsky, ed., Cahners Exposition Group, Stamford, CT, 1985, pp. 116–134.

Lynd, L. R., J. H. Cushman, R. J. Nichols, and C. E. Wyman, "Fuel Ethanol from Cellulosic Biomass," *Science* **251**(4999), 1318–1323 (1991).

Lynd, L. R., M. S. Laser, D. Brandsby, B. E. Dale, B. Davison, R. Hamilton, M. Himmel, et al., "How Biotech Can Transform Biofuels," *Nature Biotechnol.* **26**(2), 169–172 (2008).

Lynd, L. R., W. H. van Zyl, J. E. McBride, and M. Laser, Consolidated Bioprocessing of Cellulosic Biomass. An Update, *Current Opinion in Biotechnology*, **16**, 577–583 (2005).

Marshall, R. O. and E. R. Kooi, *Science* 25, 648–649 (1957).

Mascoma, http://www.mascoma.com/, March, 2009.

McLaughlin, J. L., "Annonaceous Acetogenins: Antitumor and Pesticidal Principles of the Paw Paw Tree," *Abstract, Purdue Univ. Chromatography Workshop '95*, W. Lafayette, IN, 1995, p. 47.

Mohagheghi, A., N. Dowe, D. Schell, Y. C. Chou, C. Eddy, and M. Zhang, "Performance of a Newly Developed Integrant of Zymomonas mobilis for Ethanol Production on Corn Stover Hydrolysate," *Biotechnol. Lett.* **26**(4), 321–325 (2004).

Mosier, N., C. Wyman, B. Dale, R. Elander, Y. Y. Lee, M. Holtzapple, and M. Ladisch, "Features of Promising Technologies for Pretreatment of Lignocellulosic Biomass," *Bioresource Technol.* **96**(6), 673–686 (2005a).

Mosier, N., R. Hendrickson, N. Ho, M. Sedlak, and M. R. Ladisch, "Optimization of pH Controlled Liquid Hot Water Pretreatment of Corn Stover," *Bioresource Technol.* **96**, 1986–1993 (2005b).

Mosier, N. S., C. Wyman, B. Dale, R. Elander, Y. Y. Lee, M. Holtzapple, and M. R. Ladisch, "Features of Promising Technologies for Pretreatment of Cellulosic Biomass," *Bioresource Technol.* **96**, 673–686 (2005c).

Mosier, N. S., R. Hendrickson, M. Brewer, N. Ho, M. Sedlak, G. Welch, B. Dien, A. Aden, and M. R. Ladisch, "Industrial Scale-up of pH-Controlled Liquid Hot Water Pretreatment of Corn Fiber for Fuel Ethanol Production," *Appl. Biochem. Biotechnol.* **125**, 77–85 (2005d).

Natake, M. and S. Yoshimura, , *Agric. Biol. Chem.* **28**(8), 505–509 (1964).

Natake, M., *Agric. Biol. Chem.* **30**(9), 887–895 (1966).

Natake, M., *Agric. Biol. Chem.* **32**(3), 303–313 (1968).

Patil, D. R., D. G. Rethwisch, and J. S. Dordick, "Enzymatic Synthesis of a Sucrose-Containing Linear Polyester in Nearly Anhydrous Organic Media," *Biotechnol. Bioeng.* **37**, 639–646 (1991).

Petsko, G. A., "Protein Engineering," in *Biotechnology and Materials Science—Chemistry for the Future*, M. L. Good, ed., *American Chemical Society*, Washington, DC, 1988, pp. 53–60.

Polastro, E. T., A. Walker, and H. W. A. Teeuwen, "Enzymes in the Fine Chemicals Industry: Dreams and Realities," *Bio/Technology*, 7(12), 1238–1241 (1989).

Pollack, A., Redesigning Crops to Harvest Fuel, *New York Times*, September 8, 2006 (see nytimes.com).

RFA (Renewable Fuels Association), http://www.ethanolrfa.org/industry/statistics/ (Nov. 13, 2007).

Rouhi, A. M., "Researchers Unlocking Potential of Diverse, Widely Distributed Saponins," *Chem. Eng. News* 73(37), 28–35 (1995).

Ryu, K., D. R. Stafford, and J. S. Dordick, "Peroxidase-Catalyzed Polymerization of Phenols: Kinetics of p-Cresol Oxidation in Organic Media," in *Biocatalysis in Agricultural Biotechnology*, ACS Symp. Ser. 389, J. R. Whitaker, ed., 1989.

Schoemaker, H. E., D. Mink, and M. G. Wubbolts, "Dispelling the Myths: Biocatalysis in Industrial Synthesis," *Science* 299 (5613), 1694–1697 (2003).

Schultz, P. G., R. A. Lerner, and S. J. Benkovic, "Catalytic Antibodies," *Chem. Eng. News* **68**(22), 1–56 (1990).

Schwietzke, S., Y. Kim, E. Ximenes, and N. Mosier, "Ethanol Production from Maize," in *Biomass for Maize Biotechnology*, B. Larkins, ed., Springer (in press, 2008).

Sedlak, M. and N. W. Y. Ho, "Production of Ethanol from Cellulosic Biomass Hydrolysates Using Genetically Engineered Saccharomyces Yeast Capable of Cofermenting Glucose and Xylose," *Appl. Biochem. Biotechnol.* **113–116**, 403–416 (2004).

Shapouri, H., J. A. Duffield, and M. Wang, *The Energy Balance of Corn Ethanol: An Update*, Agricultural Economic Report 814, 2002.

Shaw, A., K. K. Podkaminer, S. G. Desai, J. S. Bardsley, S. R. Rodgers, P. G. Thorne, D. A. Hogsett, and L. R. Lynd, Metabolic Engineering of a thermophilic bacterium to produce ethanol at a high yield, *PNAS*, **105**, 13769–13774 (2008).

Stone, R. "A Biopesticidal Tree Begins to Blossom, Neem Seed Oil Has Insect Toxicologists Buzzing about Its Potential as a Source of Natural Insecticides," *Science* **255**(5048), 1070–1071 (1992).

Syngenta, http://www.syngenta.com/en/about_syngenta/pop_pipline_corn.html, March, 2009.

Takasaki, Y., *Agric. Biol. Chem.* **30**(12), 1247–1253 (1966).

Takasaki, Y., K. Yoshiju, and A. Kanbagashi, in *Fermentation Advances*, D. Perlman, ed., Academic Press, New York, 1969, pp. 561–589.

Tsumura, N. and T. Sato, *Bull. Agric. Chem. Soc. Japan* **24**(3), 326–327 (1970).

Tsumura, N., M. Hagi, and T. Sato, "Enzymatic Conversion of D-Glucose to D-Fructose Part VIII. Propagation of Streptomyces phaeochromogenus in the Presence of Cobaltous Ion," *Agric. Biol. Chem.* **31**(8), 902–907 (1967).

Tsumura, N. and T. Sato, "Enzymatic Conversion of D-Glucose to D-Fructose Part V. Partial Purification and Properties of the Enzyme from Aerobacter cloncae," *Agric. Biol. Chem.* **29**(12), 1123–1128 (1965a).

Tsumura, N. and T. Sato, "Enzymatic Conversion of D-Glucose to D-Fructose Part VI. Properties of the Enzyme from Streptomyces phaeochromogenus," *Agric. Biol. Chem.* **29**(12), 1129–1134 (1965b).

Tyndall, R. M., "Application of Cellulase Enzymes to Cotton Fabrics and Garments," *AATCC Book of Papers*, 1991, pp. 259–273.

USDA Economics Research Service, Feed Grains Database, http://www.ers.usda.gov/Data/FeedGrains/, (May 19, 2009).

USDA Economic Research Service Report, *Industrial Uses of Agricultural Materials*, Report IUS-1, June 1993, pp. 10–35.

UNICA, Sugarcane Industry in Brazil, Ethanol, Sugar, Bioelectricity, www.unica.com.br., March, 2009.

Weil, J. R., A. Sarikaya, S.-L. Rau, J. Goetz, C. M. Ladisch, M. Brewer, R. Hendrickson, and M. R. Ladisch, "Pretreatment of Corn Fiber by Pressure Cooking in Water," *Appl. Biochem. Biotechnol.* **73**, 1–17 (1998).

Weil, J. R., M. Brewer, R. Hendrickson, A. Sarikaya, and M. R. Ladisch, "Continuous pH Monitoring during Pretreatment of Yellow Poplar Wood Sawdust by Pressure Cooking in Water," *Appl. Biochem. Biotechnol.* **70–72**, 99–111 (1998).

Weil, J., P. Westgate, K. Kohlmann, and M. R. Ladisch, "Cellulose Pretreatments of Lignocellulosic Substrates," *Enzym. Microbial. Technol.* **16**, 1002–1004 (1994).

Wheelwright, S. M., *Protein Purification*, Hansen Publishers, Munich, 1991.

Yamanaka, K., "Sugar Isomerases Part I. Production of D-Glucose Isomerase from Heterolactic Acid Bacteria," *Agric. Biol. Chem.* **27**(4), 265–270 (1963a).

Yamanaka, K., "Sugar Isomerases Part II. Purification and Properties of D-Glucose Isomerase from Lactobacillus brevis," *Agric. Biol. Chem.* **27**(4), 271–278 (1963b).

Yamanaka, K., "Purification, Crystallization and Properties of the D-Xylose Isomerase from *Lactobacillus brevis*," *Biochim. Biophys. Acta* **151**(3), 670–680 (1968).

CHAPTER THREE
HOMEWORK PROBLEMS

3.1. Describe what is meant by the term "industrial enzymes." What are the major uses of industrial enzymes (other than glucose isomerase), and who markets these? In what forms?

3.2. Glucose isomerase is a major industrial enzyme used in producing high-fructose corn syrup (HFCS). Give its properties:

_____ °C temperature stability

_____ pH at which it is used

_____ microorganism used to produce it

_____ cofactor(s) required for activities

_____ amount of HFCS produced annually in the United States

3.3. Explain how the development of technology that utilized enzymes to produce high-fructose corn syrup (HFCS) was able to win market share from invert (from sugarcane). Give a brief description of economic factors and trends.

3.4. How did the appearance of low-calorie sweeteners impact sugar consumption? What is the current per capita sugar consumption in the United States?

3.5. On the basis of your answers to questions 3.2–3.4, what factors might be important in the successful introduction of other types of bioproducts in the US market?

3.6. What are the key steps in transforming cellulose-containing material (referred to as *biomass*) to ethanol using bioprocessing? Give a schematic diagram and show the key steps.

3.7. How does the development of glucose isomerase provide some of the scientific foundation for the economically attractive conversion of biomass to ethanol? Give names of sugars that may have an impact. Be specific in your answer. How does genetic engineering of yeast have a possible impact?

3.8. What types of bioproducts can be derived from renewable resources, and what are their uses? Construct a table showing the bioproduct in the left column and use in the right column. How might the new biotechnology contribute to making such products more economic?

3.9. What is meant by *bioseparations*? What role does it play in obtaining bioproducts?

3.10. How does pretreatment fit into a process for converting cellulose to ethanol? What is the role of pretreatment? Why is it needed?

CHAPTER FOUR

MICROBIAL FERMENTATIONS

INTRODUCTION

The generation of products through microbial culture requires that the cells be propagated, at conditions where growth of the desired microorganism dominates in large vessels through processes that are generically referred to as *fermentations*[1]. Products are derived from the metabolism of the microorganism which can be either wild type (not genetically engineered) or recombinant. However, the presence of more than one microorganism will result in competition. An undesired microorganism is often the faster growing species and consumes fermentation media components but does not give the desired product. Like weeds in a garden, the undesired species may take over and cause lost productivity.

Control of fermentation conditions and close attention to sterile procedures minimize the chance of introducing living microorganisms that will contaminate the bioreactor. Sterilization of the fermentation medium at 121 °C for 30 min is generally sufficient to kill both living microorganisms and their spores. Sterilizing the components that make up the culture medium and then inoculating the desired microorganism into a previously sterilized, closed vessel avoids introduction of foreign microorganisms.

The term *fermentation* denotes microbial cell propagation and generation of products under either aerobic or anaerobic conditions. The secretion of metabolites from the inside of microbial cells to the surrounding medium and accumulation of the metabolite in the medium occurs as a consequence of the oxidation of monosaccharides, particularly glucose under both aerobic and anaerobic conditions. Fermentations are classified as Type I, II, or III. The classification is based on the

---

[1] The use of the term *fermentation* is inexact since fermentative conditions should technically refer to metabolism that occurs through the glycolytic pathway in the absence of oxygen as described later in this book. This footnote is intended to alert the reader to the sometimes ambiguous terminology that prevails in the field of biotechnology.

Modern Biotechnology, by Nathan S. Mosier and Michael R. Ladisch
Copyright © 2009 John Wiley & Sons, Inc.

time at which a specified metabolite (i.e., product) appears in the fermentation broth relative to the exponential growth phase of the microorganism. This naming convention was proposed by Gaden in 1959, and has found general utility in classifying industrial fermentations.[2]

The utilization of sugars in microbial culture is generally considered as the independent variable and is used to develop explanations and equations that relate substrate consumption to growth and product accumulation. While energy is released when fats and proteins are degraded by cells, monosaccharides are the major energy and carbon source in microbial cultivation, for metabolic and economic reasons (Gaden 1959). From a metabolic perspective, monosaccharides are preferred since the Embden-Meyerhoff and citric acid cycle pathways that utilize these sugars are found in almost all living organisms (discussed in Chapter 6).

The preferred fermentation substrate is glucose. Glucose is the most common carbon and energy source for fermentations since it is readily available and relatively inexpensive at 10¢/lb or less. Glucose, which is also referred to as *dextrose* or an *industrial sugar*, is derived from hydrolysis of starch from corn that has been processed by wet milling[3] or dry milling.[4] At some point in the future it is probable

---

[2] The term *fermentation* is used here to denote microbial cell propagation and generation of products under either aerobic, microaerobic, or anaerobic conditions. *Aerobic* denotes condition where air is intentionally mixed with the medium; *microaerobic* refers to air that is initially present, but is then used up or displaced as microbial growth occurs; while *anaerobic* denotes a condition where oxygen is removed and intentionally excluded from the fermentation media since it is toxic to the cells.

[3] *Corn wet milling* is a process in which the corn is steeped at about 130 °F in a solution containing 0.1–0.2% SO_2. The SO_2 inhibits microbial growth and enhances the swelling of the corn kernels. Corn entering such a process resembles the dry corn kernels found in holiday decorations, while the soaked corn is wet and resembles the kernels of sweet corn. The liquid that remains after steeping is steep water. It contains proteins, some sugars, vitamins, and minerals that were extracted from the corn. Steep water is a valuable byproduct and is used in the formulation of fermentation media (such as for ethanol production). The wet corn kernels are processed through a series of grinding mills and wet separations based on screening and centrifuge techniques. A membrane surrounds the germ that contains the corn oil. The germ is fractionated from the starch particles. The protein (gluten) holds the starch particles together, and the fiber (celluose and hemicellulose) is part of the skin that covers the kernel, or is associated with the starch. The yields on wet milling are about 67 g starch, 10 g gluten (containing protein), 6 g germ (contains oil), and 9 g fiber per 100 g corn (Ladisch and Svarczkopf 1991; Gulati et al. 1997; Hoseney 1986; Eckhoff and Tso 1991; Fox and Eckhoff 1993). Echoff and his coworkers give an excellent description, and address engineering issues of corn wet milling (Eckhoff and Tso 1991; Fox and Eckhoff 1993). The starch has many uses. If it is processed to glucose, it is first cooked in water under pressure, at about 120 °C. It is gelatinized or dissolved at this condition, so that it can be readily converted to glucose when the temperature is decreased to about 50–60 °C and enzymes are added to induce starch hydrolysis.

[4] *Corn dry milling* is a process where the corn is equilibrated to 20% moisture and then ground into small pieces that range from about 0.02 to 3 mm in size. The germ and the various sizes of particles are separated by dry sieving techniques into different size fractions, for applications related to food products. The largest sizes are pressed and converted to corn flakes (i.e., flaking grits) or used in beer brewing (i.e., brewer's grits). In dry-grind processes where the goal is ethanol production, rather than transformation of corn into specialty starch and oil products, the whole kernel is ground to a fine powder. The ground

that industrial sugars will be derived from cellulosic materials as well (Ladisch and Svarczkopf 1991; Gulati et al. 1996; Mosier et al. 2005). Advances in bioconversion technologies for fuel ethanol production from cellulose will effectively decrease costs of industrial sugars, that by virtue of their sources (wood, agricultural waste, etc.), are not suitable for food.

FERMENTATION METHODS

Fermentations Are Carried Out in Flasks, Glass Vessels, and Specially Designed Stainless-Steel Tanks

Aerated and anaerobic fermentations at the laboratory scale are typically carried out in shaken (not stirred) flasks in an incubator-shaker. There are many different types available, with a model based on temperature-controlled, heated air illustrated in Figs. 4.1 and 4.2. The air serves to both heat and cool the shake flasks that are agitated by an orbital motion of the shaker table. The flasks, nutrient medium, and caps are sterilized at 121 °C for 30 min in an autoclave. Wet heat (steam) is needed to obtain effective killing of microorganisms and spores that might be found in the flask or nutrient solution.

Shake flask studies are used early in the development of an industrial fermentation since multiple nutrient compositions and different cultures can be evaluated in parallel studies, in the same incubator (such as shown in Fig. 4.1). The pH in a shake flask is controlled using buffer, while other nutritional factors are controlled by the makeup of the broth.

The temperature range over which many types of organisms will grow are relatively narrow and will typically fall within 20–60 °C and at atmospheric pressure. Microorganisms that are psychotrophs require lower temperatures, while archaebacteria (Archaea) require higher temperatures (up to 90 °C) and sometimes, elevated pressure to grow. In this case, special incubation chambers are needed. The circulation of air in the incubator also serves to remove heat of fermentation. The oxygen requirements for aerobic microorganisms are met by the swirling action of the shake flask, accompanied by diffusion of air through the porous, sterile, cotton plug that is fitted on top of the flask.

The culturing of strict anaerobes [i.e., microorganisms that are killed by oxygen in the ppb (parts per billion) range] requires that inoculation of the microorganism be carried out in a special hood, known as an *anaerobic chamber*, so that oxygen is excluded and prevents oxygen from dissolving in the broth. This enclosed

---

corn is used for ethanol fermentation. It is either gelatinized (i.e., dissolved) and added directly to the fermentation vessel together with enzymes and yeast or saccharified by the enzyme (converted to glucose) and then added to the fermentation vessel together with yeast (Ladisch and Svarczkopf 1991; Gulati et al. 1997; Wells 1979), to produce ethanol. Ethanol production by corn dry grinding involves grinding the whole kernel into a powder. Without further fractionation, the ground corn is cooked, hydrolyzed, and fermented. After fermentation and distillation, a solid coproduct remains and is referred to as *distillers' grains*, which may be further processed into additional ethanol using cellulose conversion technology (Ladisch and Dale 2008; Kim et al. 2008a–c; Mosier et al. 2008; Dien et al. 2008; Perkins et al. 2008).

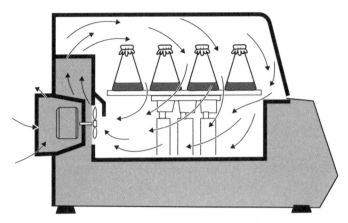

Figure 4.1. Schematic diagram of incubator-shaker used for shake flask culture of microbial cells. Cotton plugs or permeable fabric caps are sterile but allow air to pass into the flask while keeping out airborne microorganisms. The chamber is tightly constructed to attain close temperature uniformity. The lid has a large viewing window and provides full access to the chamber interior. Air is rapidly swept through a heating duct and is dispersed throughout the chamber. To achieve precise temperature and gradient, samples are uniformly bathed in a gentle flow of temperature-controlled air that circulates around, between, and over the flasks. A fresh-air intake valve permits the controlled entrance of room air when desired. Temperatures are controlled from 22.5 to 60°C. The shaker holds 21, 125-mL Erlenmeyer flasks containing 30 mL fluid. [Courtesy of New Brunswick Scientific, Inc., *General Catalogue 1028* (1982), p. B-18.]

chamber resembles a glovebox. Traces of oxygen are scavenged from the N_2 gas by passing the gas over hot (200–300°C) copper filings that react with trace oxygen to form copper oxides, and this oxygen-free nitrogen is used to fill the chamber.

Controlled laboratory fermentations at a larger scale are typically carried out in sealed glass vessels of the type shown in Fig. 4.2. Volumes of these vessels range from 1 to 20 L, with the working volume (broth) typically being about 70% of the fermentor volume. At the laboratory, scale, the entire vessel, including headplates and fermentation media, can be placed into an autoclave for sterilization. Such a fermentor contains sealed ports for pH and O_2 probes, as well as ports for addition of acid or base for purposes of pH control. Aeration of the fermentation broth is carried out by bubbling compressed air that has been sterile-filtered through a 0.2-μm filter, into the bottom of the vessel. The 0.2-μm cutoff of such a filter is small enough to block any airborne spores or microorganisms from entering the fermentor with the air. Ports enable samples to be taken for intermittent analysis by chromatography, mass spectroscopy, or other methods. These data are then used to generate graphs of the fermentation timecourse. Heating and cooling of the fermentor is achieved by passing thermostat water through the cooling coil (i.e., hollow baffle) or a jacket (Fig. 4.3).

Industrial-scale fermentation for manufacture of high-value products is carried out in large, 316 stainless-steel vessels, of the type shown in Fig. 4.3. The principle on which this equipment is based is similar to that of the laboratory-scale fermentor, although numerous and complex mixing, aeration, and scaleup issues must be

A   *Air Exhaust*
B   *Water Circulation Lines*
C   *Exhaust Gas Condenser*
D   *pH Electrode Port*
     (Capped)
E   *pH Electrode Port*
     (Capped)
F   *Circulating Water Out*
     Fitting
G   *Fittings for Acid and Base*
H   *Inoculation/Filling Port*
I   *Rubber Diaphragm*
J   *Thermowell*
K   *Sampling Line*
L   *Sparger*
M   *Magnetic Disc (Foam*
     Breaker)
N   *Auxiliary Addition Ports*
O   *Nutrient Feed Line*
P   *Harvest Line Plugged*
Q   *Circulating Water-in Fitting*
R   *Thermistor Well*
S   *Level Probe Fitting*
     (Plugged)
T   *Foam Breaker*
U   *Impeller*
V   *Vessel Retaining Ring*
     (Lower)
W   *Base Plate*
X   *Hollow Baffle*
Y   *Cylindrical Tube Glass*
     Vessel
Z   *O_2 Electrode Port*
     (Capped)

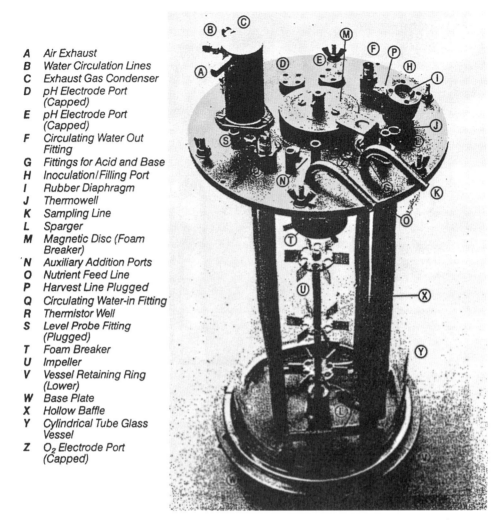

Figure 4.2. Picture of a laboratory fermentor showing major components, except for an agitation motor that is fitted to the impeller shaft that protrudes from the center of the headplate [courtesy of New Brunswick Scientific, Inc., *General Catalogue 1028* (1982), p. C-22].

addressed. Sterilization is carried out in place, and special designs are needed to minimize the risk of holdover of small amounts of liquid that could serve as reservoirs of microbial contamination from one batch to the next. Polished surfaces minimize the occurrence of small pores or pockets in the metal that might retain microbial spores. Layout of the piping must ensure that there are no low spots where liquid could pool and be a source of contamination.

This brief overview of fermentor design is intended to help the reader understand how the products discussed in this chapter are obtained at either a bench or industrial scale. Further background on fermentor design, fermentation system layout, and scaleup can be found in several biochemical engineering textbooks (Aiba et al. 1973; Bailey and Ollis 1977; Shuler and Kargi 1992; Blanch and Clark 1996).

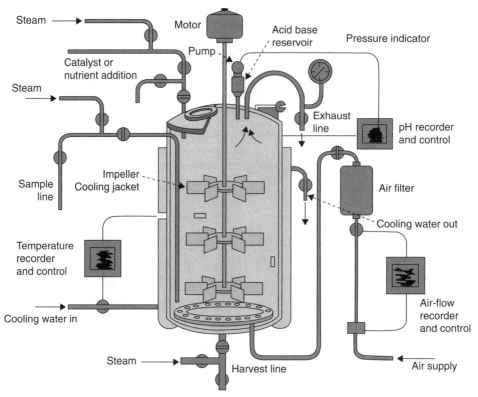

Steam

Motor

Acid base reservoir

Pump

Catalyst or nutrient addition

Pressure indicator

Steam

Exhaust line

pH recorder and control

Impeller
Cooling jacket

Air filter

Sample line

Temperature recorder and control

Cooling water out

Air-flow recorder and control

Cooling water in

Steam

Harvest line

Air supply

Batch-stirred reactor, with peripherals; workhorse of the fermention industry

Figure 4.3. Diagram of an instrumented fermentor for aerated fermentation of products generated under sterile conditions in a closed, agitated vessel. This type of fermentor may range in volume from 50 to 200,000 L. Metal construction usually consists of polished type 316 stainless steel. [From Bjurstrom (1985), Fig. 4.3, p. 128.]

Humphrey (1998) retrospectively still reminds bioengineers that "scale-up is still an art not a science." Unlike a chemical reactor, a biological reactor contains living entities that respond to externally applied changes in their environment by generating new reaction pathways, new enzymes, and new products. These changes are propagated and copied through cell division in a manner that is neither possible nor observed in chemical reactors that contained a fixed amount of a defined catalyst.

MICROBIAL CULTURE COMPOSITION AND CLASSIFICATION

Microbial Cells: Prokaryotes versus Eukaryotes

Microbial cells are either prokaryotes or eukaryotes. Microbial organisms are important organisms with a long history of human uses for everything ranging from the production of bread, beer, wine, and cheese to more recent uses for generating biofuels, biopolymers, and pharmaceuticals. Microbial cells fall into both primary categories of living things on Earth first proposed by Edouard Chatton (1883–1947), who in the 1930s classified all organisms as either prokaryotes or eukaryotes.

Prokaryotes are characterized by a single-cell wall beneath which there is a membrane enclosing the cytoplasm containing DNA, RNA, proteins, and other cell contents. A prokaryotic cell does not have an obvious organized internal structure (Alberts et al. 1989). *Eukaryotes* have a nucleus that contains most of the cell's DNA, as well as distinct structures within the cytoplasm. Eukaryotic cells have structural characteristics that allow communication between the cell and its environment, and with other cells. This enabled development of cell aggregates and multicellular life. It is believed that prokaryotic cells were present on Earth 3.5–4 billion years ago; anaerobic eukaryotes, at 1.4–1.5 billion years; and aerobic eukaryotic cells containing mitochondria, at about 850 million years.

Eukaryotes and prokaryotes are similar in some aspects of their chemistry, physiology, and metabolism (Kabnick and Peattie 1991). These similarities include

1. Cell membranes consisting of lipids
2. DNA that stores code for all the proteins in the cell
3. Transfer of the DNA's information via RNA
4. Protein synthesis carried out via structures called *ribosomes*

Both prokaryotes and eukaryotes may have either anaerobic or aerobic metabolism. Some have both anaerobic and aerobic metabolic pathways. These represent facultative microbes that can adapt to either aerobic or air-limited (fermentative) conditions. Some prokaryotes are obligate anaerobes that cannot survive in the presence of oxygen.

Differences between prokaryotic and eukaryotic cells include size, the nature of cell division, and transmembrane movement. Prokaryotes tend to be smaller, primarily unicellular, and exhibit either anaerobic or aerobic metabolism or both (facultation). In comparison, eukaryotes are larger, mainly multicellular, and have aerobic metabolism (Alberts et al. 1989).

Procaryotic and eukaryotic cells have different structures as indicated in Fig. 4.4a. The bacterium, *Escherichia coli*, is a single-cell microorganism with a characteristic dimension of about 1 μm. *E. coli* does not make spores, and reproduces by cell division. Gram negative *E. coli* was first isolated by the pediatrician Escherich from the feces of a cholera patient. Subsequent studies showed this microorganism to inhabit the intestines of humans and other vertebrates (Koch and Schaechter 1985). *E. coli* is relatively easy to grow, and its physiology, genetics, and biochemistry have been thoroughly studied. Consequently, it is not surprising that *E. coli* became the first microorganism used as the host cell for recombinant plasmids containing the genes for the insulin A chain, insulin B chain, or proinsulin, and ultimately became the vehicle for manufacture of the first commercial recombinant product, insulin, beginning in about 1980. Chromosomes are unraveled by attachments in the inner membrane of prokaryotes compared to cytoskeletal structures in eukaryotes.

Transmembrane movement in eukaryotes can occur through membrane-bound vesicles in a process known as *endocytosis* (components entering the cell) or exocytosis (exiting the cell). Both prokaryotes and eukaryotes secrete metabolic products through the cell membrane, directly, and are of industrial significance for production of antibiotics, amino acids, and carboxylic acids via fermentation.

(a)

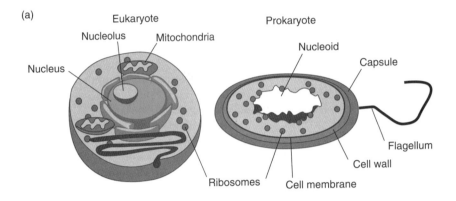

(b)

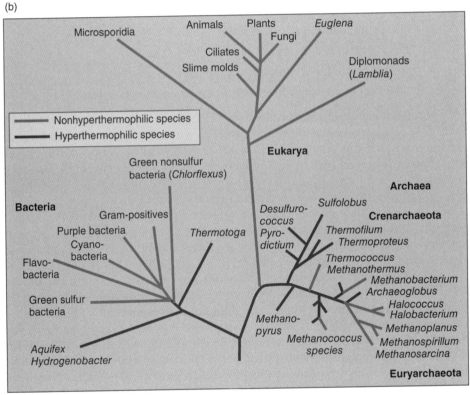

Figure 4.4. (a) Schematic representations of a eukaryote and a prokaryote. [*Note*: The prokaryote is not to scale with the eukaryote. Prokaryotic cells are approximately $\frac{1}{100}$ th the length of a eukaryotic cell. This image is from the Science Primer (http://www.ncbi. nlm.nih.gov/About/Primer), a work of the National Center for Biotechnology Information, part of the National Institutes of Health. As a work of the US federal government, the image is in the public domain (see http://www.ncbi.nlm.nih.gov/About/disclaimer.html)]. (b) Woese family tree showing relationship between one-celled life and higher organisms [reprinted from Morrell (1997), p. 701].

Classification of Microorganisms Are Based on Kingdoms

With the advent of DNA sequencing techniques, the "five-kingdom classification" system [Plantae, Fungi, Animalia, Protista (unicellular eukaryotes), and Monera (the prokaryotes)] has been amended to include three "superkingdoms" or "domains," first proposed by Carl Woese (Morrell 1997). These domains include Eukarya (all eukaryotes), Bacteria (all blue-green algae and true bacteria), and Archaea (archaebacteria) (Fig. 4.4b). This split of prokaryotes into two domains (Bacteria and Archaea) was required because of the striking physiological and genetic differences between bacteria and archeabacteria, many of which are adapted to extreme living conditions (hot springs, hot vents on the ocean floor, etc.). Much less is currently known about the archaebacteria, but active research is investigating the genomes of these organisms for enzymes that operate at extreme temperature with higher reaction rates than exist in bacteria, as well as for novel enzyme catalysts that may have industrial applications. Also among the archaebacteria are the organisms classified as *methanogens*, which are anaerobic microorganisms that produce methane (natural gas) as a byproduct of metabolism. Diversa (http://www.diversa.com/) is an example of a company that is screening genomic data from bacteria and archaebacteria from around the world, including extreme environments like hot springs, for genes that have novel or improved functionality for industrial applications. Methanogens are also being studied for their potential use for generating renewable natural gas from human and animal waste. Anaerobes that ferment CO and H_2 to ethanol are being developed as biocatalysts for producing ethanol from synthesis gas (Henstra et al. 2007; Datar et al. 2004).

The taxonomy of microorganisms and division of such organisms into classes of increasing complexity—archaebacteria, eubacteria, and eukaryotes—helps to differentiate both their structure and associated functions in a meaningful way. Differences include the organization of the nuclear material, cell membranes, and complexity of structures internal to the cell. A feature common to all cells is the structure of DNA, and its function as a repository of information for the biosynthesis of proteins, the catalysts that form the essence of life in the cell's cytoplasm.

Prokaryotes Are Important Industrial Microorganisms

Two important prokaryotes for bacterial fermentations are *Bacillus* spp. and *Escherichia coli*. *Bacillus* is a source of the enzymes amylase and glucose isomerase, proteases used in detergents, and antibiotics. Amylase is an enzyme (i.e., biocatalyst) used to hydrolyze starch to maltodextrins. The maltodextrins are subsequently converted to glucose by another enzyme, glucoamylase. Glucose isomerase converts glucose to a mixture of 42–44% fructose and glucose as described in Chapter 3 (bioproducts and biofuels). The fructose–glucose mixture is further processed by liquid chromatography to obtain a 55% fructose product known as *high-fructose corn syrup* (HFCS). Annual US production of HFCS since the beginning of the twenty-first century has been 18 to 24 billion lb/year and forms the basis of the wet-milling industry in the United States.[5]

---

[5] Information regarding US agricultural commodity and food production has been published by the US Department of Agriculture Economic Research Service (http://www.ers.usda.gov/).

Escherichia coli is the microorganism instrumental in the development of bacterial recombinant biotechnology with products ranging from therapeutic proteins to biofuels and monomers. Other examples of procaryotic cells are mycoplasmas, *Staphylococcus*, *Anabaena* (a cyanobacterium), and *Spirillius*. Mycoplasmas are small (0.3-µm) organisms that resemble bacteria and normally lead a parasitic existence within plant or animal cells (Alberts et al. 1989). *E. coli* has been genetically modified by insertion of the genes for the reduction of pyruvate to ethanol, thus enabling this bacterium to ferment both glucose and xylose to ethanol (Jarboe et al. 2007).

The first microorganism used to obtain the commercial recombinant product, insulin, was *E. coli* containing recombinant plasmids with genes for insulin A and insulin B chains. Insulin derived by recombinant technology was first sold commercially in 1982, and accounted for an estimated $170 million in sales in 1990 (Thayer 1991). In 2000, retail sales of insulin in the United States were $1.3 billion (Mendosa 2001). By 2005, only 2000 of the 3.5 million insulin users purchased an Iletin™ brand insulin (derived from pork pancreas), and these sales may have been for cats or dogs. The sale of pork-derived insulin was stopped by Lilly in 2005, while sales of beef–pork-derived insulin was discontinued in 1998. Total insulin sales by Lilly in 2005 were $2 billion/year, and these are now all derived fermentation-based processes (Swiatek 2005).

Escherichia coli K12 cells are well characterized. A detailed composition was published by Battley (1991). The unit carbon formula representing the structural components of the cell is

$$CH_{1.595} O_{0.374} N_{0.263} P_{0.023} S_{0.006} \tag{4.1}$$

with a corresponding unit carbon formula weight of 24.190 Da. When glycogen is included as part of the cellular mass, the unit carbon formula becomes

$$CH_{1.597} O_{0.385} N_{0.257} P_{0.022} S_{0.006} \tag{4.2}$$

with a corresponding unit carbon formula weight of 24.253 Da. The distribution of the key cellular constituents, excluding glycogen, is summarized in Table 4.1, and shows that proteins, RNA, and DNA make up almost 90% of the dry weight of the cell. Combustion of dry *E. coli* cells gives a weight of ash (i.e., metal oxides and salts) equivalent to about 12% of initial dry cell weight. This corresponds to 3.38% metal ions, of which P, S, and K^+ are the major ions, and Fe^{3+}, Mg^{2+}, Ca^{2+}, and Cl^- are present at much smaller concentrations. The percentage of ions in the solids that make up the total ash is reported to be about 28% (Battley 1991).

Eukaryotes Are Used Industrially to Produce Ethanol, Antibiotics, and Biotherapeutic Proteins

Eukaryotic cells, unlike prokaryotic cells, have distinctive organelles, and a nucleus that contains most of the cells' DNA. The nucleus is enclosed by a double layer of membrane, which separates its contents from the cytoplasm, where most of the cell's metabolic reactions occur. Organelles in the cytoplasm include mitochondria and chloroplasts. Mitochondria are found in almost all eukaryotic cells, and contain the

Table 4.1.

Total Weights of Monomer Constituents that Make Up
Macromolecular Components in 100 g Dry Weight of *E. coli* K-12 Cells

Monomers from	Weight g/100 g Cells						
	Total[a]	C	H	O	N	P	S
Proteins	64.18	29.25	4.95	19.48	9.75	—	0.75
RNA nucleic acids	21.58	7.25	0.88	8.02	3.49	1.95	—
DNA nucleic acids	3.27	1.17	0.14	1.12	0.53	0.31	—
Lipids	9.17	5.92	0.96	1.75	0.14	0.40	—
Lipopolysaccharides	4.03	1.89	0.33	1.63	0.08	0.08	—
Peptidoglycans	2.84	1.23	0.20	1.15	0.27	—	—
Polyamines	0.40	0.22	0.05	—	0.13	—	—
Subtotal	105.47	46.93	7.51	33.15	14.39	2.74	0.75
Water eliminated by polymerization	−10.97		−1.23	−9.74			
Total polymers	94.5	46.93	6.29	23.41	14.39	2.74	0.75

[a]Does not include glycogen.
Source: Adapted from E. H. Battley, "Calculation of Heat of Growth of *Escherichia coli* K-12 on Succinic Acid,"*Biotechnology Bioengineering* 37, 334–343 (1991), Table II, p. 337. Copyright 1991, John Wiley & Sons, Inc.

enzymes that carry out metabolism based on oxygen. Chloroplasts are found in plant cells that are capable of photosynthesis (Alberts et al. 1989).

Eukaryotes that are important to bioprocessing include yeast (*Saccharomyces cerevisiae*) for ethanol production and the fungi *Penicillium* spp. for streptomycin and tetracycline production (Evans 1965). Mammalian cells are another type of eukaryote used in industrial manufacture of therapeutic proteins. These cells include recombinant Chinese hamster ovary (CHO) cells for monoclonal antibodies and tissue-type plasminogen activator (tPA). For example, tissue-type plasminogen activator (tPA) is a protein that activates plasminogen to form plasmin. The plasmin rapidly hydrolyzes fibrin, before the plasmin itself is inactivated by antiplasmin. A significant infusion of tPA can dissolve a blood clot quickly enough to prevent tissue damage that would otherwise occur during a heart attack. This protein has numerous disulfide bonds and is glycosylated. Consequently, tPA is produced by culture of a mammalian cell line (i.e., CHO) into which the tPA gene has been cloned. These cells are capable of glycosylating the protein, unlike the bacterium *E. coli*, which cannot (Builder and Grossbard 1986; Builder et al. 1988).

Wild-Type Organisms and Growth Requirements in Microbial Culture

Wild-Type Organisms Find Broad Industrial Use

Future developments in microbial fermentations are likely to utilize engineered yeast, bacteria, or mammalian cells. Nonetheless, there is a broad array of wild-type organisms that are not genetically modified that generate an equally broad array of microbial products. From the domains of Eukarya and Bacteria, it is estimated that there are several hundred microorganisms that make commercially useful products out of the millions of species which exists in nature. These products include principally antibiotics and food products as shown in Table 4.2 [from Phaff (1981)].

Table 4.2.

Prokaryotes and Eukaryotes Used as Microbial Industrial Organisms as Defined by Phaff (1981, p. 88)

Name	Organism Type	Product
Industrial Chemicals		
Saccharomyces cerevisiae	Yeast	Ethanol (from glucose)
Kluyveromyces fragilis	Yeast	Ethanol (from lactose)
Clostridium acetobutylicum	Bacterium	Acetone and butanol
Aspergillus niger	Mold	Citric acid
Xanthomonas campestris	Bacterium	Polysaccharides
Entomopathogenic Bacteria		
Bacillus thuringiensis	Bacterium	Bioinsecticides
Bacillus popilliae	Bacterium	Bioinsecticides
Amino Acids and Flavor-Enhancing Nucleotides and Acids		
Corynebacterium glutamicum	Bacterium	L-Lysine
Corynebacterium glutamicum	Bacterium	5′-Inosinic acid and 5′-guanylic acid; glutamate
Single-Cell Proteins		
Candida utilis	Yeast	Microbial protein from paper-pulp waste
Saccharomycopsis lipolytica	Yeast	Microbial protein from petroleum alkanes
Methylophilus methylotrophus	Bacterium	Microbial protein from growth on methane or methanol
Foods and Beverages		
Saccharomyces cerevisiae	Yeast	Baker's yeast, wine, ale, sake
Saccharomyces carlsbergensis	Yeast	Lager beer
Saccharomyces rouxil	Yeast	Soy sauce
Candida milleri	Yeast	Sour French bread
Lactobacillus sanfrancisco	Bacterium	Sour French bread
Streptococcus thermophilus	Bacterium	Yogurt
Lactobacillus bulgaricus	Bacterium	Yogurt
Propionibacterium shermanii	Bacterium	Swiss cheese

Table 4.2. *(Continued)*

Name	Organism Type	Product
Gluconobacter suboxidans	Bacterium	Vinegar
Penicillium roquefortii	Mold	Blue-veined cheeses
Penicillium camembertii	Mold	Camembert and Brie cheeses
Aspergillus oryzae	Mold	Sake (rice-starch hydrolysis)
Rhizopus	Mold	Tempeh
Mucor	Mold	*Sufu* (soybean curd)
Monascus purpurea	Mold	Ang-Kak (red rice)
Vitamins		
Eremothecium ashbyi	Yeast	Riboflavin
Pseudomonas denitrificans	Bacterium	Vitamin B_{12}
Propionibacterium	Bacterium	Vitamin B_{12}
Polysaccharides		
Leuconostoc mesenteroides	Bacterium	Dextran
Xanthomonas campestris	Bacterium	Xanthan gum
Pharmaceuticals		
Penicillium chrysogenum	Mold	Penicillins
Cephalosporium acremonium	Mold	Cephalosporins
Streptomyces	Bacterium	Amphotericin *B*, kanamycins, neomycins, streptomycin, tetracyclines, and others
Bacillus brevis	Bacterium	Gramicidin *S*
Bacillus subtilis	Bacterium	Bacitracin
Bacillus polymyxa	Bacterium	Polymyxin *B*
Rhizopus nigricans	Mold	Steroid transformation
Arthrobacter simplex	Bacterium	Steroid transformation
Mycobacterium	Bacterium	Steroid transformation
Hybridomas	Mammal	Immunoglobulins and monoclonal antibodies
Mammalian cell lines	Mammal	Interferon
Escherichia coli (via recombinant-DNA technology)	Bacterium	Insulin, human growth hormone, somatostatin, interferon
Carotenoids		
Blakesiea trispora	Mold	β-Carotene
Phaffia rhodozyma	Yeast	Astaxanthin

Microbial Culture Requires that Energy and All Components Needed for Cell Growth Be Provided

Basic requirements for microbial growth must be satisfied by the growth medium (Aiba et al. 1973). The media's constituents must provide sources of energy, carbon, nitrogen, and minerals. Amino acids and vitamins, in addition to sugars, must be part of the media. The development of an economic medium for a particular bioreaction process requires trial-and-error experimentation. Selection, discovery, and/or engineering of organisms that are able to utilize inexpensive media components, and subsequent optimization of culture conditions including pH, temperature, and/or the rate at which oxygen is supplied (i.e., aeration rate) must be carried out.

Adenosine triphosphate is the source of energy used by living cells that must come from the oxidation of medium components or from light. The role of ATP is succinctly summarized by Aiba et al. (1973): "The coupling of ATP to thermodynamically unfavorable reactions allows them to proceed at a useful rate. Photosynthetic bacteria and algae can utilize the energy from light for ATP formation; autotrophic bacteria can generate ATP from the oxidation of inorganic compounds; while heterotrophic bacteria, yeasts, and fungi form ATP while oxidizing organic compounds." In this context microorganisms can be divided into three categories (Aiba et al. 1973):

Autotrophic bacteria, which can generate ATP from oxidation of inorganic compounds

Heterotrophic bacteria, *yeasts*, and *fungi*, which utilize organic compounds (glucose, fats, alcohol, alkanes, aromatics) to generate ATP

Photosynthetic bacteria and *algae*, which utilize energy from light to generate ATP

MEDIA COMPONENTS AND THEIR FUNCTIONS (COMPLEX AND DEFINED MEDIA)

The media used in fermentations may be defined or complex. What do defined media, an Indianapolis 500 race car driver's dining preference, and the care and feeding of industrial microbes have in common? The driver stated it very succinctly. In response to a reporter's request to name his favorite dining experience, he said "home"—since he got "what he wanted, and he knew what was in it." Defined media fit this definition, but sometimes people do not know exactly what microbes want. In this case, complex media are used. Examples of defined and complex media are given in the next several pages.

Carbon Sources Provide Energy, and Sometimes Provide Oxygen

Carbon is a major component of cell biomass, since 50% of the dry weight of cells is carbon. Facultative organisms are organisms that can grow under either aerobic or anaerobic conditions, and incorporate about 10% of the carbon in the substrate into cell material under conditions of anaerobic metabolism. In the presence of oxygen, up to 55% of the substrate carbon is converted into carbon incorporated into the cell biomass. CO_2 is the source of carbon for autotrophs (e.g., alga) and autolithotrophs (some types of archaebacteria found at the bottom of the ocean). In this case the energy (light) and carbon sources are separate entities.

Heterotrophs obtain both carbon and energy from sugar. The oxygen and carbon are both derived from the sugar itself under anaerobic conditions. The glycolytic pathway (Fig. 4.10b) results in 10% of the carbon in the substrate being incorporated into cell mass. Under aerobic conditions, however, the carbon is obtained from the sugar, while oxygen is derived from both the sugar as well as dissolved oxygen that is transferred from the air to the culture media (i.e., fermentation broth) and from the media to the cells.

The initial concentration of a sugar in the fermentation broth may be based on the expected fraction of carbon in the substrate that will be incorporated into the cells to give a specified amount of microbial biomass.[6] Aiba et al. (1973) give the example of a fermentation where the goal is to obtain 40 g of cells (dry weight basis)/L of broth. For aerobic conditions, and if oxygen transfer is not limiting, the total carbon that would need to be initially present in the medium is:

$$\frac{\left|\dfrac{40\,\text{g cells (dry weight)}}{\text{L broth}}\right| \dfrac{0.5\,\text{g carbon}}{\text{g cells}}}{\left|\dfrac{1\,\text{g glucose in broth}}{0.5\,\text{g glucose incorporated into cells}}\right| \dfrac{180\,\text{g glucose}}{72\,\text{g carbon}}} = 100\,\frac{\text{g glucose}}{\text{L broth}}$$

The molecular formula for glucose is $C_6H_{12}O_6$ [molecular weight (MW) = 180]. Hence there is 72 g of carbon (= 6 × 12) in 180 g of glucose.

A Nitrogen Source Is Needed for Protein Synthesis. Nitrogen is about 10% of the weight of most organisms, and is supplied to microbial cultures through a variety of compounds, depending on the organism's ability to metabolize different sources of nitrogen, and its ability to synthesize the basic amino acids required for growth. The most direct source of nitrogen is inorganic, and may consist of ammonia or ammonium salts. However, this is insufficient if a specific amino acid needed for growth cannot be made by the organism. The amino acid must then be supplied in the broth. Other nutrient requirements include vitamins and nucleic acids.

Proteins (i.e., polymers of amino acids), nucleic acid bases (purine, pyrimidines), and vitamins, as well as individual amino acids are recognized as organic sources of nitrogen since they contain N, as well as C, H, and O. These sources also sometimes contain S in their structure. Costs of nitrogen sources for culture media range from $0.20 to $1.00/lb and can be expensive relative to the carbon source. Complex sources of nitrogen include soybean, peanut, fish meals, yeast extract, whey, casein, and hydrolyzed proteins. Complex sources of nitrogen do not have a clearly defined composition, and often contain other growth factors that are defined on the basis of their effect, rather than by identification of their chemical structure. Byproducts from sugar processing, such as molasses from sugarcane or

---

[6] *Microbial biomass* refers to the cell mass that is generated when microorganism grow in fermentation media. This term is distinct from cellulosic biomass which denotes plant matter.

corn steep liquor from corn wet milling, contain such nutrients. Corn steep liquor or molasses is therefore used as a supplement or replacement for water in making up a major part of the liquid fraction of the starting culture media.

Defined or Synthetic Media Are Made from Known Reagents. A culture medium made up by weighing out and adding together distinct individual chemical components is referred to as a "chemically defined" or "synthetic medium" (Shuler and Kargi 1992; Goodhue et al. 1986). A chemically defined medium has a typical compositional ratio of about 2 : 1 : 1 (by weight) carbon source, ammonium sulfate, and dipotassium phosphate, in 10 mL of a salt solution and 990 mL distilled water (Goodhue et al. 1986), although there is significant variation from the mean. When 0.05–0.5% of a single vegetable or meat extract or preparation is added to the chemically defined medium, a semidefined medium results. Small amounts of yeast extract, meat or soy peptones (hydrolyzed proteins), malt extract, casein hydrolysates, or corn steep liquor often enhance growth since they contain growth factors and vitamins that promote cell growth and may be added to fermentation media as well (Goodhue et al. 1986). This type of medium is then referred to as being "complex."

The inorganic components are essential for the growth of the organisms, since they provide cofactors required for numerous reactions in the metabolic pathways discussed later in this chapter (Fig. 4.9), as well as the chemical structure and function of proteins required for cell growth. The need for small amounts of inorganic constituents is also reflected by the composition of inorganic constituents of different microorganisms in Table 4.3.

Table 4.3.

Inorganic Constituents of Bacteria and Yeast

Element	Bacteria	Fungi (g/100 g Dry Weight)	Yeast
Phosphorus	2.0–3.0	0.4–4.5	0.8–2.6
Sulfur	0.2–1.0	0.1–0.5	0.01–0.24
Potassium	1.0–4.5	0.2–2.5	1.0–4.0
Magnesium	0.1–0.5	0.1–0.3	0.1–0.5
Sodium	0.5–1.0	0.02–0.5	0.01–0.1
Calcium	0.01–1.1	0.1–1.4	0.1–0.3
Iron	0.02–0.2	0.1–0.2	0.01–0.5
Copper	0.01–0.02	—	0.002–0.01
Manganese	0.001–0.01	—	0.005–0.007
Molybdenum	—	—	0.0001–0.0002
Total ash	7–12	2–8	5–10

Source: From Aiba et al. (1973), Table 2.5, p. 29.

Complex Media Have a Known Basic Composition but a Chemical Composition that Is Not Completely Defined

A complex medium utilizes naturally occurring components whose exact composition is not known and that may vary from one fermentation batch to the next. Constituents in complex media might include corn steep liquor, yeast extracts or hydrolysate, or molasses (see Tables 4.4 and 4.5 for approximate compositions). The typical composition of molasses (Table 4.4) illustrates how the basic composition of a complex medium is known, but its chemical composition is not completely defined (Atkinson and Mavituna 1983). Approximately 20% of the constituents are known only to be organic components. Numerous examples of complex media are given in the *Biochemical Engineering and Biotechnology Handbook* (Atkinson and Mavituna 1983). The media are often formulated for specific microorganisms, and are based on years of trial-and-error experimentation in various research laboratories and industrial facilities. For example, the formulation of *Escherichia* media might consist of 10 g tryptone, 1.0 g yeast extract, 8 g NaCl, and 1 g glucose, in 1 L of deionized water to which 2 mL of 1 M $CaCl_2$ and 1.0 mL of thiamin has been added (Atkinson and Mavituna 1983). Shuler and Kargi (1992) give an example of a complex medium used in penicillin fermentation in which an antifoaming[7] agent consists of lard.

Table 4.4.

Composition of Molasses from Sugar Beet and Cane Processing

Component	% (by weight)	
	Beet Molasses	Sugarcane Molasses
Sucrose	48.5	33.4
Raffinose	1.0	—
Hydrolyzed sucrose (invert)	1.0	21.2
Organic components (that are not sugars)	20.7	19.6
Nitrogen (protein)	2.0	—
Ash (minerals)	10.8	9.8
Water (= 100 − sum of other constituents)	16.0	16.0
Total	100	100

Source: Atkinson and Mavituna (1983), Table A1, p. 1104, based on data of Rhodes and Fletcher.

---

[7] An antifoaming agent, which is usually referred to as *antifoam*, reduces the surface tension and inhibits formation of foam that would otherwise fill up the headspace of the bioreactor, and could literally cause a significant fraction of the bioreactor's content to overflow through a relief port (designed for this purpose). Foaming is caused by the agitation and passage of air through the broth that has protein in it. The protein is generated as a consequence of cell growth. Air is required so that oxygen is provided for the growth of an aerobic microorganism.

Table 4.5.

Comparison of Major Components in Selected Fermentation Media Components

Components	Brewer's Yeast	Yeast Hydrolysate	Blackstrap Molasses	Corn Steep Liquor
Overall (%)				
Dry matter	95.0	94.5	78.0	50.0
Protein	43.0	52.5	3.0	24.0
Carbohydrate	39.5	—	54.0	5.8
Fat	1.5	0.0	0.4	1.0
Fiber	1.5	1.5	—	1.0
Ash	7.0	10.0	9.0	8.8
Minerals (%)				
Ca	0.1	—	0.74	—
Mg	0.25	—	0.35	—
Phosphorous	1.40	—	0.08	—
Available phosphorus	1.40	—	—	—
Potassium	1.48	—	3.67	—
Sulfur	0.49	—	—	—
Vitamins (mg/kg)				
Biotin	—	—	—	0.9
Choline	4840	—	660	—
Niacin	498	—	47	—
Pantothenic acid	121	—	43	—
Pyridoxine	50	—	44	19.4
Riboflavin	35	—	4.4	—
Thiamine	75	—	0.9	0.88
Amino acids (%)				
Arg	2.2	3.3	—	0.4
Cys	0.6	1.4	—	0.5
Gly	3.4	—	0.10	1.1
His	1.3	1.6	—	0.3
Ile	2.7	5.5	—	0.9
Leu	3.3	6.2	0.01	0.1
Lys	3.4	6.5	0.01	0.2
Met	1.0	2.1	—	0.5
Phe	1.8	3.7	—	0.3
Thr	2.5	3.5	0.06	—
Trp	0.8	1.2	—	—
Tyr	1.9	4.6	—	0.1
Val	2.4	4.4	0.02	0.5

Source: Adapted from Miller and Churchill (1986), Table 2, Appendix A, pp. 130–131.

Industrial Fermentation Broths May Have a High Initial Carbon (Sugar) Content (Ethanol Fermentation Example)

An example of a defined (synthetic) medium for alcohol production is given in Table 4.6. The concentration of glucose is 100 g/L and would result in 51.1 g ethanol if the fermentation efficiency were 100%. It is unlikely that industrial fuel alcohol fermentation would use a synthetic or defined medium as given in Table 4.6, since it is more expensive than a glucose syrup or hydrolyzed starch from corn wet or dry milling plants. These processes would use corn steep liquor together with starch or dextrose (glucose) and mineral supplements. Corn steep liquor contains some of the vitamins and amino acids required for the fermentation (Table 4.4) (Miller and Churchill 1986). Nonetheless, the composition given in Table 4.6 is useful as an indication of the components required for ethanol production by yeast.

If starch is used as the carbon source and the glucose is produced by the action of starch-hydrolyzing enzymes in the same tank in which the fermentation

Table 4.6.

Composition of a Defined (Synthetic) Medium for Ethanol Production

Component	Amount (g/L)
Glucose	100.0
Ammonium sulfate	5.19
Potassium dihydrogen phosphate	1.53
Magnesium sulfate 7H$_2$O	0.55
Calcium chloride 2H$_2$O	0.13
Boric acid	0.01
Cobalt sulfate 7H$_2$O	0.001
Copper sulfate 5H$_2$O	0.004
Zinc sulfate 7H$_2$O	0.010
Manganous sulfate 7H$_2$O	0.003
Aluminum sulfate	0.001
Biotin	0.002
Pantothenate	0.003
Inositol	0.000125
Thiamin	0.00625
Pyridoxine	0.125
p-Aminobenzoic acid	0.005
Nicotinic acid	0.00625
	0.001
	0.005

Source: Atkinson and Mavituna (1983), Table 27, p. 1018.

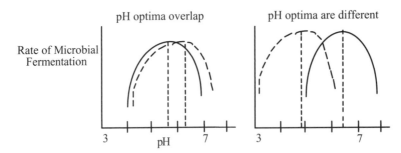

Figure 4.5. Overlap of pH optima for hydrolysis and fermentation are needed for efficient simultaneous saccharification and fermentation (SSF). If the pH optima are different, SSF may not be practical because of the mismatch between rates of enzyme-catalyzed hydrolysis and microbial fermentation.

is carried out, the fermentation is referred to as *simultaneous saccharification and fermentation* (SSF). The enzymes are amylases (degrading soluble starch or amylose principally to the disaccharide, maltose) and glucoamylase (hydrolyzes maltose to glucose). The enzymes may either be added to the fermentation broth or generated and secreted by the yeast that also converts the glucose to ethanol. If cellulose is the carbon source, and celluloses are generated, such as by a bacterium, that produces ethanol, the fermulation is referred to as *consolidated bioprocessing*.

Simultaneous saccharification and fermentation simplifies the equipment required to carry out the conversion of starch from corn to ethanol, since both hydrolysis and fermentation are carried out in the same tank, rather than in separate vessels. However, the success of this approach requires that the pH optima of the enzymes and the microorganism overlap (Fig. 4.5). If the pH optima are different, or if the temperature required for enzyme hydrolysis is significantly higher than the temperature range for yeast growth, the simultaneous saccharification and fermentation may not be practical. A mismatch between rates of hydrolysis and fermentation, and/or stability of the microorganism relative to that of the enzyme at a higher temperature, will result in an inefficient process.

The Accumulation of Fermentation Products Is Proportional to Cell Mass in the Bioreactor

Frequently the product of industrial interest is secreted by the cell into the fermentation broth. Products include low-molecular-weight components such as ethanol, lactic acid, amino acids, antibiotics, and some types of proteins. The product may also be intracellular, such as the enzyme glucose isomerase,[8] that is so large (MW > 100,000) that it can not pass through the cell's membrane. Hence, it is an intracellular product. Proteins produced using recombinant *E. coli* are another form of intracellular product. Bacteria to which a gene for expressing a foreign protein or peptide such as human insulin has been inserted produce such a high concentration of a particular protein that it coagulates within the cell and forms an inclusion

---

[8] Glucose isomerase is a commercially important enzyme that converts glucose in corn syrup to an equilibrium mixture of fructose and glucose, as described in Chapter 3. The equilibrium mixture, when processed further, yields high-fructose corn syrup (HFCS), which is used as a sweetener in a wide range of foods.

body (i.e., a body included within the cell). Other types of recombinant cells will secrete the protein product. Sometimes, the product of interest is the cell itself, such as baker's yeast. In all cases, it is useful to present the timecourse of the fermentation in terms of cell mass that has accumulated as a function of the time over which the fermentation has been carried out.

A Microbial Fermentation Is Characterized by Distinct Phases of Growth

Figure 4.6 illustrates several phases of growth. Initially there is a lag phase, during which the inoculum of microbial cells acclimates to its environment and cell growth is small. The accumulation of cell mass then increases, and enters an exponential growth phase. Finally, cell growth levels off because of depletion of substrate in the culture media, and attains the stationary phase. The stationary phase is followed by the death phase. Fermentations are usually stopped, and the broth processed to recover the product, shortly before the stationary phase is attained, as indicated by point B in Fig. 4.6. When viable (living) cells in a smaller fermentation vessel are used to inoculate a larger fermentation vessel, the broth is harvested and transferred to the larger vessel in midexponential phase (denoted by x in Fig. 4.6).

The average rates shown in Fig. 4.6 represent productivities that are measured between point A where the fermentor is inoculated and points B or C, where the fermentation is stopped and the broth is pumped out. The largest average productivity of cells coincides with the line having the greatest slope. These productivities are higher than those measured in the context of process economics where the time required to carry out a fermentation includes the time of the fermentation itself plus the turnaround time. As pointed out by Gaden (1959), operational factors involved in turnover of a tank include cleaning, sterilization, filling, inoculating, cell growth, and emptying the vessel. However, the modeling or quantitation of the fermentation

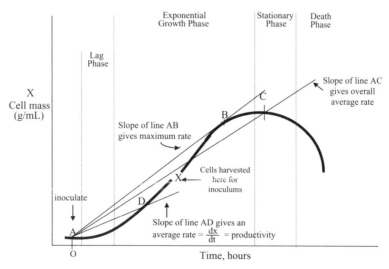

Figure 4.6. Schematic illustration of several phases of growth [adapted from Gaden (1959), Fig. 1, p. 417 and Monod (1942), Fig. 1, p. 7]. The cell mass concentration (y axis) typically varies between 2 and 50 g/L; time (x axis), between 4 h and 5 days.

itself is most meaningful during the times when cell growth is occurring. Consequently, expressions for rates of cell growth and product formation are based on the exponential growth phases, with the lag phase handled separately (Bailey and Ollis 1977; Basak et al. 1995). The rate of disappearance of nutrients during the exponential growth phase is based on a limiting substrate that is dissolved in the liquid phase, and represents the dominating source of carbon and energy. This is often the sugar, glucose.

Expressions for Cell Growth Rate Are Based on Doubling Time

The generation of cells or cell mass X is given by:

$$X = X_0 2^n \tag{4.3}$$

where the initial cell mass is given by X_0 and the number of generations, by n. If the doubling of cell mass ($n = 2$) occurs over a time period t_d, then the growth rate is given by $\mu_d = n/t_d$. Equation (4.3) can be rewritten as

$$X_d = X_0 2^{\mu_d t_d} \tag{4.4}$$

where X_d is the cell mass at time t_d and $n = \mu_d t_d$. Rearrangement of Eq. (4.4), and taking the logarithm, gives

$$\log \frac{X_d}{X_0} = \mu_d t_d \log 2 \tag{4.5}$$

and with further rearrangement an expression for a growth rate μ_d in terms of doubling time t_d is obtained (Monod 1942):

$$\mu_d = \frac{\log X_d - \log X_0}{t_d \log 2} \tag{4.6}$$

A different form of Eq. (4.6) enables a definition of the specific growth rate, used in unstructured models, to be rationalized on the basis of a semilog plot of cell mass as a function of time (Fig. 4.7). If the natural logarithm ($\ln X = 2.303 \log X$)

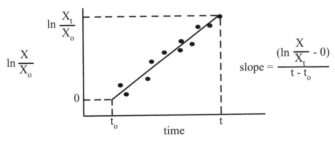

Figure 4.7.   Linearized (semilog) plot of cell mass (X) as a function of time.

of cell mass is plotted as a function of time during the exponential growth phase, the data should fall on a straight line as shown in Fig. 4.7.

The dots (•) represent data points. The fractional cell mass at $t = t_0$ is defined by $X/X_0 = 1$ so that $\ln(X/X_0) = 0$. Since the slope over a small time interval is defined by $d(\ln(X/X_0))/dt$, the expression for the specific growth rate μ is

$$\frac{d\ln(X/X_0)}{dt} = \frac{1}{X}\frac{dX}{dt} = \mu \tag{4.7}$$

The slope of the graphical representation in Fig. 4.6 corresponds to the definition of specific growth rate in Eq. (4.7) for an initial population of cells (X_0). Unstructured models[9] describe the time-course of cell accumulation during the exponential growth phase. Such models are useful in addressing the industrially relevant exponential growth phase. This is illustrated in Fig. 4.8 with a linear scale on the left and a logarithmic plot on the right.

The goal in a practical fermentation is to maximize the amount of product obtained per unit volume of the fermentation vessel. This is achieved in many fermentations when the lag and stationary phases are as short as possible, and the timecourse of the overall batch process coincides with the exponential growth phase. The correlation of cell population to the amount of product formed requires that yield coefficients be defined that have the following general form:

$$Y = \frac{\text{increase in product}}{\text{increase in cell mass}} = \frac{\Delta P}{\Delta X} \tag{4.8}$$

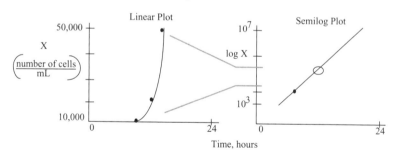

Figure 4.8. Comparison of linear and semilog plots of cell mass versus time from fermentation.

---

[9] An *unstructured model* is a mathematical representation of cell growth as a function of changes in broth composition. Since the model is not based on a mechanism (such as the kinetics of a metabolic pathway) but only an observation of changes in composition, it is termed *unstructured*. Unstructured models quantify the increase of cell population only in terms of population number or mass and do not include expressions that are written in terms of the cell's metabolic pathways. Such models do not reflect the physiological state of the population (Bailey and Ollis 1977). Hence, an unstructured model does not distinguish the differences in cell metabolism that may occur between lag and exponential, or stationary and exponential, growth phases.

Yield coefficients relate concentration of product to the amount of cell mass as well as the pattern of the cell mass accumulation during the course of the fermentation.

Products of Microbial Culture Are Classified According to Their Energy Metabolism (Types I, II, and III Fermentations)

Products of microbial metabolism are measured in terms of weight of product per weight of cell mass formed, or weight of product per weight of a substrate consumed. Fermentation efficiencies are calculated from the ratio of product formed to the amount of total product that theoretically could be formed. The correlation of the amount of a metabolite generated followed by excretion into the medium, or accumulation inside the cell, is the basis of a classification scheme that reflects the role of the metabolite with respect to the cell's growth.

There are three major fermentation rate patterns Type I, II, or III that help to correlate yield to the rate of cell mass accumulation. These are illustrated in Figs. 4.9, 4.10, and 4.11, respectively, and were proposed by Bailey and Ollis (1977) and by Gaden (1959). Depending on the type of fermentation (I, II, or III), different proportions of the glucose will be assimilated into cell mass, extracellular product, growth energy (heat), and maintenance energy. *Maintenance energy* is the energy required to maintain cellular functions, that is, the biochemical reactions that define a living cell when the cell is in a resting state.

Type I fermentations (either aerobic or anaerobic) represent processes where the main product is a coproduct of the cell's primary energy metabolism (Fig. 4.9). Examples include the aerobic propagation of cells, the accumulation of ethanol under nonaerobic conditions, and the production of lactic acid. The formation of these products coincides with the glycolytic and/or citric acid cycle pathways illustrated in Fig. 4.10.

A *Type II* fermentation defines processes in which product accumulation is an indirect consequence of the cell's energy metabolism. The product is not directly obtained from oxidation of the carbon source. Rather, it results from transformation of or between metabolites, such as citric acid derived through the citric acid cycle or the synthesis of amino acids by the cell. There is a delay between onset of growth and accumulation of the product in the broth.

Note that the concentrations of cell mass, product, and substrate in Fig. 4.9 are scaled to facilitate comparison between the curves, for purposes of explanation.

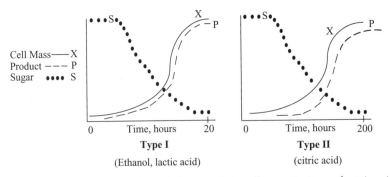

Figure 4.9. Schematic representation of characteristic cell mass (—), product (- - -), and sugar (···) accumulation for Types I and II fermentations. (Gaden 1959; Bailey and Ollis 1977).

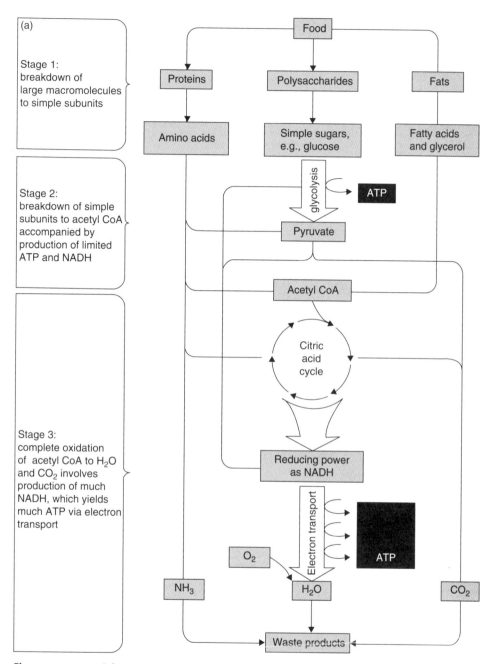

Figure 4.10. (a) Schematic representation of the three stages of catabolism;

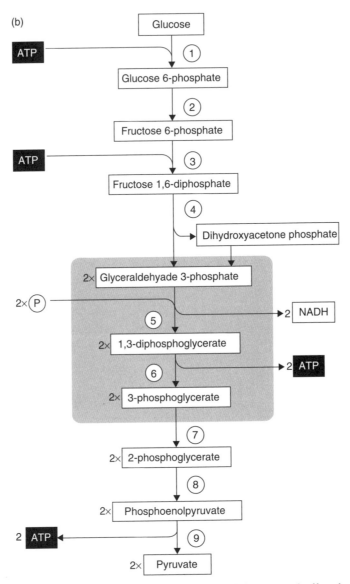

Figure 4.10. *(Continued)*   (b) glycolysis (glycolytic or Embden–Meyerhoff pathway);

The concentration of cells, product, and substrate are actually quite different. For example, in an ethanol fermentation, if the initial substrate concentration were 100 g/L, the final cell and ethanol concentrations might be on the order of 10 and 45 g/L, respectively. In the case of citric acid, the concentration of mycelial cell mass is about 20 g/L, compared to 160 g/L citric acid for an initial sugar concentration of 240 g/L (Bailey and Ollis 1977).

Type III fermentations characterize processes in which the product appears or is accumulated independently of the cell's energy metabolism, that is, not directly derived from the glycolytic or citric acid cycle pathways shown in Fig. 4.10c (Gaden

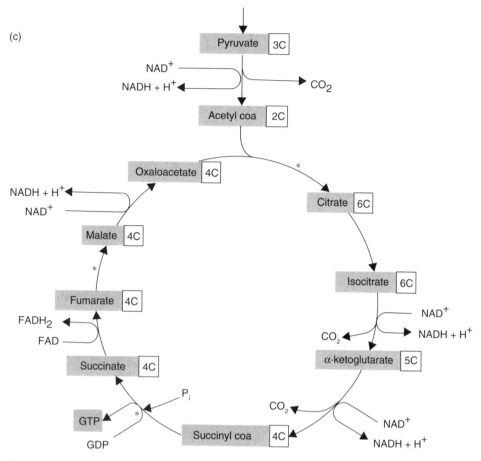

Figure 4.10. *(Continued)*   (c) citric acid cycle, with the number of carbons in each chemical species are indicated in the box [from Alberts et al. 1989, Fig. 4.18, p. 64; Fig. 4.20, p. 66; and Fig. 4.23, p. 68];

1959). The product accumulation pattern is illustrated in Fig. 4.11. The synthesis of a secondary metabolite is usually initiated late in the growth cycle since this metabolite could also be toxic to the cell that makes it. Nonetheless, these metabolites may have survival value for the organism that produces them, and are produced as mixtures of closely related chemical compounds. Wang et al. (1979) give the example of antibiotics where at least 20 penicillins, 8 alfatoxins, and 10 bacitracins are produced by a narrow taxonomic range of organisms. Mycotoxins and pigments are also secondary metabolites.

Type III fermentations (which are usually aerated) have at least three subtypes: a, b, and c as illustrated in Figs. 4.11 and 4.12. Subtype III (a) is a pattern where cell accumulation is followed by product generation. As stated by Gaden (1959, p. 422): "all aspects of energy metabolism are maximized with virtually no accumulation of desired product; in the second phase oxidative metabolism is practically over and product accumulation is maximum." Penicillin and streptomycin

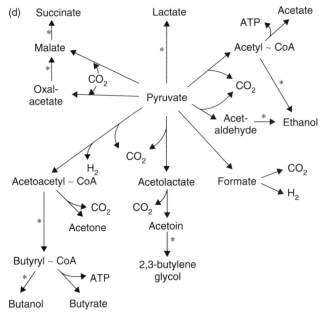

Anaerobic metabolism of pyruvate by different organisms.
*=Reactions that oxidize NADH$_2$ → NAD$^+$

Organism	Major Products
Clostridia	Butyric, acetic acids, butanol, ethanol, acetone, CO$_2$, H$_2$.
Enteric bacteria	Acetic acid, ethanol, CO$_2$,H$_2$ (or formic acid), lactic acid, succinic acid, 2,3-butylene glycol.
Yeast	Ethanol, CO$_2$.
Homofermentative lactobacilli	Lactic acid.

Figure 4.10. *(Continued)* (d) products from pyruvate anaerobic metabolism of pyruvate by different microorganisms that do not involve the citric acid cycle [from Aiba et al. (1973), Fig. 3.7, p. 67].

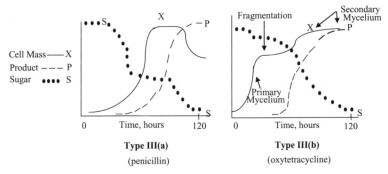

Figure 4.11. Schematic representation of curves for characteristic cell mass (—), product (- - -), and sugar (···) accumulation. Two examples of Type III fermentations are given in this figure.

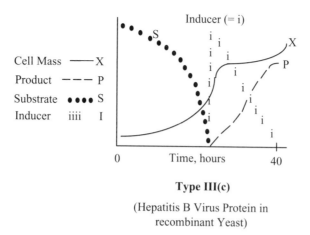

Type III(c)

(Hepatitis B Virus Protein in
recombinant Yeast)

Figure 4.12. Characteristic cell mass (—), product (- - -), and sugar (···) accumulation for Type III fermentation where the product is not produced until an inducer is added.

fermentations are examples of Type III (a). The penicillin concentration after 80 h of fermentation might be 0.8 g/L, compared to a mycelial cell mass of 20 g/L. Type III (b) consists of two growth phases, with the second phase resulting in formation of the product as illustrated in the right-hand graph of the schematic in Fig. 4.11 (Goodhue et al. 1986). This behavior is characteristic of fermentations in which the microorganism initially undergoes mycelial growth, with the microbial structures consisting of multiple cells that exhibit a high level of metabolism. These mycelia slow in their growth rate, fragment, and start to grow again, on which significant antibiotic production occurs. Examples of Type III (b) fermentations are the production of oxytetracycline and chloramphenicol (Gaden 1959).

The classification of fermentations was proposed long before genetically engineered microorganisms were developed or gene regulation and protein expression were understood. The growth and induction of a protein product in recombinant bacteria, such as *E. coli*, also fits the classification of a Type III fermentation as is illustrated in Fig. 4.12. In this case, propagation of the microorganism is carried out to maximize cell mass, and then expression (i.e., production) of the protein is initiated by a change in composition in the broth or addition of an inducer. This approach has been used in both prokaryotes [*E. coli* for the production of insulin as discussed by Ladisch and Kohlmann (1992)], and eukaryotes (yeast for the production of the envelop protein of hepatitis B virus).[10]

---

[10] Hepatitus B virus (HBV) causes several types of liver disease and infects 350 million people worldwide—mostly without producing symptoms. A fraction of the cases lead to liver failure or liver cancer and kill an estimated 1 million people worldwide and 4000 in the United States (Marshall 1998). Many people who are infected do recover. However, some do not. This has resulted in cases where people are chronically infected and suffer from further long-term complications, including cirrhosis and/or hepatocellular carcinoma. These people are also carriers of the virus, which can infect others. This has motivated research on development of vaccines. The first-generation vaccine, made from 22-nm particles or HBV surface antigen particles purified from the plasma of chronic carriers, has proved safe and effective.

The production of the envelop protein of hepatitis B used an ingenious approach based on a regulated galactose-1-phosphate uridyltransferase promoter (abbreviated Gal-10). The cells are grown in a medium containing glucose. When the glucose is nearly depleted, galactose is added. The presence of galactose initiates a sequence of events that activates the promoter so that the gene with which it is associated is transcribed, with the resulting mRNA translated into protein. This chain of events results in secretion of the desired product (Kniskern et al. 1991). This approach enables significant cell mass to be propagated before the fermentation generates the product, which is toxic to the host cell and inhibits the cell's growth. Since the yield of product is proportional to the number of cells, an objective of this approach is to maximize product accumulation by maximizing cell mass[11] before inducing expression of the protein.

Product Yields Are Calculated from the Stoichiometry of Biological Reactions (Yield Coefficients)

The yield of products is related to the number of cells as defined by Eq. (4.8) and illustrated in Figs. 4.8, 4.10, and 4.11. The stoichiometry between substrates and products can be related to yield by atomic formulas for the substrates, products, and cells that are present during a given time interval of the fermentation. The stoichiometric formula for cell growth can be represented by the method of Wang et al. (1979) using the following reaction:

$$A (C_a H_b O_c) + B (O_2) + D (NH_3)$$

$$\text{carbon source} \quad + \quad \text{oxygen} \quad + \quad \text{nitrogen source}$$

$$\longrightarrow M (C_\alpha H_\beta O_\gamma N_\delta) + N (C'_\alpha H'_\beta O'_\gamma N'_\delta) + P (CO_2) + Q (H_2O) + \text{heat} \quad (4.9)$$

$$\text{cell mass} \qquad \text{product}$$

$$E (C_\alpha H_\beta O_\gamma N_\delta)$$

$$\text{inoculum}$$

The carbon source (usually a sugar), oxygen, and nitrogen source are as indicated. The distribution of cell mass and the metabolites (product, water, and CO_2) on the right side of Eq. (4.9) are represented by the stoichiometric coefficients M, N, P, and Q. The starting amounts of carbon, oxygen, nitrogen (represented as NH_3) and inoculum are given on the left side of Eq. (4.9) in terms of coefficients A, B, D, and E. Since the inoculum results in only a small initial cell mass (i.e., $X_0 << X$), the

---

Larger quantities were needed, and hence recombinant organisms for the production of the antigen were pursued as a potential source of the vaccine (Kniskern et al. 1991). However, since the first vaccine was licensed in 1981 and 200 million people have been immunized, approximately 20,000 adverse reactions have been reported. Some of these resulted in immune disorders that resemble multiple sclerosis. This has caused concern about government policies that mandate immunization of healthcare workers and newborn infants (Marshall 1998).

[11] Note that schematic (Miller and Churchill 1986) is a representation of a possible timecourse for explanation purposes. The authors of this work did not present time course data (Kniskern et al. 1991).

contribution of E ($C_\alpha H_\beta O_\gamma N_\delta$) due to initial cell mass is ignored and does not appear on the right side of Eq. (4.9).

The yield coefficient is given the name "coefficient" rather than "constant" since it represents an averaged behavior of the cell. It does not represent changes in the cell's metabolism that may occur between the beginning and the end of the fermentation. A change in the proportion of substrate that is directed to cell mass, products, or heat will occur depending on the nature of the fermentation (Shuler and Kargi 1992; Bailey and Ollis 1977), and is discussed in Chapters 8 and 9, on metabolism and metabolic engineering.

Yield coefficients may be defined on the basis of the stoichiometry of Eq. (4.9) when prior research or experimental data have shown that this accurately represents the elemental balance in the cell. The measured values are empirical since they depend on culture conditions. Hence, the application of the yield coefficient needs to be considered on a case-by-case basis. The biology, biochemistry, and metabolism of the living cell must be understood, so that the yield coefficient is properly used and interpreted.

The yield coefficient $Y_{X/S}$, for the amount of cell mass that appears for a given disappearance of the carbon energy source (i.e., substrate), is given in Eq. (4.10). The subscripts for the yield coefficient are shown in the order of the terms in the numerator and denominator, respectively. The parameters α, β, γ, δ correspond to the subscripts in the molecular formulas in Eq. (4.9), while the numerical coefficients for each of the subscripts represent the molecular weights of carbon, C (= 12), hydrogen, H (= 1), oxygen, O (= 16), and nitrogen, N (= 14):

$$Y_{X/S} = \underbrace{\frac{\text{mass of cells produced}}{\text{mass of substrate consumed}}}_{\text{Experimentally measured}} = \frac{X - X_0}{S_0 - S} = \frac{\Delta X}{\Delta S} \overset{?}{=} \underbrace{\frac{\left[\dfrac{M(12\alpha + 1\beta + 16\gamma + 14\delta)}{(1 - r)}\right]}{A(12a + 1b + 16c)}}_{\text{Elemental balance}}$$

(4.10)

The correspondence between measurement ($\Delta X/\Delta S$) and theory is not exact as indicated by the question mark over the equal sign. The parameter r represents the fraction of cell mass that corresponds to ash (Table 4.2), so that the balance represented in Eq. (4.10) is on a C, H, O, and N basis, when divided by $(1 - r)$. This is inexact, for example, when a balance is calculated for ATP, since the inorganic component, phosphorous, is part of its structure and represents a part of its molecular weight. Consequently, the reaction in Eq. (4.10) must show additional terms that explicitly account for phosphorous, if a theoretical estimate were to be made.

The yield coefficient, Y_{PS}, is based on the amount of product (as C, H, O, and N) that appears relative to substrate (carbon energy source) that disappears. The calculated values of $Y_{X/S}$, $Y_{P/S}$, and other yield coefficients represent approximations.

$$Y_{P/S} = \frac{(\Delta P)}{\Delta S} = \frac{P - P_0}{S_0 - S} \overset{?}{=} \underbrace{\frac{N}{A} \frac{(12\alpha' + 1\beta + 16\gamma' + 14\delta')}{(12a + 1b + 14c)}}_{\text{Elemental balance}} \left(\frac{\text{g product}}{\text{g substrate}}\right) \quad (4.11)$$

Another yield coefficient gives the amount of cells propagated per amount of oxygen consumed, where ΔO_2 represents the cumulative difference between oxygen entering and leaving the bioreactor as part of the gas stream.

$$Y_{X/O_2} = \frac{X - X_0}{(O_2)_0 - O_2} = \frac{(\Delta X)}{\Delta O_2} = \frac{\text{g cells generated}}{\text{g oxygen consumed}} \qquad (4.12)$$

The ratio of the weight of cells synthesized (dry basis) to mole ATP synthesized per mole energy substrate (such as glucose) consumed is given by (Aiba et al. 1973):

$$Y_{X/ATP} = \frac{X - X_0}{\left(\dfrac{ATP}{S_0 - S}\right)} = \frac{(\Delta X)}{\left(\dfrac{ATP}{\Delta S}\right)} \qquad (4.13)$$

The amount of ATP generated relative to the carbon substrate consumed (such as glucose) depends on whether the fermentation or culture is carried out under anaerobic or aerobic conditions. The metabolic pathway for anaerobic growth proceeds through the glycolytic pathway (Fig. 4.10b) and results in a predictable 2 mol ATP formed (from ADP) per mole glucose consumed for a wide range of microorganisms. Aiba et al. (1973) present the analysis of Bauchop and Elsden, supported by data of other researchers, which shows that the value of $Y_{X/ATP}$ is constant for anaerobic metabolism and averages 10 ± 2 (g cells generated)/(mole ATP generated) if the energy substrate is the only factor that limits cell growth (Aiba et al. 1973). If inhibitors of cell growth are present, the yield will fall below this range.

If the energy substrate is present at an excess concentration (i.e., it is no longer the limiting substrate), part of the substrate might be stored as a polysaccharide (such as glycogen). The yield of cells (measured on a dry basis) can then appear to be larger, with values of $Y_{X/ATP}$ exceeding 10. A significant variation in Y_{ATP} is possible with values ranging from 6 to 29 g cells/[(mol ATP)/(g energy substrate)] (Wang et al. 1979). Some of this variation can result from differences in the maintenance requirements of the cells, and the conditions for growth.

Yield coefficients are useful and widely used despite their inherent approximations. The rates of cell growth and product accumulation in Type I fermentations parallel each other as conceptually illustrated in Fig. 4.9. Hence $Y_{X/P}$ is constant and can be used in formulating an unstructured model. Similarly, the decrease in substrate concentration and increase in cell mass and product are inversely related, and the magnitude of the slopes is similar. A constant yield coefficient $Y_{X/S}$ results. Expressions for modeling the timecourse of Type I fermentations can be developed in which the concentrations of cell mass, dissolved products, and substrates are related to one another through the yield coefficient. Similarly the relation between product and cell mass in Type II fermentations can be related, if product accumulation and substrate depletion parallel cell growth, even though there is a temporal offset between cell and product accumulations.

The Embden–Meyerhof Glycolysis and Citric Acid Cycles Are Regulated by the Relative Balance of ATP, ADP, and AMP in the Cell

The amount of high-energy phosphate bonds is given by the energy charge ratio in Eq. (4.14) (Cooney et al. 1969), as proposed by Atkinson in 1969 (Wang et al. 1979):

$$\text{Energy charge} = \frac{[\text{ATP}] + \frac{1}{2}[\text{ADP}]}{[\text{ATP}] + [\text{ADP}] + [\text{AMP}]} \tag{4.14}$$

In vitro studies show that catabolic enzymes such as phosphofructose kinase in the glycolytic pathway and isocitrase dehydrogenase in the citric acid cycle (Fig. 4.10c) are inhibited by a high energy charge. These enzymes lead to production of energy as ATP and NADH, respectively (Wang et al. 1979). The control of the activity of these enzymes is inversely proportional to the amount of chemical energy in the cell, and hence is part of the mechanism that regulates the cell's level of activity. Conversely, pyruvate carboxylase is activated by high energy levels, since this enzyme catalyzes the carboxylation of pyruvate to give oxaloacetate for the citric acid cycle as shown Fig. 4.10b and c (Aiba et al. 1973; Alberts et al. 1989; Murray et al. 1990). This balances the generation of ATP by an increased rate of its use. The other enzyme that catalyzes a reaction leading to the citric acid cycle is coenzyme A carboxylase, which forms citric acid. It is also activated by a high level of ATP.

The cell yields of aerobically grown bacteria, and other organisms, are difficult to classify with respect to ATP generation. A narrow range of values for $Y_{X/\text{ATP}}$ can not be assigned (Aiba et al. 1973; Shuler and Kargi 1992; Wang et al. 1979). However, the maximum amount of ATP formed can be estimated. Under aerobic conditions, it is as large as 38 mol ATP per mole of glucose for eukaryotes, where oxidation of glucose occurs in the mitochondria. The maximal yield is 16 mol ATP/mole of glucose for prokaryotes (Aiba et al. 1973). These represent maxima, since the citric acid cycle is catalytic, with at least five points of entry and exit, and an ill-defined stoichiometry with respect to the acetyl-CoA that marks the transition between the Embden–Meyerhoff–Parnas pathway and the citric acid cycle (Fig. 4.10c).

REFERENCES

Aiba, S., A. E. Humphrey, and N. F. Millis, *Biochemical Engineering*, 2nd ed., Academic Press, New York, 1973, pp. 28–32, 56–91, 195–345.

Alberts, B., D. Bray, J. Lewis, M. Raff, K. Roberts, and J. D. Watson, *Molecular Biology of the Cell*, 2nd ed., Garland Publishing, New York, 1989, pp. 58, 65, 78, 205–216, 341, 345–348, 353–354, 359, 362–363, 373, 381–383, 601.

Altaras, N. E., D. C. Cameron, "Metabolic Engineering of a 1,2-Propanediol Pathway in *Escheria coli*," *Appl. Environ. Microbiol.*, **65**(3), 1180–1185 (1999).

Atkinson, B. and F. Mavituna, *Biochemical Engineering and Biotechnology Handbook*, Nature Press, Macmillan, 1983, pp. 51–62, 320–325, 1018, 1058, 1004.

Bailey, J. E. and D. F. Ollis, *Biochemical Engineering Fundamentals*, McGraw-Hill, New York, 1977, pp. 3, 14–15, 221–254, 338–348, 357, 371–383.

Basak, S., A. Velayudhan, and M. R. Ladisch, "Simulation of Diauxic Production of Cephalosporin C by *Cephalosporium acremonium*: Lag Model for Fed-Batch Fermentation," *Biotechnol. Progress* **11**, 626–631 (1995).

Battley, E. H., "Calculation of Heat of Growth of *Escherichia coli* K-12 on Succinic Acid," *Biotechnol. Bioeng.* **37**, 334–343 (1991).

Bjurstrom, E., "Biotechnology: Fermentation and Downstream Processing," *CEP* **92**(4), 126–158 (1985).

Blanch, H. W. and D. S. Clark, *Biochemical Engineering*, Marcel Dekker, New York, 1996, pp. 276–330.

Builder, S. E., R. van Reis, N. Paoni, and J. Ogez, "Process Development and Regulatory Approval of Tissue-Type Plasminogen Activator," *Proc. 8th Intnatl. Biotechnology Symp.*, Paris, July 7–22, 1988.

Builder, S. E. and E. Grossbard, "Laboratory and Clinical Experience with Recombinant Plasminogen Activator," in *Transfusion Medicine: Recent Technological Advances*, Alan R. Liss, New York, 1986, pp. 303–313.

Cameron, D. C., N. E. Altaras, M. L. Hoffman, A. J. Shaw, "Metabolic Engineering of Propanediol Pathways," *Biotechnol. Progr.*, **14**(1), 116–125 (1998).

Cooney, C. L., D. I. C. Wang, and R. I. Mateles, "Measurement of Heat Evolution and Correlation of Oxygen Consumption during Microbial Growth," *Biotechnol. Bioeng.* **11**, 269–281 (1969).

Datar, R. P., R. M. Shenkman, B. G. Cateni, R. L. Huhnke, and R. S. Lewis, "Fermentation of Biomass-Generated Producer Gas to Ethanol," *Biotechnol. Bioeng.* **86**(5), 587–594 (2004).

Eckhoff, S. R. and C. C. Tso, "Starch Recovery From Steeped Corn Grits as Affected by Drying Temperature and Added Commercial Protease," *Cereal Chem.* **68**(3), 319–320 (1991).

Fox, E. J. and S. R. Eckhoff, "Wet-Milling of Soft Endosperm, High-Lysine Corn Using Short Steep Times," *Cereal Chem.* **70**(4), 402–404 (1993).

Gaden, E. L. Jr., "Fermentation Process Kinetics," *J. Biochem. Microbiol. Technol. Eng.* **1**(4), 413–429 (1959).

Goodhue, C. T., J. P. Rosazza, and G. P. Peruzzotti, "Methods for Transformations of Organic Compounds," in *Manual of Industrial Microbiology and Biotechnology*, A. L. Demain and N. A. Solomon, eds., American Society for Microbiology, Washington, DC, 1986, pp. 108–110.

Gulati, M., K. Kohlmann, M. R. Ladisch, R. Hespell, and R. J. Bothast, "Assessment of Ethanol Production Options for Corn Products," *Bioresource Technol.* **58**, 253–264 (1996).

Henstra, A. M., J. Sipma, A. Rinzema, and A. J. M. Stams, "Microbiology of Synthesis Gas Fermentation for Biofuel Production," *Current Opin. Biotechnol.* **18**(3), 200–206 (2007).

Hoseney, R. C., *Principles of Cereal Science and Technology*, American Association of Cereal Chemists, St. Paul, MN, 1986, pp. 153–163.

Humphrey, A., "Shake Flask to Fermentor: What Have We Learned?" *Biotechnol. Progress* **14**, 3–7 (1998).

Jarboe, L. R., T. B. Grabar, L. P. Yomano, K. T. Shanmugan, and L. O. Ingram, "Development of Ethanologenic Bacteria," *Adv. Biochem. Eng. Biotechnol.* **108**, 237–261 (2007).

Kabnick, K. S. and D. A. Peattie, "*Gardia*: A Missing Link between Prokaryotes and Eukaryotes," *Am. Scientist* **79**(1), 34–43 (1991).

Kniskern, P. J., A. Hagopian, D. L. Montgomery, C. E. Carty, P. Burke, C. A. Schulman, K. J. Hofmann, et al., "Constitutive and Regulated Expression of the Hepatitis B Virus

(HBV) Pre SZ + S Protein in Recombinant Yeast," in *Expression Systems and Products for rDNA Products*, R. T. Hatch, C. Goochee, A. Moreira, and Y. Alroy, eds., ACS Symp. Ser. 477, 1991, pp. 65–75.

Koch, A. L. and M. Schaechter, "The World and Ways of *E. coli*," in *The Biology of Industrial Microorganisms*, A. L. Demain and N. A. Solomon, eds., Benjamin/Cummins, Menlo Park, CA, 1985, pp. 1–25.

Ladisch, M. R. and K. Kohlmann, "Recombinant human insulin," *Biotechnol. Prog.* 8(6), 469–478 (1992).

Ladisch, M. R. and K. A. Svarczkopf, "Ethanol Production and the Cost of Fermentable Sugars from Biomass," *Bioresource Technol.* 36, 83–95 (1991).

Marshall, E., "A Shadow Falls on Hepatitis B Vaccination Effort," *Science* 28(5377), 630–631 (1998).

Mendosa, D., "Is the Cost of Insulin Skyrocketing?" *Diabetes Wellness Newslett.* 1, 7 (April 2001).

Miller, T. L. and B. W. Churchill, "Substrates for Large-Scale Fermentations," in same reference as for 44, 1986, pp. 129–136.

Monod, J., *Recherches sur La Croissance des Cultures Bacteriennes*, Actualities Scientifiques et Industrielles 911, Microbiologie—Exposés Publiés sous la Direction de J. Bordet (Director de L'Institut, Pasteur de Bruxelles, Prix Nobel), Hermann and Cie, Paris, 1942, pp. 16–17.

Morrell, V., Scarred Revolutionary, *Science* 276(5313), 699–702 (1997).

Mosier, N. S., C. Wyman, B. Dale, R. Elander, Y. Y. Lee, M. Holtzapple, and M. R. Ladisch, "Features of Promising Technologies for Pretreatment of Cellulosic Biomass," *Bioresource Technol.* 96, 673–686 (2005).

Murray, R. K., P. A. Mayes, D. K. Granner, and V. W. Rodwell, *Harper's Biochemistry*, 22nd ed., Appleton & Lange, Norwalk, CT, 1990, pp. 10–14, 58, 68–69, 102–106, 146, 160–161, 165, 199–200, 318–322, 336–338, 549–550, 555.

New Brunswick Scientific, Inc., *General Catalogue 1028*, 1982, pp. 13–18, C-22.

Perkins, D., W. Tyner, and R. Dale, "Economic analysis of a modified dry grind ethanol process with recycle of pretreated and enzymatically hydrolyzed distillers' grains," *Biores. Tech.* 99, 5243–5249 (2008).

Phaff, H. I., "Industrial Microorganisms," *Sci. Am.* 245(3), 77–89 (1981).

Shuler, M. L. and F. Kargi, *Bioprocess Engineering: Basic Concepts*, Prentice-Hall, Englewood Cliffs, NJ, 1992, pp. 48–56, 59, 158–159.

Swiatek, J., "Eli Lilly Will End Its Sale of 4 Insulins," *The Indianapolis Star*, Business Section (July 6, 2005).

Thayer, A. M., "Biopharmaceuticals Overcoming Market Hurdles," *Chem. Eng. News* 69(8), 27–48 (1991).

Tong, I-T., D. C. Cameron, "Enhancement of 1,3 Propanediol Production by Co-fermentation in *Escherichia coli*, Expressing Klebsiella pneumonia dha regulon genes," *Appl. Biochem. Biotechnol.*, 34–35(1), 149–159 (1992).

Wang, D. I. C., C. L. Cooney, A. L. Demain, P. Dunnill, A. E. Humphrey, and M. D. Lilly, *Fermentation and Enzyme Technology*, Wiley, New York, 1979, pp. 10–11, 14–36, 75–77.

Wells, G. H., "The Dry Side of Corn Milling," *Cereal Foods World* 24(7), 333 (1979).

4.1. For each term, check only *one* answer:

Yeast is a

_____ prokaryote

_____ eukaryote

E. coli is a bacterium that

_____ has a nuclear membrane that surrounds its DNA

_____ does *not* have a nuclear membrane that surrounds its DNA

E. coli is a

_____ procaryote

_____ eucaryote

A *type I fermentation* is a fermentation where an extracellular product is the direct consequence

_____ of the microbes energy metabolism

_____ of the microbe's response to environmental stress

Ethanol fermentation is:

_____ Type I

_____ Type II

_____ Type III

Which one is more resistant to being killed during *sterilization* and therefore requires that a higher temperature be used?

_____ vegetative cells

_____ spores

Ethanol fermentation is

_____ anaerobic

_____ microaerobic

_____ aerobic

A product of *glycolysis* is:

_____ ADP

_____ pentose phosphate

_____ ATP

4.2. Fermentation conditions result, at most, in production of 2 mol of ethanol from 1 mol of glucose. Sugarcane molasses is being evaluated as a potential substrate for ethanol production. The composition of the molasses is given in Table 4.4. Carry out a preliminary evaluation of the substrate costs for such a process.

    (a) What is the theoretical yield of ethanol (weight basis) per gram glucose? (20 points)

    (b) If molasses is used as a fermentation substrate, what is the minimum substrate cost for ethanol production? (20 points)

(c) If the target price is $1.10/gal (reflects operating and capital costs) and capital cost is equivalent to $2.50/annual gallon installed capacity, does molasses appear to be a good substrate for this process? Show your analysis and explain. Why or why not? (20 points)

4.3. The calculation of theoretical yields of propanediol from a fermentable substrate enables estimation of substrate costs for the best-case scenarios. Assume that glucose costs 5¢/lb (calculated on a dry-weight basis). (Tong and Cameron 1992; Cameron et al. 1998; Altaras et al. 1999).

(a) What is the cost of substrate if a hypothetical electrochemical route is able to split water and *all* of the glucose is converted to a mixture of 1,2-propanediol and 1,3-propanediol? The hypothetical reaction sequence for producing the propanediol is as shown below, with the [H] requirement obtained from the dissociation of the water:

$$C_6H_{12}O_6 + 8[H] \rightarrow 2C_3H_8O_2 + 2H_2O \qquad \text{(HP4.3)}$$

(b) What is the maximum yield of propanediol per pound glucose if the glucose provides both the carbon that is incorporated into the propanediol, and sufficient hydrogen (reducing equivalents) needed for formation of the propanediol, as well as energy requirements of the cell (results in formation of CO_2)? Explain your rationale and your work.

(c) What is the best possible yield under anaerobic fermentation conditions? Explain your rationale, and show your work.

(d) What is the best possible yield under aerobic conditions (involving the TCA cycle)?

$$\text{If } Y_{ATP} = \left(\frac{\text{mole ATP}}{\text{mole glucose}} \right), \text{ the}$$

$$\text{yield of propanediol is } 1.42 \frac{\text{mol}}{\text{mol}} \text{ or } 0.58 \frac{\text{g}}{\text{g}}$$

The calculation of theoretical yields of propanediol from a fermentable substrate enables estimation of substrate costs for the best-case scenarios.

4.4. Your company, which is located in Indiana, is considering producing fuel ethanol from waste cellulose. The microorganism to be used is *Saccharomyces cerevisiae* var. anamensis. Estimate the fermentation volume and supply of municipal solid waste (containing 50% cellulose, dry-weight basis) required to produce 20 million gal per year of fuel-grade alcohol containing 0.6% water. A concentration of 16.7% glucose is used as the initial concentration in the feed. Give your solution in terms of tons per year of municipal solid waste (dry basis) and liters of fermentation volume. Assume the turnaround time between batches to be 8 h, with an additional 4 h for the lag phase. The working volume of the fermentor, which is agitated, is 70% of the total volume.

CHAPTER FIVE

MODELING AND SIMULATION

INTRODUCTION

The calculation and visualization of timecourses of cell mass accumulation, substrate depletion, and product formation during microbial fermentation or cell culture depends on simultaneous solution of first-order differential equations. Once the equations are developed, and the values of constants and yield coefficients are defined, time-dependent concentration profiles may be generated. We use the Runge–Kutta method for this purpose since it is easy to set up and can be solved using a Visual Basic™ program embedded in an Excel™ spreadsheet or using software such as Matlab™, MathCAD™, Maple™, or other software capable of numerically solving systems of ordinary differential equations.

Simulations are readily carried out, and results plotted in an Excel spreadsheet for batch and fed-batch fermentations. Batch fermentations occur when a microorganism, inoculated into a broth containing the substrate and other nutritional factors, grows until all of the substrate is depleted. No carbon substrate is added after the fermentation begins. In the case of fermentations that require air, the air will be bubbled through the broth on a continuous basis.

In contrast to the differential equations required for a batch fermentation, a continuous culture may be modeled using algebraic equations. An Excel spreadsheet may be used to carry out material balances or solve algebraic equations. As in the batch case, the goal is to calculate cell mass, substrate utilization, and output of product. Unlike the case of batch fermentation, the calculated values are constants and do not change with time.

This chapter gives a brief background of the numerical techniques that are used to simulate the performance of microbial bioreactors. This chapter follows an outline of particularly lucid lecture notes that were originally developed by Dr. Ajoy Velayudhan, who worked in LORRE at Purdue University in 1989–90. The authors thank Dr. Velayudhan for his contribution, as well as that of Professor Henry

Modern Biotechnology, by Nathan S. Mosier and Michael R. Ladisch
Copyright © 2009 John Wiley & Sons, Inc.

Bungay of Rensselaer Polytechnic Institute, who pioneered methods of numerical solution of modeling microbial fermentations using the PC.

THE RUNGE–KUTTA METHOD

Simpson's Rule

Simpson's rule is a special case of the Runge–Kutta method. Simpson's rule is applied to the numerical integration of a single separable First-order differential equation of the form $dx/dt = f(t)$ with the initial condition of $x = x_0$ at $t = 0$. For an nth-order equation, this would be expressed as

$$\frac{d^n x}{dt^n} = f(t) \tag{5.1}$$

where the initial conditions are given by $x = x_0$ at $t = 0$, such that

$$\frac{dx}{dt} = \frac{d^2 x}{dt} = \frac{d^3 x}{dt} = \cdots \frac{d^{n-1} x}{dt^{n-1}} = 0$$

The solution is given by

$$x = x_0 + \int_{t_0}^{t} f(t)\,dt + \int_{t_0}^{t} f(t)\,dt^n \tag{5.2}$$

The numerical solution of the first-order differential equation

$$\frac{dx}{dt} = f(t) \tag{5.3}$$

is given by Simpson's rule for solution of directly integrable ordinary differential equations, where the initial condition x_0 at $t = t_0$ is known. This enables calculation of the first step of a solution that has the following form:

$$x = x_0 + \int_{t_0}^{t} f(t)\,dt \tag{5.4}$$

The first step gives the value x_1, and is calculated from the equation

$$x_1 = x_0 + \frac{h}{6}(f_0 + 4f_{1/2} + f_1) \tag{5.5}$$

where h denotes the step size (Fig. 5.1), and

$$\begin{aligned} t_1 &= t_0 + h \\ x_1 &= x(t_1) \\ f_1 &= f(t_1) \end{aligned} \tag{5.6}$$

The calculated value for x_1 becomes the starting value for the second step in this calculation. This calculation is repeated n times for the independent variable t, that

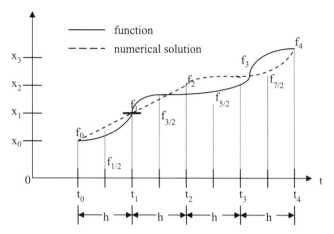

Figure 5.1. Schematic diagram of numerical integration by Simpson's rule.

is, increased by step h, to obtain the value of x_{n+1}, where n is the number of steps into which the solution procedure has been divided. The solution process is represented in the schematic diagram of Fig. 5.1.

Once x_1 has been calculated for step size h, x_2 can be obtained:

$$x_2 = x_1 + \frac{h}{6}(f_1 + 4f_{3/2} + f_2) \tag{5.7}$$

This process is repeated to calculate subsequent values of x through step $n + 1$:

$$x_{n+1} = x_n + \frac{h}{6}(f_n + 4f_{n+1/2} + f_{n+1}) \tag{5.8}$$

Fourth-Order Runge–Kutta Method

The differential equation $(dx/dt) = f(x,t)$ with the initial condition $x = x_0$ at $t = t_0$ is solved numerically by the Runge–Kutta method. As before, $t_n = t_0 + nh$, $x_n = x(t_n)$, and $f_n = f(t_n)$. The Fourth-order Runge–Kutta method describes the value of x_{n+1} in a manner similar to that described for Simpson's rule [Eq. (5.5)]:

$$x_{n+1} = x_n + \frac{h}{6}(k_0 + 2k_1 + 2k_2 + k_3) \tag{5.9}$$

where

$$k_0 = f(x_n, t_n) \tag{5.10}$$

$$k_1 = f\left(x_n + \frac{k_0}{2}, t_n + \frac{h}{2}\right) \tag{5.11}$$

$$k_2 = f\left(x_n + \frac{k_1}{2}, t_n + \frac{h}{2}\right) \tag{5.12}$$

$$k_3 = f(x_n + k_2, t_n + h) \tag{5.13}$$

Since the value of the initial condition, x_0, is known, x_1 may be calculated. Using x_1, the value of x_2 may be found. This is repeated until x_{n+1} is calculated. When $f(x,t)$ is independent of x, that is, $f(x,t) = g(t)$, the Runge–Kutta method gives

$$x_{n+1} = x_n + \frac{h}{6}(g_n + 2g_{n+1/2} + 2g_{n+1/2} + g_{n+1}) = x_n + \frac{h}{6}(g_n + 4g_{n+v_2} + g_{n+1}) \quad (5.14)$$

where $g_n = g(t_n)$; this form is the same as used for Simpson's rule [Eq. (5.5)].

An example of the solution process using the Fourth-order Runge–Kutta equation is given by calculation of microbial growth [example from Bungay (1993), p. 3.11]. Exponential growth is given by

$$f(x_0, t_0) = \frac{dx}{dt} = \mu x \quad (5.15)$$

where a fermentation medium is inoculated with a microbial culture, and the broth is subject to the initial conditions of $x = x_0$ at $t = t_0 = 0$, where x is the concentration of living cells in g/L. The definition of μ is developed in equations 4.3 to 4.7 (Chapter 4).

We assume here that there is sufficient substrate to rapidly promote growth. In other words, the nutrients are in excess. The constants are given below:

$$\mu = 0.2\,h^{-1} \quad \text{(specific growth rate)}$$

$$h = 0.25\,h \quad \text{(step size)}$$

$$x_0 = 10\,g/L \text{ at } t = 0 \quad \text{(initial cell concentration)}$$

The value of x_1 may be calculated from x_0, which is the concentration of the cells. From Eqs. (5.10)–(5.13) the calculated results are

$$k_0 = f(x_0, t_0) = \mu x_0 = (0.2)(10) = 2$$

$$k_1 = f\left(x_0 + \frac{k_0}{2}, t_0 + \frac{h}{2}\right) = \mu\left(x_0 + \frac{k_0}{2}\right) = (0.2)\left(10 + \frac{2}{2}\right) = 2.2$$

$$k_2 = f\left(x_0 + \frac{k_1}{2}, t_0 + \frac{h}{2}\right) = \mu\left(x_0 + \frac{k_1}{2}\right) = (0.2)\left(10 + \frac{2.2}{2}\right) = 2.22$$

$$k_3 = f(x_0 + k_2, t_0 + h) = \mu(x_0 + k_2) = (0.2)(10 + 2.22) = 2.444$$

so

$$x_1 = x_0 + \frac{h}{6}(k_0 + 2k_1 + 2k_2 + k_3)$$

$$= 10 + \frac{0.25}{6}(2 + 2 \times 2.2 + 2 \times 2.22 + 2.444)$$

$$= 10.5535$$

Using a hand calculator, a similar solution is obtained (with much less work) from

$$x_1 = 10 \exp(0.2 \cdot 0.25) = 10.5127$$

The integration using the Runge–Kutta method gave an answer that was off by about 0.4%. This illustrates the utility and accuracy of the Runge-Kutta technique.

Ordinary Differential Equations (ODEs)

Runge–Kutta Technique Requires that Higher-Order Equations Be Reduced to First-Order ODEs to Obtain Their Solution

An nth-order ordinary differential equation is represented by

$$\frac{d^n x}{dt^n} = f\left(x, \frac{dx}{dt}, \cdots, \frac{d^{n-1}x}{dt^{n-1}}, t\right) \qquad (5.16)$$

where initial conditions at $t = t_0$ are specified for each equation:

$$x = x_0 \qquad (5.17)$$

$$\frac{dx}{dt} = x_0' \qquad (5.18)$$

$$\frac{d^{n-1}x}{dt^{n-1}} = \frac{dx_0^{n-1}}{dt^{n-1}} \qquad (5.19)$$

First-order ordinary differential equations are constructed from the nth-order one as represented in

$$\begin{aligned}
y_1 &= \frac{dx}{dt} \\[6pt]
y_2 &= \frac{dy_1}{dt} = \frac{d^2 x}{dt^2} \\[6pt]
y_3 &= \frac{dy_2}{dt} = \frac{d^3 x}{dt^3} \\[6pt]
&\ \ \vdots \\[6pt]
y_n &= \frac{dy_{n-1}}{dt} = \frac{d^n x}{dt^n}
\end{aligned} \qquad (5.20)$$

where y_1 through y_n represent a series of first-order differential equations.
An example given by Bungay (1993, p. 3.5), namely

$$\frac{d^2 x}{dt^2} = -32.2 \, \text{ft/s}^2 \qquad (5.21)$$

reduces to two ordinary differential equations:

$$y_1 = \frac{dx}{dt} \tag{5.22}$$

$$y_2 = \frac{d^2x}{dt^2} = -32.2 \tag{5.23}$$

The input variables to the integration are y_1 and y_2 [corresponding to $I(1)$ and $I(2)$] with output variables of $O(1)$ for dx/dt and $O(2)$ for X. Since y_2 [corresponding to $I(2)$] is a constant $(=-32.2)$, Eq. (5.33) becomes

$$I(1) = O(2) \tag{5.24}$$

$$I(2) = -32.2 \tag{5.25}$$

where $O(2)$ is the output variable for the integration of Eq. (5.25).

Systems of First-Order ODEs Are Represented in Vector Form

The example of Eqs. (5.23) and (5.24) illustrated the solutions of two differential equations. The number of equations will be larger as the complexity of the microbial system increases and other parameters such as product and substrate inhibition are considered. In this case, a system of first-order ODEs will result, and economy of expression of these equations can be obtained by using vector notation. For a condition of first-order equations represented by

$$\frac{dx_1}{dt} = f_1(x_1, x_2, \ldots, x_n, t)$$

$$\frac{dx_2}{dt} = f_2(x_1, x_2, \ldots, x_n, t) \tag{5.26}$$

$$\vdots$$

$$\frac{dx_n}{dt} = f_n(x_1, x_2, \ldots, x_n, t)$$

initial conditions at $t = t_0$ are given by $x_1 = x_{1,0}, \ldots, x_n = x_{n,0}$. A more concise representation is given in vector form, where

$$\frac{d\tilde{x}}{dt} = \tilde{f}(\hat{x}, t); \quad \text{initial condition} \quad \tilde{x} = \tilde{x}_0 \quad \text{at} \quad t = t_0 \tag{5.27}$$

where

$$\tilde{x} = \begin{bmatrix} x_1 \\ x_2 \\ x_3 \\ \vdots \\ x_n \end{bmatrix} \tag{5.28}$$

and

$$\tilde{f} = \begin{bmatrix} f_1(\tilde{x}, t) \\ f_2(\tilde{x}, t) \\ f_3(\tilde{x}, t) \\ \vdots \\ f_n(\tilde{x}, t) \end{bmatrix} \qquad (5.29)$$

The vector form of the Runge–Kutta equations follows

$$\frac{d\tilde{x}}{dt} = f(\tilde{x}, t) \qquad (5.30)$$

where

$$\tilde{x} = \tilde{x}_0 \quad \text{at} \quad t = t_0 \qquad (5.31)$$

and

$$\tilde{x}_{n+1} = \tilde{x}_n + \frac{h}{6}\left(\tilde{k}_0 + 2\tilde{k}_1 + 2\tilde{k}_2 + \tilde{k}_3\right) \qquad (5.32)$$

where

$$\tilde{k}_0 = \tilde{f}(\tilde{x}_n, t_n)$$

$$\tilde{k}_1 = \tilde{f}\left(\tilde{x}_n + \frac{\tilde{k}_0}{2}, t_n + \frac{h}{2}\right)$$

$$\tilde{k}_2 = \tilde{f}\left(\tilde{x}_n + \frac{k_1}{2}, t_n + \frac{h}{2}\right) \qquad (5.33)$$

$$\tilde{k}_3 = \tilde{f}\left(\tilde{x}_n + \tilde{k}_2, t_n + h\right)$$

As long as initial conditions are given for each output variable ($\tilde{x}$ in this case), the solution will be obtained in the same manner as illustrated by solution of Eq. (5.15). There is a set of distinct k values for each ordinary differential equation. The vector notation of Eqs. (5.30)–(5.33) provides a concise manner in which to write these equations.

KINETICS OF CELL GROWTH

The kinetics of cell growth may be described through equations that are based on balanced growth and utilize an unstructured model. Balanced growth assumes that the properties of the cells are independent of their age. Only the number of cells changes, while the characteristics of the individual cells are the same for all cells, regardless of their age. The model is unstructured since it does not relate specific changes in rates of metabolic pathways to changes in cell division or production of given metabolites. These assumptions define Monod kinetics as represented in Eqs. (5.16)–(5.18):

$$\frac{1}{N}\frac{dN}{dt} = \mu \quad \text{or} \quad \frac{1}{x}\frac{dx}{dt} = \mu \tag{5.34}$$

where N represents concentration in cell number and x represents concentration on a weight basis. We will use the cell weight basis and designate cell mass in terms of x. In the case where there is no inhibition, the growth rate is given by

$$\mu = \frac{1}{x}\frac{dx}{dt} = \frac{\mu_{max}S}{K_s + S} \tag{5.35}$$

Rapid growth is represented by

$$\mu = \frac{1}{x}\frac{dx}{dt} = \frac{\mu_{max}S}{K_{s,1} + K_{s,2}S_0 + S} \tag{5.36}$$

Rapid growth given in Eq. (5.36) refers to conditions within the cell system (consisting of a large population of cells). Rapid growth is defined by:

1. Rapid nutrient consumption
2. Internal cell nutrients that are far less than nutrients in the external medium
3. Absence of a truly exponential phase so that Eq. (5.34) is not directly applicable

Determining constants for Eq. (5.35) with an initial rate approach would result in the type of plot shown in the schematic diagram given by Fig. 5.2. The measurements of data points would consist of carrying out fermentations (refer to Chapter 4), using different initial concentrations of the carbon source (glucose in this case) and measuring the rates of cell growth over relatively short time increments on the order of an hour. Plotting the inverse of the specific growth rate $(1/\mu)$ versus the inverse of the starting substrate concentration $(1/S)$ results in a graph similar to Fig. 5.2, where the line that best fits the data intercepts the y axis at the value for the inverse of the maximum specific growth rate $(1/\mu_{max})$.

The plot in Fig. 5.2 arises from inverting Eq. (5.35) to give

$$\frac{1}{\mu} = \frac{K_s + S}{\mu_{max}S} = \frac{K_s}{\mu_{max}} \cdot \frac{1}{S} + \frac{1}{\mu_{max}} \tag{5.37}$$

The simulation of cell growth may be represented in formulas for solution by the Runge–Kutta method. In this case, the Monod-type equation [Eq. (5.35)] becomes

$$\frac{dx}{dt} = \frac{\mu_{max}S}{K_s + S} \cdot X \tag{5.38}$$

where the change in substrate concentration is related to cell growth by the yield coefficient [refer to Chapter 4, Eqs. (4.9)–(4.13)]:

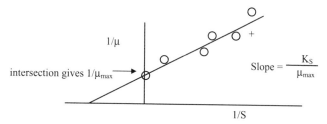

Figure 5.2. Schematic representation of inverse plot of Monod equation that may be used to represent microbial growth data.

$$\frac{ds}{dt} = -Y_{s,x} \frac{dx}{dt} \tag{5.39}$$

In order to solve these equations and obtain a plot that shows the timecourse of substrate and cell mass, these equations are expressed in terms of input (I) and output (O) variables, where I represents a derivative that is integrated and O represents a dependent variable that is an output of the integration step:

$$O = S_I$$
$$I = \frac{dx}{dt} \tag{5.40}$$

For multiple equations, an index number denotes the derivative or variable number, and thereby keeps track of the different equations (Bungay 1993). In the case of cell growth and substrate depletion in the absence of inhibition, Eqs. (5.38) and (5.39) represent cell growth, and the formalism for numerical solution is given by Eq. (5.38) as

$$I(1) = \left(\frac{\mu_{max} O(2)}{K_s + O(2)} \right) O(1) \tag{5.41}$$

where $I(1) = dx/dt$ and $O(1) = x$. The equation for the change in substrate concentration is given by Eq. (5.42)

$$I(2) = -Y_{s/x} I(1) \tag{5.42}$$

where $I(2) = dS/dt$. The output variable for substrate concentrations is therefore $O(2)$, and is used in Eq. (5.41). Solution of these equations gives the timecourse shown in Fig. 5.3, and was obtained for the initial condition of

$$\text{Cell mass: } O(1) = 1\,\text{g/L}$$

$$\text{Glucose: } O(2) = 200\,\text{g/L}$$

with the constants of

$$K_s = 0.315 \frac{\text{g substrate}}{\text{L}}$$

Yeast Fermentation

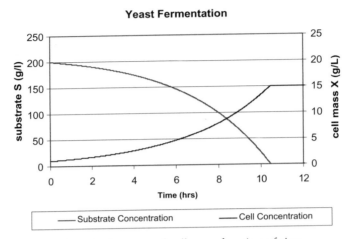

Figure 5.3.   Concentration of substrate and cells as a function of time.

$$\mu_{max} = 0.25\,h^{-1} \ \text{at} \ 35\,^{\circ}C$$

$$Y_{s/x} = 14.3 \frac{\text{consumed g substrate}}{\text{g cells generated}}$$

These constants are in a range that is typical for yeast (Wilke et al. 1981; Maiorella et al. 1983, 1984a,b). The solution of these two differential equations is given in Fig. 5.3.

The reader should note that the initial condition for each variable must be specified if a solution is to be obtained.

When cells are inoculated into a sterilized fermentation medium, there is a lag between inoculation and initiation of cell growth. The lag phase may last from 30 min to a half a day, and represents a time period that is required for the cells to become acclimated to their new environment. The solutions represented by Eqs. (5.41) and (5.42) do not account for a lag phase, with time zero denoting the time at which growth begins, after the lag phase is complete. The lag phase may be modeled as described by Basak et al. (1995) in the case study for the cephalosporin C fermentation.

K_s Represents Substrate Concentration at Which the Specific Growth Rate Is Half of Its Maximum

The specific growth rate is defined by

$$\frac{1}{x}\frac{dx}{dt} = \mu \qquad (5.43)$$

and represents growth rate of an initial (living) inoculums at concentration X. Values for K_s are generally less than 0.25 g/L, and are a function of substrate as well as microorganism (Table 5.1). The value of K_s is in the range of 250 mg/L for yeast

Table 5.1.

Kinetic Constants for Ethanol Fermentation

Microorganism	Substrate Concentration (g/L)	Temperature (°C)	K_s (mg/L)	μ_{max} (h^{-1})	$Y_{P/S}$ (g/g)	$Y_{X/S}$ (g/g)	Reference
Saccharomyces cerevisiae							
Industrial	100 glucose	30	—	0.27	0.46	0.055	Davis et al., (2006)
424A (LNH-ST)	20 glucose	—	—	0.29	—	—	Govindswamy and Vane, (2007)
ATCC 4226	—	—	315	—	—	—	Bazua and Wilke, (1977)
Bacteria							
Zymomonas mobilis	ATCC 31821	30	—	0.37	0.48	0.028	Lawford and Rousseau, (1999)
E. coli	95 xylose	35	2–4	0.23	0.47	—	Tao et al., (2001)
T. saccharo-lyticum	70 xylose	55	—	0.37	0.46	0.12	Shaw et al., (2008)

and 2–4 mg/L for *E. coli*. However, industrial fermentations generally have substrate concentrations exceeding 50 g/L (50,000 mg/L). Consequently, Eq. (5.17) simplifies to the form of Eq. (5.44) since $K_s \ll S$:

$$\mu = \frac{\mu_{max}S}{K_s + S} \cong \mu_{max} \qquad (5.44)$$

The specific growth will be near the maximum value of μ_{max} during the initial stages of many practical fermentations, but will drop in later stages of a batch fermentation as the substrate concentration (S) drops. A growth rate that is half of its maximal value is represented by

$$\mu = \frac{\mu_{max}}{2} = \frac{\mu_{max}S}{K_s + S} \qquad (5.45)$$

and when solved for K_s gives $K_s = S$. Consequently, the value of K_s represents the concentration at which the specific growth rate is half of its maximal value (μ_{max}). This is illustrated schematically in Fig. 5.4.

Inhibition of cell growth may occur if the substrate concentration in the fermentation media becomes too large. In this case the water activity (proportional to

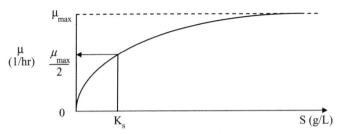

Figure 5.4. Schematic representation of definition of K_s.

osmotic pressure) of the media exceeds a critical level, and the membranes of the cells are no longer sufficient to overcome the concentration gradient of media components (such as salts) into the cell. The cell must utilize significant energy to pump out salts so that internal and external osmotic pressures are balanced. If the concentration is high enough, the cell can no longer grow. Thus at very high substrate concentrations the growth of bacteria and yeast is inhibited.

The equations for cell growth have assumed that all of the carbon-containing substrate is transformed into cell mass or metabolites or products associated with cellular activity and energetics. A fraction of the substrate is also used to maintain function of the cell. This becomes noticeable, from a modeling perspective, at very low substrate (carbon source) concentration where there is minimal cell growth and the substrate is used principally to maintain cell viability. The Monod equation can be used to correct for this effect through a maintenance coefficient as shown in

$$\mu = \mu_{max} \frac{S}{K_s + S} - k_e \tag{5.46}$$

where k_e = a maintenance constant. A continuous bioreactor system, discussed later in this chapter, may be used to determine values of k_e.

SIMULATION OF A BATCH ETHANOL FERMENTATION

The fermentation of glucose to ethanol has been thoroughly studied. The metabolism, physiology, nutritional requirements, and other yeast fundamentals are given by Russell (2003). From the perspective of fermenting glucose from cellulose, the physiology of ethanol-fermenting yeasts and the production of coproducts and operational parameters of the ethanol fermentation are thoroughly reviewed and critically analyzed in a biotechnology report by Maiorella et al. (1984a). This work is widely cited and continues to be practically relevant to this day.

The fermentation of glucose to ethanol is represented by the following sequence:

$$\text{Glucose} + \text{nutrients} \rightarrow \text{cells} + \text{ethanol} + \text{byproducts} \tag{5.47}$$

Small amounts of oxygen are needed as an essential growth factor for the synthesis of unsaturated fatty acids and steroids (Bazua and Wilke 1977). However, this is not an aerated fermentation; rather, it is microaerobic. Once the exponential growth phase begins, oxygen is depleted and vigorous generation of CO_2 occurs with one mole of CO_2 generated for each mole of ethanol produced. The experimental conditions used to generate the constants given below were at a constant temperature of 30 °C, a pH ranging from 3.9 to 4.15, and a limiting amount of substrate (1% glucose). The composition of the fermentation medium consisted of one liter of tap water; 10 g glucose; 1.5 g yeast extract; 2.5 g NH_4Cl; 5.5 g $Na_2HPO_4 \cdot 7H_2O$; 3 g KH_2PO_4; 0.25 g $MgSO_4$; 0.01 g $CaCl_2$; and citric acid/sodium citric at approximately 5 and 2.5 g, respectively, to control the pH at about 4. Optimal compositions for ethanol fermentation are discussed by Russell (2003).

The cells catalyze the biotransformation of glucose to both ethanol as well as to additional cells. The glucose that forms the carbon source for this fermentation is taken as the limiting substrate, and hence is used as the substrate variable S in the growth and ethanol production equations. The ethanol fermentation itself is a Type I fermentation (Chapter 4). The product ethanol is formed as a direct consequence of the cell's energy metabolism. Ethanol production is therefore directly proportional to cell mass as defined by the product yield coefficient.

The effect of ethanol accumulation on the fermentation kinetics of the yeast *Saccharomyces cerevisiae* ATCC 4126 were studied by Bazua and Wilke (1977) using a continuous fermentation system. Their data were fit by three constant equations:

$$\mu_{max} = \frac{\mu_0 - aP_{av}}{b - P} \tag{5.48}$$

$$v_{max} = \frac{v_0 - a'P_{max}}{b' - P} \tag{5.49}$$

where a, a' = empirical constants (h^{-1})

b, b' = empirical constants (g/L)

P = product concentration (g/L)

P_{av} = average product concentration in continuous fermentor runs (g/L)

μ_0 = specific growth rate (h^{-1}) in absence of ethanol

μ_{max} = maximum specific growth rate (h^{-1})

v_0 = specific ethanol production rate (h^{-1}), in absence of initial ethanol

v_{max} = maximum specific ethanol production rate (h^{-1})

v_0 = maximum specific ethanol production rate at P = 0 (h^{-1})

The specific ethanol production rate describes the amount of ethanol that is generated for a given amount of living cells per hour. Hence, this is defined as g ethanol per g cells per hour, and is equivalent to $(1/x)\,(dP/dt)$.

Ethanol Case Study

The values of the constants in Eqs. (5.48) and (5.49) were obtained by a least-squares fit of the data for cell mass and ethanol concentration as a function of substrate concentrations. The values of the constants for cell growth were

$$\mu_0 = 0.428$$
$$a = 0.182 \qquad (5.50)$$
$$b = 133.78$$

while for ethanol production these were

$$v_0 = 1.802$$
$$a' = 0.506 \qquad (5.51)$$
$$b' = 119.25$$

The values of μ_0 and v_0 correspond to the values of μ_{max} and v_{max} when $P_{av} = 0$. For values of μ_{max} and v_{max} equal to zero, the value of P_{max} was found to be between 93.8 and 93.1 g ethanol/L. In this manner the toxic concentrations of ethanol were defined for this particular strain of cells. Current industrial practice has achieved maximal ethanol concentrations of 15% or higher. Improvements in tolerance to ethanol are important to improving the economics of producing ethanol for fuel via fermentation (Alper et al. 2006).

Analysis of the data of Bazua by Maiorella et al. (1983) showed a linear relationship between the specific growth rate and ethanol productivity under conditions where ethanol is an inhibitor (Fig. 5.5). Ethanol inhibition was observed to begin at about 25 g/L ethanol in the fermentation beer and peaked at 95 g/L.

This result was interpreted in the context of ATP metabolic pathway. The ethanolic pathway generates ATP for cell maintenance and growth, which, when inhibited, gives a constant proportional decrease in cell growth rate and ethanol productivity. Ethanol productivity and available ATP for biosynthesis decrease with increasing inhibition. Maiorella et al. (1983) also noted that there was no apparent change in cell morphology associated with ethanol inhibition.

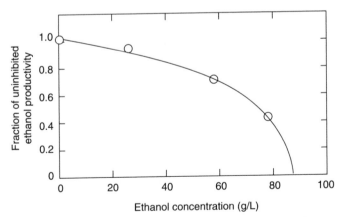

Figure 5.5. Inhibitory effect of ethanol on specific ethanol production by *Saccharomyces cerevisiae* [data are from B. L. Maiorella, H. W. Blanch, and C. R. Wilke, "Feed Component Inhibition in Ethanolic Fermentation by *Saccharomyces cerevisiae*," *Biotechnology Bioengineering* **26**, 1155–1166 (1984), Fig. 1, p. 1004, figure reprinted with permission. Copyright 1984, John Wiley & Sons, Inc.]

Table 5.2.

Byproduct Inhibition Summary

Product	Concentration at High Inhibition[a] (g/L)	Inhibition Mechanism
Ethanol	70	Direct inhibition of ethanol production pathway
Fumaric acid	2.7	Chemical interference with cell maintenance functions
Acetic acid	7.5	Chemical interference
Lactic acid	38	Chemical interference
Propanol	12	Chemical interference
Methyl-1-butanol	3.5	Chemical interference
3-Butanediol	90	Chemical interference
Acetaldehyde	5.0	Chemical interference

Source: Reprinted with permission from B. Maiorella, H. W. Blanch, and C. R. Wilke, "By-Product Inhibition Effects on Ethanolic Fermentation by *Saccharomyces cerevisiae*," *Biotechnology Bioengineering* **25**, 103–121 (1983), Table III, p. 110. Copyright 1983, John Wiley & Sons, Inc.

Inhibition can be caused by other byproducts of the ethanol fermentation as well. These are characterized as chemical interference where cell membrane disruption is involved (as in acid attack), and the cell inhibitors that act in this manner are summarized in Table 5.2. Other inhibitors include minerals brought in as part of the fermentation media components (Maiorella et al. 1984b) (Fig. 5.6).

These early studies led to a thorough analysis of the ethanol fermentation and the definition of a series of equations and constants that are useful for simulating an ethanol fermentation by *Saccharomyces cerevisiae*. The equation for the specific ethanol production rate as a function of substrate and product (ethanol) accumulated is given by

$$v = v_{max} \left[\frac{S}{S + K_m} \right] \left[1 - \frac{P}{P_{max}} \right]^n \tag{5.52}$$

where v = specific ethanol production rate [g ethanol produced/(g cells·h)]

S = substrate (glucose) concentration (g/L)

v_{max} = maximum specific production rate [1.15 g ethanol/(g cells·h)]

K_m = Monod constant (0.315 g/L)

n = toxic power constant (0.36)

P_{max} = maximum product concentration (87.5 g/L)

The inhibitory effect of ethanol concentration is pronounced, with a rapid decrease occurring above 60 g/L (Fig. 5.5) as modeled by the inhibition term $[1 - P/P_{max}]^n$ (Levenspiel 1980).

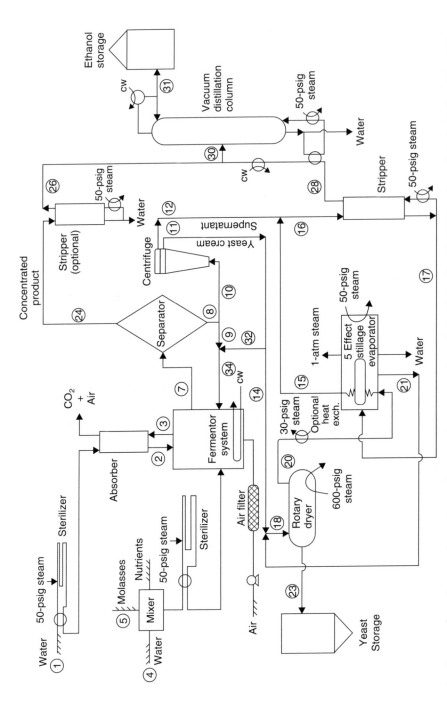

Figure 5.6. Process flow diagram for molasses fermentation system (from B. L. Maiorella, H. W. Blanch, and C. R. Wilke, "Feed Component Inhibition in Ethanolic Fermentation by *Saccharomyces cerevisiae*," *Biotechnology Bioengineering* 26, 1155–1166 (1984), Fig. 10, p. 1163, figure reprinted with permission. Copyright 1984, John Wiley & Sons, Inc.)

The cell growth rate μ [g new cells formed/(g cells·h)] was proposed to be directly related to the ethanol production rate by

$$\mu = Ev \qquad (5.53)$$

The constant E denotes the efficiency of substrate utilization for cell mass production. For nonstressed cells, in an optimal growth medium, E has a value of 0.249. The rate of glucose utilization is related to the ethanol production rate by the yield factor:

$$Y_{P/S} = 0.434 = \frac{\Delta P}{\Delta S} \qquad (5.54)$$

As pointed out by Maiorella et al. (1983), this value is less than the theoretical yield of 0.51 since a portion of the glucose is utilized for cell mass and secondary product formation, including organic acids and glycerol.

LUEDEKING–PIRET MODEL

Another type of unstructured model is given by the Luedeking–Piret model, which was developed for the lactic acid fermentation (Luedeking and Piret 1959). This model is plotted in Fig. 5.7 and has both growth and non-growth-associated terms, as indicated in the following equation:

$$\underset{\substack{\text{Product} \\ \text{accumulation}}}{\frac{dP}{dt}} = \underset{\substack{\text{Growth-} \\ \text{associated}}}{\alpha \frac{dx}{dt}} + \underset{\substack{\text{Non-growth-} \\ \text{associated}}}{\beta x} \qquad (5.55)$$

This equation was developed for use with *Lactobacillus debruickii* in which the production of lactic acid was found to occur semiindependently of cell growth. The metabolic pathway that produces lactic acid is discussed in Chapter 9. Aiyer and Luedeking (1996) also applied this type of model to batch fermentation of ethanol over a range of yeast concentrations (0.1 to 6 g/L) and pH's (4 to 6.8).

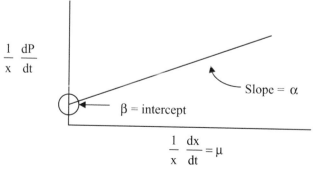

Figure 5.7. Graphical representation of Luedeking–Piret model.

CONTINUOUS STIRRED-TANK BIOREACTOR

The practice of chemical engineering utilizes continuous stirred-tank reactors (CSTRs) in which reactants are continuously added to a tank containing a catalyst to form and continuously remove a desired product (Levenspiel 1972). In biological systems there is a CSTB or continuous stirred-tank bioreactor, which is also referred to as a *chemostat* or *turbidostat*. Among the early researchers to work with continuous systems for monocultures were Novick and Szilard (1950a,b). The term *turbidostat* probably arose from the observable changes in turbidity that occurred when the rate of addition of substrate (liquid) changed, causing the steady-state population of growing cells to change in number. As the number of cells increased, the turbidity also increased. The principle of the chemostat is based on the assumption that a drop of liquid added to a stirred vessel is uniformly and instantaneously distributed throughout the vessel. This is analogous to the assumption that underlies the model of a CSTR. The time required to mix a small volume of medium should be small relative to the replacement time t_r, given by V/F, where V denotes culture volume and F is the flow rate at which the medium is added to the fermentor. The dilution rate is inversely proportional to the replacement time.

A major difference between a CSTR and CSTB is the occurrence of cell growth in the CSTB, and the difficulty in stoichiometrically accounting for the disappearance of substrate directly into products in a CSTB, since the product includes living cell mass, CO_2 and individual biochemical species.

Uses of a CTSB for research purposes may include

1. Study of biomass growth rate with respect to changes in substrate concentration
2. Study of biomass growth rate in response to environmental parameters such as pH and temperature
3. Substrate-limited growth with a constant flow rate
4. Study of metabolic flux for systems biology models of cellular metabolism.

Theories attributed to Monod (1942) and Novick and Szilard (1950a,b) first showed that the specific growth rates could be fixed at any values ranging from zero to the maximum specific growth rate μ_{max}.

The microbial biomass balance for a culture volume V and flow rate of medium added F is given by

$$Vd\bar{x} = V\mu\bar{x}dt - F\bar{x}\cdot dt \qquad (5.56)$$

When divided by $V\cdot dt$, Eq. (5.56) becomes

$$\frac{d\bar{x}}{dt} = (\mu - D)\bar{x} \quad \text{where} \quad D = \frac{F}{V} \qquad (5.57)$$

Equation (5.57) represents a cell balance across the continuous stirred-tank bioreactor (Fig. 5.8) with $d\bar{x}/dt$ denoting cell concentration in the reactor which is distinct from cell mass, x, in the expression for specific growth rate μ. In the special case when cell mass, $\bar{x}$, is constant:

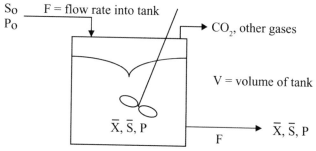

Figure 5.8. Schematic representation of a continuous stirred-tank bioreactor (CSTB). S_0 and P_0 represent initial or inlet substrate and product concentrations. Outlet concentrations for cell mass ($\bar{X}$), substrate ($\bar{S}$), and product (P) are the same as their concentrations in the tank.

$$\frac{d\bar{x}}{dt} = 0 \quad \text{then} \quad \mu = D \tag{5.58}$$

where $1/D$ is residence time. The balance for the growth limiting substrate (e.g., glucose) is

Net change = input − output = (substrate used for growth, product formation or maintenance)

$$V ds = FS_0 dt - F\bar{S} dt = V\left(\left[\frac{\mu x}{Y_{X/S}}\right] + [\mu X Y_{P/X}] + [kX] + \mu X Y_{CO_2/X} + \frac{kX}{Y_{S/CO_2}}\right) dt \tag{5.59}$$

where F =   flow rate of the culture medium into the tank (L/h)
      V =   volume, culture medium in tank (L)
      $Y_{X/S}$ =   yield coefficient (g cells formed/g substrate utilized)
      $Y_{P/X}$ =   yield coefficient (g product formed/g cells formed)
      $Y_{CO_2/X}$ =   yield coefficient (g CO_2 formed/g cell mass formed)
      Y_{S/CO_2} =   yield coefficient (g substrate utilized/g CO_2 formed)
      μ =   specific growth rate (h^{-1})
      X =   cell mass (g/L)
      dt =   small time interval (h)
      $\bar{S}$ =   substrate concentration in vessel (g/L) = $S_0 - S_r$
      S_0 =   substrate concentration in feed (g/L)
      k =   maintenance coefficient (g substrate/g cells) needed to maintain metabolic functioning of the cell in the absence of growth

The yield coefficient (discussed in Chapter 4) $Y_{x/s}$ denotes the ratio $\Delta X/\Delta S$, but is only an approximation. Substrate may also be converted to product P, or to CO_2, which must be accounted for separately as indicated in Equation (5.59).

The equations for the CSTB are based on average concentrations. For the substrate concentration $\bar{S}$ resulting only in cell growth we obtain:

$$\frac{d\bar{S}}{dt} = D(S_0 - \bar{S}) dt - \frac{\mu x}{Y_{X/S}} \tag{5.60}$$

where S_0 is the inlet substrate concentration.

At steady state, the $\overline{S}$ substrate and cell mass $\overline{x}$ concentrations in the tank are constant:

$$\frac{d\overline{x}}{dt} = \frac{d\overline{S}}{dt} = 0 \tag{5.61}$$

At constant cell mass in the CTSB, $d\overline{x}/dt = 0$ and Eq. (5.57) becomes

$$(\mu - D)\overline{x} = 0 \tag{5.62}$$

and since $d\overline{S}/dt = 0$, then Eq. (5.60) becomes

$$D(S_0 - \overline{S}) - \frac{\mu\overline{x}}{Y_{x/s}} = 0 \tag{5.63}$$

The expression for the specific growth rate is given by

$$\mu = \frac{\mu_{max}\overline{S}}{\overline{S} + K_s} \tag{5.64}$$

If the specific growth rate, $\mu = D$, as in Eq. (5.58), then Eq. (5.64) is:

$$D(\overline{S} + K_s) = \mu_{max}\overline{S} \tag{5.65}$$

Then the average substrate concentration in the CSTB follows from Eq. (5.65):

$$(\mu_{max} - D)\overline{S} = K_s D$$

or

$$\overline{S} = \frac{K_s D}{\mu_{max} - D} \tag{5.66}$$

Similarly, when $\mu = D$, the average concentration of cell mass follows from Eq. (5.63):

$$\overline{x} = Y_{x/s}(S_0 - \overline{S}) \tag{5.67}$$

When Eq. (5.67) is combined with Eq. (5.66), an expression for average cell mass $\overline{x}$ in terms of the yield coefficient $Y_{X/S}$ and maximum specific growth rate μ_{max} results:

$$\overline{x} = Y_{x/s}\left(S_0 - \frac{K_s D}{(\mu_{max} - D)}\right) \tag{5.68}$$

Equation (5.69) gives the definition of the critical dilution rate D_c, where the value of the steady-state biomass concentration approaches zero (Fig. 5.9):

$$\mu = D_c = \frac{\mu_{max}S_0}{S_0 + K_s} \tag{5.69}$$

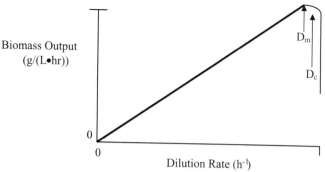

Figure 5.9. Biomass as a function of dilution rate. D_c denotes the critical dilution rate at which cell growth can no longer keep up with rate at which cells are removed from CSTB.

Further, when $S_0 \gg K_s$, Eq. (5.69) becomes

$$\mu = D_c = \mu_{max} \tag{5.70}$$

$1/D_c$ is the residence time at which the cells wash out of the CTSB.

The rate of output of cell mass per unit volume of a CSTB is given by

$$R = D\bar{x} \tag{5.71}$$

combining Eqs. (5.71) and (5.68) gives

$$R = DY_{x/s}\left[S_0 - \frac{K_s D}{\mu_{max} - D}\right] \tag{5.72}$$

The steady-state rates of output are thus seen to increase linearly with respect to D.

The maximum biomass output occurs at dilution rate D_m for substrate concentration utilized, S_r as derived from Eq. (5.72)

$$\frac{dR}{dD} = 0 = \frac{d}{dD}\left[DY_{x/s}\left(S_r - \frac{K_s D}{\mu_{max} - D}\right)\right] \tag{5.73}$$

where differentiating and setting the derivative to zero gives

$$D_m = \mu_{max}\left[1 - \left(\frac{K_s}{S_0 + K_s}\right)^{1/2}\right] \tag{5.74}$$

The steady-state cell mass concentration at this condition is then obtained from Eq. (5.67) and D_m is defined by Eq. (5.74):

$$\bar{x}_m = Y_{x/s}\left[S_0 + K_s - \{K_s(S_0 + K_s)\}^{1/2}\right] \tag{5.75}$$

When $S_0 \gg K_s$, Eq. (5.75) becomes

$$\bar{x}_m \cong Y_{x/s}S_0 \tag{5.76}$$

BATCH FERMENTOR VERSUS CHEMOSTAT

The maximum biomass output rate of a chemostat exceeds that of a batch reactor. If x_m = maximum biomass concentration, and x_0 = inoculum concentration, then for a given maximum specific growth rate μ_{max}, attained in a batch culture, we obtain

$$\int_{x_0}^{x_m} \frac{dx}{x} = \mu_{max} \int_{t_{lag}}^{t_c} dt \qquad (5.77)$$

The time t_c needed to attain a maximum cell mass concentration is given by

$$t_c = \frac{1}{\mu_{max}} \ln\left(\frac{x_m}{x_0}\right) + t_{lag} \qquad (5.78)$$

where t_{lag} = lag period for growth and turnaround time.

The output of microbial biomass of a batch culture is

$$R_{m,batch} = \frac{Y_{X/S} S_0}{t_c} \qquad (5.79)$$

where t_c is as defined by Eq. (5.78). The maximum output rate of the continuous bioreactor is given by

$$R_{m,CTSB} = D_m \bar{x}_m \cong D_m Y_{x/s} S_0 \qquad (5.80)$$

with $D_m = \mu_{max}$. Consequently the ratio of the output of a CTSB to a batch reactor is

$$\frac{R_{m,CSTB}}{R_{m,batch}} = \frac{D_m Y_{x/s} S_0}{\dfrac{1}{t_c} \cdot Y_{x/s} S_0} = \mu_{max} t_c = \ln\left(\frac{x_m}{x_0}\right) + \mu_{max} t_{lag} \qquad (5.81)$$

The doubling time t_d of an exponentially growing batch culture at its maximal rate is

$$t_d = \frac{0.693}{\mu_{max}} \qquad (5.82)$$

Consequently, Eq. (5.81) becomes

$$\frac{R_{m,chemostat}}{R_{m,batch}} = \ln\left(\frac{x_m}{x_0}\right) + \frac{0.693 t_{lag}}{t_d} \qquad (5.83)$$

If t_{lag} were zero, the ratio of outputs would be given by $\ln(x_m/x_0)$, with $x_m > x_0$. Hence, it is shown that the output of the chemostat will be greater than that of a batch reactor.

An example is given by continuous ethanol fermentation as reported by Xu et al. (2005) with flocculating yeast and Nagashima et al. (1984a,b) using yeast entrapped in calcium alginate beads. Xu et al. showed that flocculating yeast self-aggregate and settle to the bottom of the tank. This yeast could achieve ethanol concentrations of about 60–120 g/L with corresponding cell mass of 5–15 g/L. However, the culture tended to oscillate at dilution rates of 0.04 (h^{-1}), causing ethanol concentration and productivity to drop dramatically for several days at a time, during a 60-day run. The productivity of CTSB was at 3.44 g L^{-1}h^{-1} at its maximum.

The immobilized cell reactor of Nagashima et al. (1984a,b) showed ethanol productivities based on bioreactor volume of 20 g L^{-1}h^{-1} at 85–90 g/L ethanol with a biocatalyst lifetime in excess of 6 months. Bacterial contamination of the bioreactor was avoided by keeping the pH of the feed (glucose) solution at pH 4, and by adding some unspecified bacteriocidal compounds. The concentration of the yeast was equivalent to 250 g cells/L gel, or about 100 g cells/L of reactor. Both Xu and Nagashima achieved continuous ethanol production with productivities much larger than a batch reactor at 1–2 g L^{-1}h^{-1}.

Continuous fermentation has the potential of reducing the volume of the bioreactors needed for a given level of production. However, this type of bioreactor requires that the inlet sugars be soluble in the aqueous media. Simultaneous hydrolysis and fermentation of a solid substrate, such as cellulose, may need to be handled differently.

REFERENCES

Aiyar, A. S. and R. Luedeking, "A Kinetic Study of Alcoholic Fermentation of Glucose by Saccharomyces Cereviseae," *Chemical Eng. Progress Symp. Ser. 69*, **62**, 55–59 (1966).

Alper, H., J. Moxley, E. Nevoigt, G. R. Fink, and G. Stephanopoulos, "Engineering Yeast Transcription Machinery for Improved Ethanol Tolerance and Production," *Science* **314**(5805), 1565–1568 (2006).

Bailey, J. E. and D. F. Ollis, *Biochemical Engineering Fundamentals*, McGraw-Hill, New York, 1977, pp. 3, 14–15, 221–254, 338–348, 357, 371–383.

Basak, S., A. Velayudhan, and M. R. Ladisch, "Simulation of Diauxic Production of Cephalosporin C by Cephalosporium acremonium: Lag Model for Fed-Batch Fermentation," *Biotechnol. Progress* **11**(6), 626–631 (1995).

Bazua, C. D. and C. R. Wilke, "Ethanol effects in the Kinetics of a continuous fermentation with *Saccharomyces Cereviseae*," *Biotechnol Bioeng Symp.* 7, 105–118 (1977).

Bungay, H. R., *Basic Biochemical Engineering*, 2nd ed., Biline Associates, 1993, pp. 3.4–3.5, 6.14–6.27.

Davis, L., P. Rogers, J. Pearce, and P. Peiris, "Evaluation of Zymomonas-based Ethanol Production from a Hydrolyzed Waste Starch Stream," *Biomass Bioenergy* 30, 809–814 (2006).

Govindaswamy, S. and L. M. Vane, "Kinetics of Growth and Ethanol Production on Different Carbon Substrates Using Genetically Engineered Xylose Fermenting Yeast," *Bioresource Technol.* 98, 677–685 (2007).

Lawford, H. G. and J. D. Rousseau, "The effect of glucose on high-level xylose fermentations by recombinant Zymomonas in batch and fed-batch fermentations," *Applied Biochemistry and Biotechnology* 77(1–3), 235–249 (1999).

Levenspiel, O., *Chemical Reaction Engineering*, 2nd ed., J Wiley, New York, 1972, pp. 125–155.

Levenspiel, O., "The Monod Equation: A Revisit and a Generalization to Product Inhibition Situations," *Biotechnol. Bioeng.* **22**, 1671–1687 (1980).

Luedeking, R. and E. L. Piret, "A Kinetic Study of the Lactic Acid Fermentation: Batch Process at Controlled pH," *J. Biochem. Microbiol. Technol. Eng.* **1**, 393–412 (1959).

Maiorella, B. L., H. W. Blanch, and C. R. Wilke, "Economic Evaluation of Alternative Ethanol Fermentation Processes," *Biotechnol. Bioeng.* **26**(9), 1003–1025 (1984a).

Maiorella, B. L., H. W. Blanch, and C. R. Wilke, "Feed Component Inhibition in Ethanolic Fermentation by *Saccharomyces cerevisiae*," *Biotechnol. Bioeng.* **26**, 1155–1166 (1984b).

Maiorella, B. L., H. W. Blanch, and C. R. Wilke, "By-Product Inhibition Effects on Ethanolic Fermentation by *Saccharomyces cerevisiae*," *Biotechnol. Bioeng.* **25**(1), 103–121 (1983).

Monod, J., *Recherches sur La Croissance des Cultures Bacteriennes*, Actualities Scientifiques et Industrielles 911, Microbiologie—Exposés Publiés sous la Direction de J. Bordet (Director de L'Institut Pasteur de Bruxelles, Prix Nobel), Hermann and Cie, Paris, 1942, pp. 16–17.

Nagashima, M., M. Azuma, S. Noguchi, K. Inuzuka, and H. Samejima, "Continuous Ethanol Fermentation Using Immobilized Yeast Cells," *Biotechnol. Bioeng.* **26**, 992–997 (1984a).

Nagashima, M., M. Azuma, S. Noguchi, and K. Inuzuka, "Continuous Alcohol Production with Immobilized Growing Microbial Cells," *Arab Gulf J. Sci. Res.* **2**(1), 299–314 (1984b).

Novick, A. and L. Szilard, "Description of the Chemostat," *Science* **112**, 715–716 (1950a).

Novick, A. and L. Szilard, "Experiments with the Chemostat on Spontaneous Mutations of Bacteria," *Proc. Natl. Acad. Sci. USA* **36**, 708–719 (1950b).

Russel, I., "Understanding Yeast Fundamentals," in *The Alcohol Textbook*, 4th ed., K. A. Jacques, T. P. Lyons, and D. R. Kelsall, eds., Nottingham Univ. Press, 2003, pp. 85–119.

Shaw, A. J., K. K. Podkaminer, S. G. Desai, J. S. Bardsley, S. R. Rogers, P. G. Thorne, D. A. Hogsett, and L. R. Lynd, "Metabolic engineering of a thermophilic bacterium to produce ethanol at high yield," *Proceedings of the National Academy of Sciences of the United States of America* **105**(37), 13769–13774 (2008).

Tao, H., R. Gonzalez, A. Martinez, M. Rodriguez, L. O. Ingram, J. F. Preston, and K. T. Shanmugam, "Engineering a homo-ethanol pathway in escherichia coli: increased glycolytic flux and levels of expression of glycolytic genes during xylose fermentation," *Journal of Bacteriology*, **183**(10), 2979–2988 (2001).

Wang, D. I. C., C. L. Cooney, A. L. Demain, P. Dunnill, A. E. Humphrey, and M. D. Lilly, *Fermentation and Enzyme Technology*, Wiley, New York, 1979, pp. 10–11, 14–36, 75–77.

Wilke, C. R., R. D. Yang, A. F. Sciamanna, R. P. Freitas, "Raw Materials Evaluation and Process Development Studies for Conversion of Biomass to Sugars and Ethanol, *Biotechnol. Bioeng.*, **23**, 163–183 (1981).

Xu, T. J., X. Q. Zhao, and F. W. Bai, "Continuous Ethanol Production Using Self-Flocculating Yeast in a Cascade of Fermentors," *Enzym. Microbial Technol.* **37**, 634–640 (2005).

5.1. Biotin is added in trace amounts (0.1 mg/L) to a microbial fermentation that transforms glucose into glutamic acid. The productivity, which is related to cell mass, increases. When the biotin concentration is increased further to (1 mg/mL) in a second experiment, cell growth slows and glutamic acid production is also lower. When biotin is not added, the cells do not grow, and glutamic acid is not produced. Substrate depletion is not a factor since $S \gg K_s$ in all three experiments, where

$$\mu = \frac{\mu_{max}[S]}{K_s + [S]}$$

While accumulation of glutamic acid in the fermentation broth does not cause inhibition of growth, intracellular accumulation of glutamic acid does cause inhibition. Explain the role of biotin in achieving maximal glutamic acid accumulation in the fermentation broth.

5.2. The sterilization of a solution containing glucose is carried out at 121 °C for 30 min. Heatup to sterilization temperature is achieved within 10 s using a heat exchanger. The initial cell mass (expressed as the number of viable cells) is 10 cells/mL, while the initial concentration of glucose is 100 g/L. Set up the equations for solution (using Matlab) that will give the change in glucose, hydroxymethyl furfural, and levulinic acid concentrations, as well as viable cell count as a function of time during the sterilization. The glucose is subject to degradation via the pathway

$$G \xrightarrow{\;K_1\;} D \xrightarrow{\;K_2\;} L \qquad\qquad \text{(HP5.2a)}$$

where G = glucose, D = degradation product (hydroxymethyl furfural), and L = levulinic acid and

$$X_v \xrightarrow{\;K_D\;} X_i \qquad\qquad \text{(HP5.2b)}$$

where X_v = viable cells and X_i = inactivated cells. Assume that cell growth does *not* occur during the sterilization heatup process, and that viable cells represent those cells that survive sterilization. State your assumptions.

5.3. The fermentation of xylose to ethanol is an important step in the commercial production of ethanol from cellulose. One possible process for achieving this conversion is through isomerization of xylose to xylulose, and transformation of xylulose to hexose, and hexose to ethanol by yeast, as shown by

$$A \underset{k_{-1}}{\overset{k_1}{\rightleftharpoons}} S \qquad\qquad \text{(HP5.3a)}$$

$$S \xrightarrow[\text{yeast}]{} P \qquad\qquad \text{(HP5.3b)}$$

where A = xylose (a pentose), S = xylulose (also a pentose), and P = ethanol (the product). Assume that the equation for cell growth is

$$\mu = \left(\frac{\mu_{max}[S]}{K_s + [S]}\right)\left(1 - \frac{P}{P_{max}}\right)^n$$

The yield coefficients for product and cell mass are

$$Y_{P/X} \sim \text{product/cell mass}$$

$$Y_{S/X} \sim \text{substrate/cell mass}$$

Initial concentration of xylose (A) is 100 g/L; xylulose (S), 1 g/L; cell mass (X), 10 g/L; and ethanol, 0.5 g/L. Set up the equations for accumulation of cell mass and product and concentration of fermentable substrate (S, xylulose) and starting sugar (A, xylose) as a function of time.

(a) Give the initial conditions for A, S, X, P

(b) Give the number of independent equations.

Set up the equations that give changes in A, S, X, and P as a function of time.

5.4. Simultaneous Saccharification Fermentation. The production of a biofuel using a biobased process is to be carried out by hydrolyzing cellulose (saccharification) from plant tissue and fermenting the resulting glucose to ethanol. Cellulase (an enzyme that hydrolyzes cellulose) is strongly inhibited by glucose; therefore the substrate is to be hydrolyzed in the same tank that is used to carry out the fermentation in a process called *simultaneous saccharification and fermentation* (SSF). The enzyme is added to a fermentor containing both yeast cell mass (X) and a cellulose (C) where the cellulose concentration is initially at 125 g/L. The yeast converts the glucose (S) to ethanol (P). This approach enables the cellulose to be enzymatically hydrolyzed and the glucose immediately to be fermented to ethanol at the same time, thereby avoiding the otherwise strong inhibition of the enzyme that would be caused by the accumulation of glucose due to the hydrolysis of the cellulose. Although the glucose concentration will be low, the inhibitory effect of glucose must be taken into account. The following reaction scheme shows how inhibition affects the kinetics of hydrolysis:

Cellulose (*n*) + enzyme $\rightleftarrows$ Cellulose (*n*)–enzyme $\longrightarrow$ Cellulose (*n*−1) + enzyme + glucose

$\uparrow\downarrow$

enzyme–glucose

The rate of cellulose hydrolysis can be calculated using the Michaelis–Menten equation modified to include inhibition:

$$\frac{dC}{dt} = \frac{-V_{\max} C * \frac{1000}{162}}{K_m \left(1 + \frac{C * \frac{1000}{162}}{K_m} + \frac{S * \frac{1000}{180}}{K_i}\right)}$$

Note: $\dfrac{1000}{162}$ converts cellulose concentration from g/L to mM and $\dfrac{1000}{180}$

converts glucose concentration from g/L to mM:

$$\text{Costants: } V_{\max} = k_p * \text{Et mM glucose/h}$$

$$\text{Et} \quad = 1000 \, \text{mg protein}$$

$$k_p \quad = 0.005 * 60 \, \text{mM glucose/(mg protein} * \text{h)}$$

$$K_m \quad = 4.0 \, \text{mM}$$

$$K_i \quad = 0.5 \, \text{mM}$$

The fermentation of the resulting glucose to ethanol can be modeled using the same equations as before. The cell concentration is

$$\frac{dX}{dt} = XE\nu$$

$$E\nu = \mu$$

$$\nu = \nu_m \left[\frac{S}{K_S + S}\right] \left[1 - \frac{P}{P_{\max}}\right]^n \quad \text{product (ethanol) inhibition}$$

The product (ethanol) concentration is

$$\frac{dP}{dt} = \nu X$$

The substrate (glucose) concentration equation must be slightly modified to include the appearance of glucose (from hydrolysis) as well as the disappearance (by fermentation):

$$\frac{dS}{dt} = -\frac{1}{Y_{x/s}} \frac{dX}{dt} - \frac{dC}{dt} * \frac{180}{1000}$$

Note: $\dfrac{180}{1000}$ coverts mM glucose to g/L:

$$\text{Costants: } K_S \quad = 0.315 \, \text{g/L}$$

$$P_{\max} = 87.5 \, \text{g/L}$$

$$\mu_{\max} = 1.15 \, \text{g EtOH/(g cells} \cdot \text{h)}$$

$$E \quad = 0.249$$

$$Y_{x/s} = 0.07$$

$$n \quad = 0.36$$

Initial conditions: At time $t = 0$:

$$C(0) = 125\,g/L \quad X(0) = 1\,g/L$$

$$S(0) = 0\,g/L \quad P(0) = 0\,g/L$$

(a) Construct a plot showing the concentrations of cell mass, cellulose, glucose, and ethanol versus time from the solution of the differential equations for this fermentation. Use the Runge Kutta method described in Chapter 6, together with an appropriate software package.

(b) How long does the fermentation take to complete?

(c) What are the final concentrations of the four modeled components (cell mass, etc.)?

(d) Describe how the four modeled components behave during the course of the fermentation and give rationale for this behavior.

5.5. Unstructured Inhibition Models. An increase in the concentration of substrate results in an increase in the specific growth rate as illustrated in Fig. 5.4. However, as the concentration increases even further, cell growth becomes very slow and essentially stops. For the expression for specific growth rate:

$$\mu = \frac{\mu_{max}S}{K_s + S + \dfrac{S^2}{K_i}}$$

(a) show how this expression results where K_i is a substrate inhibition constant. in a characteristic plot of:

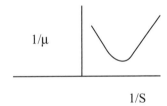

(b) K_s in the Monod equation [where $\mu = (\mu_{max}S)/(K_s + S)$] is the substrate concentration that corresponds to the point where the specific growth rate is

— 0.10 of its maximum

— half of its maximum

— 0.63 of its maximum

5.6. The anaerobic acetone/butanol/ethanol fermentation is subject to strong product inhibition because of the butanol that compromises the cell membrane. An equation that may describe this inhibition is as follows:

$$\text{Product: } \mu = \frac{\mu_{max}S}{K_S + S} \cdot \frac{K_p}{K_p + P}$$

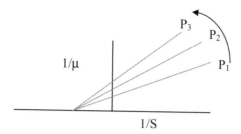

In this diagram, the concentrations of the product (also the inhibitor) are $P_3 > P_2 > P_1$. If μ_{max} and K_p are determined from fermentation data where the product concentration is close to zero, then K_p can be obtained from the data. Derive an expression, and give a graphical representation of how K_p might be determined from the measured growth data.

AEROBIC BIOREACTORS

INTRODUCTION

Aerobic bioreactors are a key component of the existing and emerging biotechnology industry. They provide numerous valuable products ranging from proteins and peptides, to antibiotics and organic acids. They also begin to motivate our desire for a more detailed fundamental knowledge of how cells use oxygen to make products that are industrially important. Chapter 6 is a bridge between the introductory material in Chapters 1 through 5, and subsequent chapters in this book that describe cellular metabolism and bioenergetics from a molecular perspective. The goal is to understand how directed (or spontaneous) changes in environmental conditions for microorganisms change the way in which the reaction pathways in a living organism adapt, thereby resulting in a system-level response that is observable as a change in cell concentration, growth rate, ATP accumulation, cell morphology, and/or secretion of an extracellular product.

A shift in metabolism is a part of the microorganism's drive to survive by providing energy needs for maintenance, and for growth. By understanding key components of the cell's metabolism and correlating changes inside of it to changes outside of the cell, we may begin to mathematically simulate, and in some cases, interpolate environmental changes to phenomenological response. This response may be at a single microorganism level, or for selective pressures that enable one microorganism to dominate others in a microbially diverse culture.

The student should not be confused by terminology that has more recently become popular such as *directed evolution* or *molecular evolution*. These are new words that describe processes that have been used by microbiologists for over 100 years to select one characteristic over another by using media that enable one microorganism, or a microorganism with desirable (to us) properties to grow more quickly. The key component that has been added since about 1976 has been the ability to introduce foreign genes into *E. coli*, yeast, and other microorganisms. While first carried out for plasmids, and later for chromosomes (Chapters 10 and 11), these applications utilized human genes for therapeutic hormones (such as for insulin or growth hormone) for transforming the bacteria *E. coli* to produce these

Modern Biotechnology, by Nathan S. Mosier and Michael R. Ladisch
Copyright © 2009 John Wiley & Sons, Inc.

human proteins (Ladisch and Kohlmann 1992; Ladisch 2001). This technique also applied to introduction of human genes into malignant cells, so that they could be grown in (aerated) bioreactors and produce therapeutically useful enzymes [such as the "clot-busting" hydrolytic proenzyme known as *tissue plasminogen activator* (tPA)] or a hormone that enabled stem cells to differentiate into blood cells [erythropoietin (EPO)].

The biotechnology associated with medical products made its way to modification of microorganisms to enhance their productivity in generating existing products. These techniques, combined with sequencing of bacterial and human genomes, could be used to both observe and alter bacterial pathways. The pathways may be directed to provide enhanced levels of desirable (commercially useful) molecules. The techniques perfected for microbially generated biopharmaceutical and pharmaceutical products have made possible a toolbox that is now more broadly applicable. These techniques, developed over a period of 30 years, enable the redirection of metabolic pathways, analysis of the role of different pathways, and insertion of new ones into yeast and bacteria for the generation of biofuels, including ethanol, butanol, and long-chain acids. These techniques enable the insertion of multiple genes into microorganisms so that they express enzymes for hydrolysis of cellulose and hemicelluloses to hexoses and pentoses. When these pathways are combined with microorganisms that are modified so that glucose, xylose, and other pentoses are fermented to ethanol, a multifunctional, self-replicating, living biocatalyst—microorganism—is obtained. This concept has been termed *consolidated bioprocessing* (CBP). CBP achieves saccharification and fermentation in a single microorganism (Hunt et al. 2006; Baum et al. 2007). Since nonfood renewable resources are the feedstocks for this type of process, and since these feedstocks contain recycled CO_2, this approach has the potential to mitigate the generation of CO_2, a greenhouse gas (Houghton et al. 2006).

The combination of genome sequencing; analytical tools developed for the human genome project for indirectly observing cellular metabolism, screening of cells with desired characteristics using molecular markers as well as selective culturing techniques, and amplification of genes from difficult-to-grow microorganisms in surrogate hosts (easier to grow and able to readily express foreign genes) are all tools that have developed since the mid-1970s. This has enabled scientists and engineers to revisit the development of living (microbial), non-living (dead whole cells), and protein (enzyme) catalysts that transform sugars or other precursors into commercially or therapeutically valuable products. This concept is extended to systems of enzymes that may be extracted from living cells, and that carry out sequences of reactions (Voloshin and Swartz 2005).

The reinvigoration of these applications has also stimulated the quest for terminology that is easy to remember. The techniques briefly summarized here are sometimes referred to as *directed evolution*, since the tools now available through catalogs of scientific supply houses contain reagents and enzymes that enable direction or redirection of a microorganism's metabolic machinery to produce one product over another. However, the process is not as easy or direct as it would at first appear, since changing several enzymes in a metabolic pathway may change other aspects of the cell's metabolism in an unanticipated manner. Hence, screening of the resulting transformed microorganisms is still needed, much as has been the case for over 100 years. The development of rapid screening methods is a critical and continuing challenge. Since selective pressures may be used to select one result (microbe) over another, this has been viewed as an evolutionary approach, where a microbe is selected, propagated, studied, modified, and then submitted to the entire process again. This has previously been described as biochemical engineering (Aiba et al. 1973) or applied bacteriology (Singleton 1995).

The development of the engineer's toolbox has been evolutionary, and it enables directed but not always predictable changes. This chapter, by way of examples, describes the characteristics of an aerated fermentation that may be anticipated but not necessarily controlled, and begins to relate these changes to the metabolism of microbial cells at industrially relevant conditions. The student will hopefully begin to develop a fundamental understanding of how things grow and how they might be improved either at a molecular level or through environmental means. As we said at the beginning of this section, the process is one of building a bridge that connects current practice to future developments in biotechnology.

The simulation and modeling of fermentations was developed in Chapter 5 without specific consideration of oxygen requirements. The assumption implicit in developing the equation for cell growth, substrate depletion, and product accumulation was that the carbon source—glucose—was the limiting substrate that controls the rate and concentrations of the cell mass and product. Oxygen, if required by the cell's metabolism, was assumed to be present in sufficient quantities to meet the needs of the microorganism's metabolic requirements, and hence oxygen concentration did not appear explicitly in any of the equations. This was the case for cell mass (biomass) accumulation in the yeast example. For ethanol production, micro-aerobic conditions were assumed in which oxygen may be present initially. It is not supplied intentionally, nor is it excluded from the fermentation. Once the fermentation is initiated, CO_2 generated as a coproduct of ethanol fermentation displaces oxygen and is removed from the fermentation vessel as it is formed.

Many of the microorganisms and products listed in Table 4.2 require that oxygen be provided by bubbling (sparging) air through the liquid in the fermentation vessel or bioreactor, in much the same way as air is filtered and bubbled through a fish tank. In the case of a bioreactor, a combination of vigorous agitation and bubbling of air through a manifold is carried out. The concentration of oxygen in the fermentation media is at most 8 ppm. Once cell growth occurs, the steady-state oxygen concentration approaches zero, even though the liquid medium in the bioreactor is being vigorously aerated. During an aerated fermentation, the cell mass may quickly increase to the point where oxygen is removed from the fermentation media as quickly as it is provided. At this point, the bioreactor is said to be at oxygen-limiting conditions. This effect is often further compounded by an increase in the viscosity of the fermentation broth due to the high cell concentration, and sometimes also due to the branched structure of product or cells and cell morphology. The higher viscosity makes it more difficult to stir and mix the broth, and results in decreased oxygen transfer into the broth. This changes the cell's metabolism, and may result in the induction or formation of a desired product. This chapter first describes an example of how an aerated fermentation is modeled as it makes a transition from oxygen sufficient to oxygen-limited growth conditions. This is followed by a discussion of $k_L a$ and the estimation of oxygen transfer. The terms aeration, dissolved oxygen, oxygen transfer, and oxygen limitation will be introduced as part of the vocabulary that describes aerated bioreactors. The example below, although published in 1984, was selected because of its completeness with respect to both the model and the accompanying rationale. This example illustrates a thought process that is as valid today as it was in 1984.

FERMENTATION PROCESS

Fermentation of Xylose to 2,3-Butanediol by *Klebsiella oxytoca* Is Aerated but Oxygen-Limited

A major fraction (~25–35%) of cellulose-containing agricultural residues, and wood chips is hemicellulose. On hydrolysis, the hemicellulose gives the monosaccharide, xylose, smaller amounts of another pentose, arabinose, and even smaller amounts of glucose or galactose. The xylose must be fermented to a value-added product to help justify the economics of cellulose conversion. The diol, 2,3-butanediol, is a potential precursor for the production of butadiene rubber or isobutanol. It can be obtained by the fermentation of the pentose, xylose, and hence offers a value-added product that can be derived from the hemicellulose portion of biomass and waste cellulose materials. This diol is formed by *Klebsiella oxytoca* when the fermentation makes the transition from an oxygen sufficient state to an oxygen-limiting state.

Jansen et al. (1984a,b) present an analysis for the production of 2,3-butanediol in an aerated fermentation by *Klebsiella oxytoca* ATCC 8724. Fermentation was carried out in a 7-L batch fermentor at pH 5.2, with 3 L of medium. Temperature was maintained at $37 \pm 0.2\,°C$. An aeration rate of 3.3 L/min, and an agitation rate of 750 rpm were used. Oxygen level was measured using a galvanic oxygen probe. Data, and a model for the metabolism of this microorganism are described. Using the two papers by Jansen et al., concentrations of cell biomass, 2,3-butanediol, and xylose as a function of time may be calculated. Assuming that the carbon source is xylose that is present at an initial concentration of 100 g/L and that the dissolved oxygen (DO) level at the beginning of the fermentation is near saturation, initial cell growth occurs quickly, but also quickly depletes dissolved oxygen. The differential equations for this fermentation can be expressed as three equation sets with one set of equations for each of the three cases below:

1. Aerobic fermentation
2. Transition from aerobic to oxygen-limiting conditions
3. Growth under oxygen-limiting conditions

The initial conditions and constants must be defined for solution using the Runge–Kutta or other numerical integration methods.

Figures 6.1, 6.2, and 6.3 summarize the key characteristics of this fermentation when carried out using the media composition of Table 6.1.

The medium was sterilized in three separate parts by autoclaving concentrated forms of the sugar solution, trace metals, and phosphate and ammonium salts, and then combining. If autoclaved together, chemical reactions between the three components or catalyzed by the salts could cause formation of products that would be toxic to the microorganism. The results of fermentation showed that xylose was consumed over a 16-h period, while the product formed starting at about 10 h (Fig.

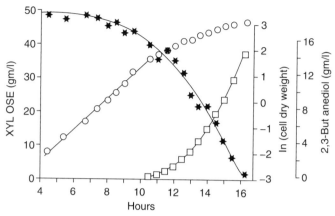

Figure 6.1. Change in xylose and 2,3-butanediol concentration as a function of time: (*) xylose; (○) cell mass (absorbance); (□) 2,3 butanediol [from Jansen et al. (1984a), Fig. 1, p. 364].

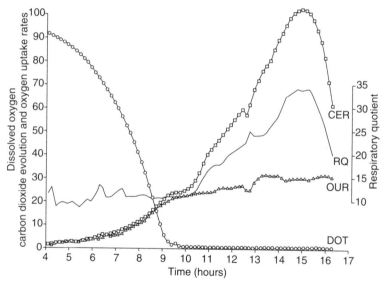

Figure 6.2. Accumulation of cell mass and protein as a function of time. (○) Optical density; (□) protein. (from Jansen et al. 1984a).

6.1). Cell protein and cell mass (proportional to optical density) increased sharply at about 10 h (Fig. 6.2), which coincided with a sharp drop and decrease of dissolved oxygen (Fig. 6.3). The transition from an oxygen-sufficient to an oxygen-limiting condition caused the cells to change their metabolism of xylose to produce energy from aerobic to anaerobic, with formation of the coproduct 2,3-butanediol. This description is consistent with the data shown in Figs. 6.1–6.3, and relates to the bioenergetics of the cell where the ATP required for growth and maintenance relative to the ATP produced by energy producing reactions balances out. This is similar

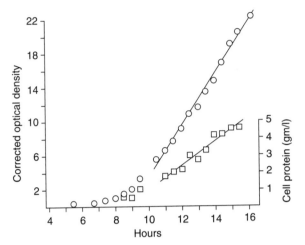

Figure 6.3. Changes in (○) dissolved oxygen as % saturation, (□) CO_2, (Δ) oxygen uptake rate, and (—) respiratory quotient. (With permission Jansen et al. 1984a.)

Table 6.1.

Composition of Media

Component	Concentration (g/L)
Xylose	5–150
$K_2HPO_4 \cdot 3H_2O$	13.7
KH_2PO_4	2.0
$(NH_4)_2HPO_4$	3.3
$(NH_4)_2SO_4$	6.6
$MgSO_4 \cdot 7H_2O$	0.25
$FeSO_4 \cdot 7H_2O$	0.05
$ZnSO_4 \cdot 7H_2O$	0.001
$MnSO_4 \cdot 7H_2O$	0.001
$CaCl_2 \cdot 2H_2O$	0.01
EDTA	0.05

Source: Jansen et al. (1984a).

to the Luedeking–Piret model in which there are growth- and non-growth-associated terms [see Eq. (5.52)] represented by parameters α and β, respectively. As pointed out by Jansen et al., the xylose metabolism may be considered as consisting of three components: "conversion to 2,3 butanediol by fermentation, oxidation to CO_2 by respiration, and assimilation to cell mass" (see Fig. 6.4).

The simulation is divided into two phases, and therefore, two sets of equations.

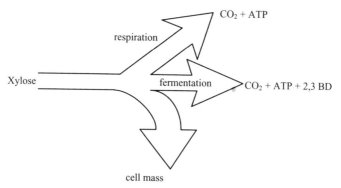

Figure 6.4. Schematic representation of xylose metabolism in *Klebsiella oxytoca* during oxygen-limited growth [from Jansen et al. (1984b), Fig. 1, p. 574].

Phase I. Oxygen-Sufficient Growth Occurs Early in the Fermentation

When oxygen is sufficient, as is the case at the initial part of the fermentation, cell mass but not 2,3-butanodiol is formed because the energy for growth and maintenance is derived from full oxidation of xylose to CO_2. The equations describing this state are

$$\frac{dx}{dt} = \mu_{max} x \tag{6.1}$$

$$\frac{dS}{dt} = -(Q_{SA} + Q_{SR} + Q_{SF}) \cdot (MW_{xylose}) \tag{6.2}$$

$$Q_{sa} = \frac{1}{120} \frac{dx}{dt} \tag{6.3}$$

$$Q_{sr} = \frac{3}{70} \left(\frac{dx}{dt} \frac{1}{Y_{ATP,1}} + m_{el} X \right) \tag{6.4}$$

$$Q_{sf} = 0 \tag{6.5}$$

$$\frac{dP}{dt} = 0 \tag{6.6}$$

Since the concentration of the substrate xylose, $S \gg K_s$, the cells grow at an exponential rate that is proportional to μ_{max} [Eq. (6.1)]. Q_{sa} denotes the rate at which xylose is assimilated into cell mass (mol xylose $L^{-1} h^{-1}$), while Q_{sr} is the rate of xylose respiration via the TCA cycle (mol xylose $L^{-1} h^{-1}$) while Q_{sf} is the rate of xylose fermentation to 2,3-butanediol (abbreviated 2,3-BD) (mol xylose $L^{-1} h^{-1}$). Q_{sf} is negligible when oxygen is sufficient. The molecular weights of xylose and 2,3-butanediol are 150 and 90, respectively. The concentration of cell mass at the transition between oxygen-sufficient and oxygen-limiting conditions is given by the solution of the equations for microbial growth under oxygen-limiting conditions, where growth is constant rather than exponential

Table 6.2.

Values of Parameters Used in Simulation of 2,3-Butanediol Production by *Klebsiella oxytoca* 8724

Parameter	Oxygen-Sufficient ($i = 1$)	Oxygen-Limiting ($i = 2$)
$Y_{ATP,i}$ (g cells/mol ATP)	11.5	10.4
μ_{max} (h^{-1})	0.6	0.6
m_{el} [mol ATP/(g cell · h)]	0.047	0.017
$k_L a C^*$ [mol oxygen/(L · h)]	—	0.027
k_d (L g^{-1} h^{-1})	—	0.0077

$$x_L (\text{g/L}) = \frac{14}{3} k_L a C^* / (m_{el} + \mu_{max}/Y_{ATP}^{max}) = 1.27$$

Source: Jansen et al. (1984a,b).

$$\frac{dx}{dt_L} = \mu_{max} X_L \tag{6.7}$$

where t_L is the elapsed time after oxygen has become limiting. Jansen et al. show that the condition for oxygen to become limiting is

$$5Q_{sr} > k_L a C^* \tag{6.8}$$

when combined with the expression for Q_{sr} [Eq. (6.4)] the growth rate for oxygen limited growth [Eq. (6.7)] is

$$\frac{dx}{dt_L} = \mu_{max} \left[\frac{\frac{14}{3} k_L a C^*}{m_{el} + \frac{\mu_{max}}{Y_{ATP}^{max}}} \right] \tag{6.9}$$

and the combination of Eqs. (6.6) and (6.8) enables calculation of X_L:

$$X_L = \frac{\frac{14}{3} k_L a C^*}{\left(m_{el} + \frac{\mu_{max}}{Y_{ATP}^{max}} \right)} \tag{6.10}$$

which for this example is (Table 6.2):

$$X_L = \frac{14}{3} \frac{0.027}{0.0047 + (0.6/11.5)} = 1.27 \, \text{g/L} \tag{6.11}$$

Phase II. A Transition to Oxygen Limitation Occurs at Low Cell Concentration (1 g/L)

Once the cell mass grows and the concentration of cell mass exceeds about 1 g/L (Fig. 6.1), the rate of oxygen dissolving into the fermentation media is less than the rate of oxygen consumption by the growing cells, so the dissolved oxygen level falls sharply as shown in Fig. 6.2. At this point, growth becomes oxygen-limited and Eqs. (6.12)–(6.16) instead of Eqs. (6.1)–(6.6) apply (Jansen et al. 1984b):

$$\frac{dx}{dt} = (\mu_{max} - k_d P)\, X_L \qquad (6.12)$$

where X_L is a constant and represents cell concentration (g/L) at the point that dissolved oxygen becomes limiting. The parameters in Eq. (6.12) are $k_d = 0.0077$ = product-induced death constant and $\mu_{max} = 0.60\,(h^{-1})$, where

$$Q_{sa} = \frac{1}{20}\frac{dx}{dt} \qquad (6.13)$$

$$Q_{sf} = \frac{18}{25}\left(\frac{dX}{dt}\frac{1}{Y_{ATP,2}} + m_{el}X - \frac{14}{3}k_L a_{eq}\right) \qquad (6.14)$$

Both Q_{sa} and Q_{sf} are expressed as moles xylose $L^{-1}h^{-1}$. In this case, the rate at which xylose is assimilated into cell mass is directly proportional to the rate of increase of cell mass. Q_{sr}, the rate of xylose respiration to CO_2 via the TCA cycle, is

$$Q_{sr} = \frac{1}{10}\left(2k_L aC^* - \frac{5}{6}Q_{sf}\right) \qquad (6.15)$$

where $k_L aC^*$ = oxygen supply rate (moles oxygen $L^{-1}h^{-1}$) and $k_L a$ = oxygen transfer coefficient. The parameter C^* denotes an equilibrium oxygen concentration in the broth.

Phase III. Butanediol Is Produced under Oxygen-Limiting Conditions

The limitation in oxygen causes growth and energy production to be more tightly coupled. Equation (6.15) assumes that the cells "regulate the amount of butanediol produced so that the total amount of ATP produced balances the amount of ATP required for growth and maintenance." The production of 2,3-butanediol is therefore

$$\frac{dP}{dt} = \frac{5}{6}Q_{sf}MW_{2,3BD} \qquad (6.16)$$

Multiplication of Eq. (6.15) by the molecular weight of butanediol (=90) gives the result on a weight (g/L) basis. Once the substrate xylose is completely depleted, the cells will be present in a constant and large concentration of 2,3-butanediol and the cells will begin dying according to the rate

$$\frac{dx}{dt} = -k_d\ \mathrm{P_{max}}\ \mathrm{X_{max}} \qquad (6.17)$$

where X_{max} represents the cell mass at the end of the growth phase when the substrate approaches zero and the maximum 2,3-butanediol concentration, P_{max}, has been obtained.

The model itself represents the metabolic activity of the cell in terms of five biochemical reactions shown in a symbolic representation in Eqs. (6.18)–(6.22) by Jansen et al. (1984b):

Oxygen-sufficient conditions:

Cell growth: $120/Y_{ATP}^{max}\,(ATP) + $ xylose $\rightarrow 120$g new cells $\qquad (6.18)$

Maintenance: $m_e\,(ATP) + 1$g cells $\rightarrow 1$g viable cells $\qquad (6.19)$

Oxidation of xylose to CO_2 via TCA cycle (respiratory pathway):

$$\mathrm{Xylose} \rightarrow 5\,CO_2 + 10\,NADH + \frac{10}{3}\,ATP \qquad (6.20)$$

$$\frac{1}{2}O_2 + NADH \rightarrow 2\,ATP \qquad (6.21)$$

Oxygen-limiting conditions:

$$\mathrm{Xylose} \rightarrow \frac{5}{3}\,CO_2 + \frac{5}{6}\,NADH + \frac{5}{3}\,ATP + \frac{5}{6}\,\mathrm{butanediol} \qquad (6.22)$$

This approach, reported in 1984, clearly illustrated how an understanding of cell energetics could be related to a quantitative model of cell growth and butanediol generation in an aerated fermentation.

The program used to carry out the simulation is shown in Table 6.3. The Runge–Kutta integration program given in the Appendix of this chapter was written by Craig Keim in 1999, while a researcher in LORRE, and modified in 2001.[1] The coding for this program is shown in Table 6.3. A code, also written by Craig Keim for solution of the differential equations by Runge–Kutta technique, is given in the Appendix of this chapter. The output of this simulation is graphed in Fig. 6.5. This figure illustrates the three regimes of oxygen-sufficient growth: 0–11 h, oxygen-limiting conditions (11–23 h), and substrate depletion accompanied by cell death (after 23 h). The references to page and table numbers in the computer code refer to the original publications of Jansen et al. (1984a,b).

---

[1] The overall approach was inspired by Harry Bungay of Rensselaer Polytechnic Institute, who recognized the value of such simulations, and cast procedures in a simplified manner so that these could be done using the early personal computers of the time (1988) and BASIC (Beginner's All-purpose Symbolic Instruction Code) programming language. Professor George Tsao was another early pioneer who showed how biochemical modeling of cellular function could be carried out with an economy of mathematical coding and experimental measurements so that phenomenological observation could be related to fundamental cellular mechanisms.

Table 6.3.

Equations for Simulation of 2,3-Butanediol Fermentation[a]

```
Public numEq As Integer            'Just some variable declarations.
Public stepSize, stopTime, startTime   'Move onto the constants subroutine below
Public O(), I()                    'and enter the relevant information
Public Qsr, Qsf

Sub constants()
    'Enter the values for the step size and the number of equations here.
    stepSize = 0.02
    numEq = 3

    '*********************************************
    ReDim I(numEq), O(numEq)       ' Just leave this line alone.   It dimensions the
                    ' input and output arrays
    '*********************************************

    ' Enter the initial conditions
    startTime = 0                  'Time value at which the intial conditions are known
    O(1) = 0.01                    'Initial cell concentration (g/L)
    O(2) = 100                     'Initial xylose concentration (g/L)
    O(3) = 0                       'Initial product concentration (g/L)

    'Enter the time at which you would like the simulation to end
    stopTime = 30

End Sub

Function inputs(eqnNumber As Integer, timeValue As Variant, outputs() As Variant)
    '*********************************************
    Dim j As Integer               'Just leave this code alone.
                    '
    For j = 1 To numEq             'Enter the input equations below...
        O(j) = outputs(j)      '
    Next j                  '

    t = timeValue                  't is the time
    '*********************************************

    'Enter the constants here
    kd = 0.0077       'Product induced death constant, Table 6.2
    yatp = 11.5       'Yatp coefficient for oxygen sufficient growth, Table 6.2
    me1 = 0.047       'Maintenance coefficient for oxygen sufficient growth, Table 6.2
    umax = 0.6        'Maximum specific growth rate, Jansen et al. (1984a,b)
    klaceq = 0.027    'Oxygen transfer rate, Table 6.2
    me2 = 0.017       'Maintenance coefficient for oxygen limited growth, Table 6.2
    Yatp2 = 10.4      'Yatp coefficient for oxygen limited growth, Table 6.2
    MWx = 150         'Molecular weight of xylose
    MWbd = 90         'Molecular weight of 2,3 butanediol
    xl = 14 / 3 * klaceq / (me1 + umax / yatp)          'equation (6.10)
```

Table 6.3. (Continued)

```
'Enter the input equations here
If O(1) <= xl Then
    'Oxygen sufficient growth
    I(1) = umax * O(1)          'eq. (6.1)
    Qsa = 1 / 120 * I(1)        'eq. (6.3)
    Qsr = 3 / 70 * (I(1) / yatp + me1 * O(1))      'eq. (6.4)
    Qsf = 0
    I(3) = 0
Else
    'Oxygen limited growth
    I(1) = (umax - kd * O(3)) * xl        'eq. (6.12)
    Qsa = 1 / 120 * I(1)                  'eq. (6.13)
    Qsf = 18 / 25 * (I(1) / Yatp2 + me2 * O(1) - 14 / 3 * klaceq)    'eq. (6.14)
    If Qsf <= 0 Then Qsf = 0
    Qsr = 1 / 10 * (2 * klaceq - 5 / 6 * Qsf)      'eq. (6.15)
    If Qsr <= 0 Then Qsr = 0
    I(3) = 5 / 6 * Qsf * MWbd             'eq. (6.16)
    If I(3) <= 0 Then I(3) = 0
End If

If O(2) <= 0 Then
    O(2) = 0
    I(2) = 0
    I(1) = -kd * O(3) * xl          'eq. (6.17)
    I(3) = 0
Else
    I(2) = -(Qsa + Qsr + Qsf) * MWx
End If

'***********************************************

inputs = I(eqnNumber)
End Function
```

[a]This program was written by Craig Keim for use in ABE580 and adapted form BasicA code developed by M. Ladisch 1/29/99. Modified 2/28/01, 12/27/05.

The connection between simulation, metabolism, and bioenergetics to make a useful product has been illustrated here. The design of fermentation vessels requires that mass transfer characteristics (especially of oxygen from air to the liquid media), combined with the cells' response to oxygen, be known or at least estimated. In most cases, maximal oxygen transfer is desired, and this requires vigorous mixing and good distribution of air (in the form of bubbles throughout the fermentor). Agitation and spargers are used for this purpose on both laboratory and production scales (refer to Fig. 4.3 in Chapter 4). In the case of microbial fermentations, the cells have membranes that resist the shear caused by mixing so that vigorous aeration is possible, although it is ultimately limited by the viscosity of the solution. In contrast, aeration of a mammalian cell culture is more challenging because the cells themselves have thin cell walls and are shear-sensitive. Mixing may cause rupture of the cell wall and lead to cell death.

The next section gives a framework for determining oxygen transfer, and introduces the principles of oxygen transfer and how the power input required for

2,3 Butanediol Production

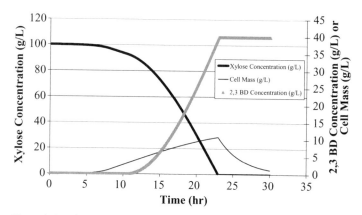

Figure 6.5. Plot of simulation of 2,3-butanediol fermentation showing cell mass, substrate concentration, and product accumulation as a function of time.

agitation may be estimated. The detailed design of bioreactors to achieve the desired level of aeration and control is the topic of another course and is not discussed.

Oxygen Transfer from Air Bubble to Liquid Is Controlled by Liquid-Side Mass Transfer

The maximum solubility of oxygen in water decreases from about 1.7 mM at 10 °C to 1.03 mM at 40 °C (Bailey and Ollis 1977). The concentration electrolytes in a fermentation medium is known to be a controlling parameter in oxygen diffusion, and also suppresses solubility. At 25 °C oxygen solubility is 1.26 mM in the absence of NaCl and 1.07 mM at 500 mM NaCl. The previous example of the 2,3-butanediol fermentation attained oxygen-limiting conditions when the dissolved oxygen reached 3% of saturation or about 0.03 mM.

A vigorously growing cell fermentation may contain 10^9 cells/mL in midexponential growth phase (discussed in Chapter 5). Bailey and Ollis (1977) calculated oxygen requirements at $375\,\text{mmol}\,L^{-1}\,h^{-1}$:

$$\frac{0.3\,\text{g O}_2}{\text{g dry-cell mass}\cdot\text{h}}\left|10^9\frac{\text{cells}}{\text{mL}}\right|10^{-10}\frac{\text{cm}^3}{\text{cell}}\left|1\frac{\text{g cell mass}}{\text{cm}^3}\right|0.2\frac{\text{g dry-cell mass}}{\text{g wet-cell mass}}$$

$$= 0.006\frac{\text{g O}_2}{\text{mL}\cdot\text{h}} = 375\frac{\text{mmol}}{\text{L}\cdot\text{h}}$$

Oxygen transfer occurs from the inside of an air bubble into the liquid that surrounds it. The major resistance is the film of liquid that surrounds each air bubble, and through which the oxygen must be transferred. The volumetric oxygen transfer, for liquid-side controlling, is given by

$$N_A = Hk_L a(\bar{p}\text{-}p^*) = k_L a(\bar{C}\text{-}C^*) \tag{6.23}$$

where N_A = moles of oxygen transferred

k_L = mass transfer coefficient on liquid side

H = Henry's constant

a = interfacial area between liquid and gas per unit volume of liquid

   = A/v = interfacial area/liquid volume

$\bar{p}$ = partial pressure of O_2 in bulk gas phase

p^* = equilibrium concentration of O_2 in gas phase with respect to $\bar{p}$

$\bar{C}$ = dissolved O_2 concentration in bulk liquid phase

C^* = equilibrium concentration of O_2 in the bulk (liquid) phase

(see also Fig. 6.6). The example of the 2,3-butanediol fermentation shows that the oxygen is used as quickly as it is supplied. This results in a bulk-phase oxygen concentration at only several percent of saturation during midexponential cell growth phase. Hence $N_A = Q_{O_2}$, where Q_{O_2} is the oxygen consumption rate in the liquid phase. The product of mass transfer rate and interfacial area is important to successful scaling up a fermentation vessel. Since C^* and $\bar{C}$ are usually known or are readily estimated, the measurement of k_L and a, separately, or $k_L a$ as a group enables Eq. (6.24) to be used.

Determination of k_L may be achieved by sparging N_2 through a tank filled with liquid media. Once this is done, air is injected and dissolved oxygen as a function of time of sparging is measured. The change in concentration of oxygen in the liquid phase is based on the (linear) difference between equilibrium (C^*) and operational or bulk-phase $\left(\bar{C}\right)$ oxygen concentrations:

$$\frac{dc_L}{dt} = k_L a\left(C^* - \bar{C}\right) \tag{6.24}$$

This equation is graphically represented in Fig. 6.7, where the slope is equivalent to $k_L a$ and $C_L = \bar{C}$.

Interfacial area is proportional to holdup in a tank. In this case

$$a = \underbrace{\frac{\text{total volume of bubbles}}{\text{total volume of broth}}}_{(holdup)} \underbrace{\frac{\text{area of bubble}}{\text{volume of bubble}}}_{(surface\ to\ volume)} = \frac{nF_0t_b}{V}\frac{\pi D^2}{\pi D^3/6} \tag{6.25}$$

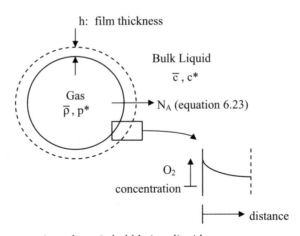

Figure 6.6.   Representation of an air bubble in a liquid.

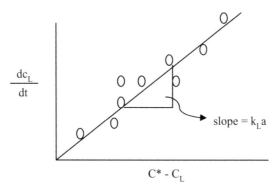

Figure 6.7. Rate of oxygen absorption as a function of concentration gradient in liquid phase.

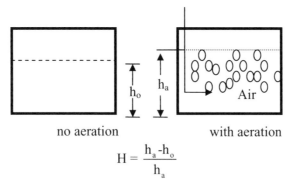

no aeration   with aeration

$$H = \frac{h_a - h_o}{h_a}$$

Figure 6.8. Schematic representation of measuring holdup H based on differences in fluid level in tanks with and without aeration.

where n = number of orifices in a sparging tube

F_0 = volumetric airflow rate per orifice

t_b = residence time of bubble in liquid

D = average diameter of air bubble

Equation (6.26) is directly expressed in terms of holdup and bubble diameter:

$$a = H\left(\frac{6}{D}\right) \tag{6.26}$$

Hence, a direct method for estimating a in a tank is enabled by measuring the increase in the liquid level when air is bubbled through the broth at a specified rate. This is illustrated in Fig. 6.8, and represents a direct method for calculating holdup H:

$$H = \frac{h_a - h_0}{h_a} \tag{6.27}$$

H also represents the volumetric void fraction of a liquid that is being aerated, that is, the ratio of air to total volume of liquid and air. This principle is illustrated by the increase in liquid level when we use a straw to blow air through a glass of water.

The interfacial area follows from Eq. (6.27), and k_L from Eqs. (6.27) and (6.25). Once k_L and a are known, the power required for scaling up oxygen transfer in an aerated bioreactor may be determined using correlations that have the form [reviewed by Ho et al. (1987)]

$$k_L a \propto \left(\frac{P_g}{V}\right)^m (v_s)^n \qquad (6.28)$$

where P_g = gassed power [horsepower (hp)]
 V = volume of gas–liquid dispersion (aerated solution) (L)
 v_s = superficial gas velocity (cm/s)

The basis of estimating power for gassed and ungassed media reflects several principles. When a liquid is aerated (i.e., "gassed"), it is less dense (refer to Fig. 6.8); hence it requires less power to mix it than does a liquid that is not aerated. Oxygen transfer is directly related to oxygen diffusion, and oxygen diffusion increases with increasing salt concentration (Fig. 6.9). Independent of diffusion, which is a molecular phenomenon, the bubble diameter also decreases with increasing salt concentration. This increases a, and through a physical phenomenon, increases the product of $k_L a$. Since the value of k_L is proportional to the oxygen diffusion coefficient, D_L (cm^2/s), we obtain

$$k_L \propto D_L^{2/3} \qquad (6.29)$$

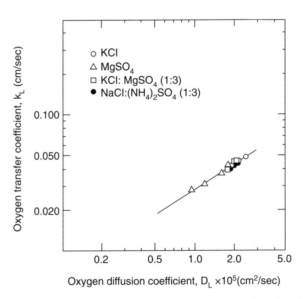

Figure 6.9. Oxygen transfer coefficient as a function of oxygen diffusion [from Ho et al. (1987), Fig. 11, p. 91].

Salt concentration has a synergistic effect. This has been shown to be the case for gas (air) in water, brine (NaCl), and solid particles in aqueous solution. An estimate of D_L enables us to obtain an estimate of k_L from Eq. (6.30).

These principles provide a framework for determining how different solution properties would effect the aeration of a fermentation broth. Ho et al. (1987) show that the power required for a gassed system follows the correlation of Michel and Miller:

$$P_g = 0.08 \left[\frac{P_0^2 ND^3}{Q^{0.56}} \right]^{0.45}$$

(6.30)

In this correlation, the different parameters are

P_g = gassed power (hp)
P_0 = ungassed power (hp)
N = rpm of impellar (min^{-1})
D = impellar diameter (ft)
Q = gas flow rate (ft^3/min)

The power number, which is proportional to the ungassed power P_0 for the Rushton impeller, was originally presented by Rushton et al. (1950) and follows the graph of Fig. 6.10. The form of Eq. (6.30) follows the curve in Fig. 6.11 and shows the proportionality between gassed and ungassed power, as well as the super-position of the data from Ho et al. on the curve for a six-blade turbine with four baffles in the tank. Figure 6.12 shows how the two are related to the density and viscosity of the media in the tank.

The power number (Rushton et al. 1950; Ho et al. 1987) is given by

$$\Phi = \frac{P_g}{\rho N^3 D^5}$$

(6.31)

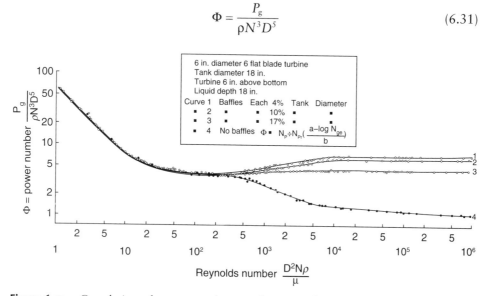

Figure 6.10. Correlation of power number as a function of Reynolds number for flat-blade turbine in a baffled reactor [from Rushton et al. (1950), Fig. 14, p. 468].

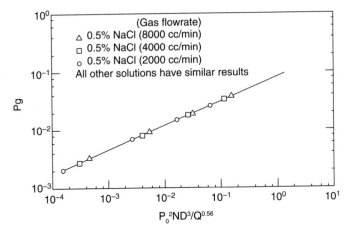

Figure 6.11.   Gassed power as a function of ungassed power, turbine configuration, and air (gas) volumetric throughput [from Ho et al. (1987), Fig. 3, p. 87, reprinted with permission from American Institute of Chemical Engineers].

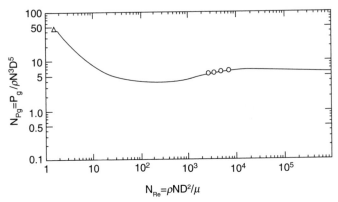

Figure 6.12.   Power number as a function of Reynolds number for an agitated tank with six-blade turbine and four baffles with data of Ho et al. (1987) superimposed [from Ho et al. (1987), Fig. 2, p. 87, reprinted with permission from American Institute of Chemical Engineers].

while the Reynolds number is

$$N_{Re} = \frac{\rho N D^2}{\mu} \tag{6.32}$$

where ρ is the density and μ is the viscosity. N is the rotation speed of the impeller (rpm), and D is the diameter of the impeller in feet. An increase in viscosity causes a decrease in N_{Re}, which, in turn, may increase the power number when $N_{Re} < 100$ (Fig. 6.10). This illustrates how rapid microbial growth, and an increase in viscosity of the broth because of cell growth, will increase (by a factor of ≤ 10) the power required to maintain a constant mixing effect (i.e., a constant Reynolds number).

CHAPTER 6 APPENDIX: EXCEL PROGRAM FOR INTEGRATION OF SIMULTANEOUS DIFFERENTIAL EQUATIONS (BY CRAIG KEIM, 1/29/99)

Runge Kutta Integration—DO NOT TOUCH

```
Sub magic()
    Static j As Integer, k As Integer
      constants         'Get all of the information about the constants
      clearData             'Clear out the old data from the spreadsheet

    iterations = Int((stopTime - startTime) / stepSize) 'Determine the number of
iterations

    t = startTime

    'Print the initial time label
    Sheets("Data").Cells(numEq + 5, 1).Value = "Time"
    Sheets("Data").Cells(numEq + 6, 1).Value = t

    'Print the time variables used (step size, start and stop times)
    Sheets("Data").Cells(2, 1).Value = "Step Size"
    Sheets("Data").Cells(2, 2).Value = stepSize
    Sheets("Data").Cells(3, 1).Value = "Start Time"
    Sheets("Data").Cells(3, 2).Value = startTime
    Sheets("Data").Cells(4, 1).Value = "Stop Time"
    Sheets("Data").Cells(4, 2).Value = stopTime

    For k = 1 To numEq
        'List the Initial Conditions
        Sheets("Data").Cells(numEq + 5, k + 1).Value = "O(" & LTrim(Str(k)) & ")"
        Sheets("Data").Cells(numEq + 6, k + 1).Value = O(k)
    Next k

    For j = 1 To iterations

            RK4 (t)
            For k = 1 To numEq
                Sheets("Data").Cells(j + numEq + 6, k + 1).Value = O(k)
            Next k
            Sheets("Data").Cells(j + numEq + 6, numEq + 2).Value = Qsr
            Sheets("Data").Cells(j + numEq + 6, numEq + 3).Value = Qsf
            t = t + stepSize
            Sheets("Data").Cells(j + numEq + 6, 1).Value = t
    Next j

End Sub

Sub RK4(timeValue As Variant)
    Static  k1(),  k2(),  k3(),  k4(),  oldOutput(),  temp1(),  temp2(),  temp3(),
newOutput()
```

```
    ReDim k1(numEq), k2(numEq), k3(numEq), k4(numEq)
    ReDim oldOutput(numEq), temp1(numEq), temp2(numEq), temp3(numEq),
newOutput(numEq)
    Dim t, j As Integer
    t = timeValue

    For j = 1 To numEq
        oldOutput(j) = O(j)
    Next j

    'Calculate the 4 Runge-Kutta constants
    For j = 1 To numEq
        k1(j) = inputs(j, t, oldOutput())
        temp1(j) = oldOutput(j) + 0.5 * stepSize * k1(j)
    Next j

    For j = 1 To numEq
        k2(j) = inputs(j, t + 0.5 * stepSize, temp1())
        temp2(j) = oldOutput(j) + 0.5 * stepSize * k2(j)
    Next j

    For j = 1 To numEq
        k3(j) = inputs(j, t + 0.5 * stepSize, temp2())
        temp3(j) = oldOutput(j) + stepSize * k3(j)
    Next j

    For j = 1 To numEq
        k4(j) = inputs(j, t + stepSize, temp3())
    Next j

    For j = 1 To numEq
        'Calculate the new y Value
        newOutput(j) = oldOutput(j) + stepSize / 6 * (k1(j) + 2 * k2(j) + 2 * k3(j)
+ k4(j))
        O(j) = newOutput(j)
    Next j
End Sub

Sub clearData()
    'Clears all of the old data prior to putting in the new data
    Static lastCell As String

    Sheets("Data").Activate
    Sheets("Data").Cells(numEq + 6, 1).End(xlToRight).Select
    ActiveCell.End(xlDown).Select
    lastCell = ActiveCell.Address()
    Sheets("Data").Range("A1:" & lastCell).Clear
    Sheets("Data").Range("A1").Select
```

End Sub

```
Sub viewCode()
    'Sheets("Module1").Activate
    Application.Goto Reference:="constants"
End Sub
```

REFERENCES

Aiba, S., A. E. Humphrey, and N. F. Millis, *Biochemical Engineering*, 2nd ed., Academic Press, New York, 1973, pp. 1–91.

Bailey, J. E. and D. F. Ollis, *Biochemical Engineering Fundamentals*, McGraw-Hill, New York, 1977, pp. 418–419, 479.

Basak, S., A. Velayudhan, and M. R. Ladisch, "Simulation of Diauxic Production of Cephalosporin C by Cephalosporium acremonium: Lag Model for Fed-Batch Fermentation," *Biotechnol. Progress* **11**, 626–631 (1995).

Baum, W., W. Provine, and C. Riva, "Making the Big Bet: Success in Developing Cellulosic Alliances," *Summary Proc. 4th Annual World Congress on Industrial Biotechnology and Bioprocessing, Linking Biotechnology, Chemistry, and Agriculture to Create New Value Chains*, Orlando, FL, March 21–24, 2007, pp. 40–41.

Bungay, H., *Basic Biochemical Engineering*, 2nd Ed. BiLine Associates, Troy, New York, 1988, 1993.

Chu, W.-B. and A. Constantinides, "Modeling, Optimization, and Computer Control of the Cephalosporin C Fermentation Process," *Biotechnol. Bioeng.* **32**, 277–288 (1988).

Ho, C. S., M. J. Stalker, and R. F. Baddour, "The Oxygen Transfer Coefficient in Aerated Stirred Reactors and Its Correlation with Oxygen Diffusion Coefficients," in *Biotechnology Processes: Scale-up and Mixing*, C. S. Ho and J. Y. Oldshule, eds., AIChE (American Institute of Chemical Engineers), New York, 1987, pp. 85–95.

Houghton, J., S. Weatherwax, and J. Ferrell, *Breaking the Biological Barriers to Cellulosic Ethanol: A Joint Research Agenda*, US Department of Energy, Washington, DC, (http://www.doegenomestolife.org/), 2006.

Hunt, S., F. Reinach, and L. Lynd, "An International Perspective on Biofuels and Cellulosic Ethanol," *Summary Proc. 3rd Annual World Congress on Industrial Biotechnology and Bioprocessing, Linking Biotechnology, Chemistry, and Agriculture to Create New Value Chains*, Toronto, Ontario, July 11–14, 2006, pp. 9–12.

Jansen, N. B., M. C. Flickinger, and G. T. Tsao, "Production of 2,3-Butanediol from D-Xylose by Klebsiella oxytoca ATCC 8724," *Biotechnol. Bioeng.* **26**, 362–369 (1984a).

Jansen, N. B., M. C. Flickinger, and G. T. Tsao, "Application of Bioenergetics to Modeling of Microbial Conversion of D-Xylose to 2,3-Butanediol," *Biotechnol. Bioeng.* **26**(6), 573–582 (1984b).

Ladisch, M. R. and K. Kohlmann, "Recombinant Human Insulin," *Biotechnol. Progress* **8**(6), 469–478 (1992).

Ladisch, M. R., *Bioseparations Engineering: Principles, Practice and Economics*, Wiley, New York, 2001, pp. 514–617.

Malmberg, L.-H. and W.-S. Hu, "Kinetic Analysis of Cephalosporin Biosynthesis in *Streptomyces clavuligerus*," *Biotechnol. Bioeng.* 38, 941–947 (1991).

Rushton, J. H., E. W. Costich, and H. J. Everett, "Power Characteristics of Mixing Impellars," *Chem. Eng. Progress* 46(9), 471–476 (1950).

Singleton, P., *Bacteria in Biology, Biotechnology and Medicine*, 3rd ed., Wiley, New York, 1995, pp. 207–245.

Voloshin, A. M. and J. R., Swartz, "Efficient and Scalable Method for Scaling Up Cell Free Protein Synthesis in Batch Mode," *Biotechnology and Bioengineering* 91(4), 516–521 (2005).

Weil, J., J. Miramonti, and M. R. Ladisch, "Literature Survey: Cephalosporin C: Mode of Action and Biosynthetic Pathway," *Enzym. Microbial Technol.* 17, 85–87 (1995a).

Weil, J., J. Miramonti, and M. R. Ladisch, "Biosynthesis of Cephalosporin C: Regulation and Recombinant Technology," *Enzym. Microbial Technol.* 17, 88–90 (1995b).

CHAPTER SIX
HOMEWORK PROBLEMS

6.1. A bacterium grown on glucose shows yield coefficients of

$$Y_{X/S} = 0.51 \frac{\text{g cell}}{\text{g substrate consumed}}$$

$$Y_{XO_2} = 1.47 \frac{\text{g cell}}{\text{g O}_2 \text{ consumed}}$$

$$Y_{X\Delta H} = 0.42 \frac{\text{g cell}}{\text{kcal heat generated}}$$

How much heat is generated for each gram of oxygen that is consumed in the fermentation? How does this impact design of the fermentation system [from Bailey and Ollis, as taken from Abbott and Clemen (1977), p. 479]?

6.2. Assume that oxygen is a limiting substrate where

$$\mu = \mu_{max} \left[\frac{\bar{c}_L}{K_{O_2} + \bar{c}_L} \right]$$

where K_{O_2} denotes the Monod-type constant for oxygen, and $\bar{c}_L$ is the bulk-phase oxygen concentration.

(a) Derive an expression for average (bulk-phase) oxygen concentration $\bar{c}_L$ as a function of c^*, Y_{O_2}, K_{O_2}, and μ_{max}.

(b) If Y_{O_2} for *K. oxytoca* is $5 \times Y_{ATP}$, and K_{O_2} is 0.10 of K_s, what is cell concentration at which metabolism of the organism changes if $k_L a C^* = 0.027$? Does this seem reasonable? (Why or why not?) State your assumptions and show your work.

6.3. The data in the table below were collected from an agitated–sparged tank (from lecture notes of George Tsao, ChE 597T, Purdue University). Determine the k_La values from the data in the table. What is the effect of the surface-active protein material on k_La?

Time (s)	Oxygen Saturation in Water	Protein Added (mg/L)
0	0	0
15	30	
32	50	
55	70	
0	0	40
20	30	
42	50	
72	70	
0	0	80
25	30	
50	50	
88	70	
0	0	160
48	30	
83	50	
140	70	

6.4. (a) If the k_L value of an aerated, bacterial bioreactor is 5×10^{-3} cm/s and the surface area of air bubbles is 3000 cm^2/L of the liquid medium, what is the expected pseudo-steady-state dissolved oxygen concentration before the exhaustion of other nutrients begins to change the rate constants k_1 and k_2 of the oxygen consumption equation given below?

$$r(c) = \frac{k_1 C}{k_2 + C}$$

where C = dissolved oxygen concentration
  k_1 = 0.09 mg oxygen/L
  k_2 = 3 mg oxygen/L
  C^* = 6.0 mg oxygen/L (at 30 °C)

(b) What is the power required for agitation if the volume of the broth in the bioreactor is 10,000 L and the superficial gas velocity is 100 cm/min with a Reynolds number of 300?

6.5. The mathematical model for simulating diauxic production of the antibiotic cephalosporin C by *Cephalosporium acremonium* is reported and reviewed by Basak et al. (1995), Malmburg and Hu (1991), and Chu and Constantinides (1988). Using this paper as a source of information and a guide to other literature as well as the reviews of Weil et al. (1995a,b), answer the following questions.

1. Specify the intermediates and major enzymes in the biosynthesis pathway of Ceph C.
2. What enzyme(s) are likely to be rate-limiting? Explain your answer.
3. How is cell morphology related to antibiotic production?
4. How will the manner in which the substrate is fed affect the final result?

6.6. Cephalosporin C is a naturally produced β-lactam antibiotic that is effective against both Gram-negative and Gram-positive bacteria. It is produced in an aerated fermentation by *Cephalosporium acremonium* on glucose and sucrose as the carbon source. Once glucose is depleted, cephalosporin C production starts, accompanied by extensive fragmentation of long, swollen hyphal fragments. Once the swollen hyphal fragments disappear, cephalosporin production stops. Write the equations for cephalosporin production, given a cell mass accumulation for thin hyphae of

$$\frac{dx_H}{dt} = \mu_H X - \mu_T X_H - \delta_H X_H \qquad \text{(HP6.6a)}$$

where the total cell mass is

$$X = X_H + X_T \qquad \text{(HP6.6b)}$$

and the cell mass concentration of the thick hyphae is

$$\frac{dx_T}{dt} = \mu_T X_H - \delta_T X_T \qquad \text{(HP6.6c)}$$

with a cephalosporin production of

$$\frac{dP}{dt} = Ex_T - \gamma P \qquad \text{(HP6.6d)}$$

and the generation of enzyme needed for this biotransformation is

$$\frac{dE}{dt} = \alpha\mu_T X_H - \beta E \qquad \text{(HP6.6e)}$$

Assume that the specific growth rates for the thick and thin hyphae (i.e., μ_T and μ_H, respectively) are constants over the range of conditions to be examined. Other fitted parameters (constants) are α and β, as well as the specific decay rate, δ_T. δ_T is also assumed to be a constant for the purposes of this problem. Set up the equations that will be needed to simulate the accumulation of the cephalosporin. Use I to denote input variables and O as output variables. Clearly identify which equation(s) represent cell mass accumulation, product concentration, and enzyme levels. Solve these equations and show the time-course for cell mass accumulation and production of cephalosporin.

ENZYMES

INTRODUCTION

Enzymes are proteins: polymers of amino acids whose structures are defined by the genetic code stored in DNA that determine the sequence of amino acids of a protein, which, in turn, defines the protein's primary structure. The primary structure helps to direct the self-assembly of the protein into secondary, tertiary, and quaternary structures that make up the protein's active, three-dimensional conformation. The self-assembly process is known as "protein folding" and brings together "amino acids from widely separated parts of the protein polymer so that a catalytically active center is formed" (Rodwell 1990).

Enzymes act as stereospecific catalysts and form the biochemical-processing systems of cells, while functioning as mediators for nearly all chemical reactions that occur in living systems. Enzymes and other proteins are involved in the repair, duplication, and expression of information contained in a cell's DNA; hydrolyze proteins and polysaccharides; and serve many catalytic functions. This chapter introduces the reader to the characteristics of biocatalysts that make up a cell's metabolic pathways. Technical principles that guide the use of enzymes for industrial bioprocessing, are discussed.

Enzymes and enzyme kinetics (discussed in Chapter 8) have been an important component of biochemistry since the nineteenth century. This discipline has now evolved to encompass modeling and characterization of enzyme function on the basis of interactions of multiple enzymes, and mathematical molding of dynamic metabolic pathways.

ENZYMES AND SYSTEMS BIOLOGY

Systems biology and directed evolution have been added to the lexicon of enzyme applications since about 2000. Systems biology evolved from earlier studies of metabolic pathway engineering (Dhurjati and Mahadevan 2008). Metabolic engineering analyzes up- and downregulation or the overexpression of proteins that

---

achieve catalytic activity that enhances the flow of substrate to a desired end product. An understanding of a microorganism's metabolic pathways and identification of a missing or weak link between substrate and desired product could lead to the insertion of genes, which, when expressed, generate a missing enzyme or enzyme pathway. An example is the insertion of the xylose kinase gene into yeast, or the alcohol dehydrogenase pathway into *E. coli*, to obtain organisms capable of fermenting both xylose and glucose to ethanol. This development increases yields of ethanol (used as a fuel) from grasses, trees, or crops by up to 50%. Alternately, carbon from a substrate could be directed to a specific product by blocking or knocking out pathways that direct some of the substrate to other products. An example would be a bacterium that produces two types of organic acids—lactate and acetate—where only one of them is the desired product. Knockout of the acetate pathway would lead to more lactate.

For the most part, however, gains are incremental [e.g., cephalosporin C; see literature survey by Weil et al. (1995a,b)], and the self-regulating characteristics of the pathways in living organisms prove challenging. This has opened a new field for modelers who utilize computers and textbook biology to complement the intellectually and operationally challenging benchwork of transforming, growing, and analyzing products from living cells. The term *in silico*, which is the application of linear algebra and numerical methods to the modeling of nonlinear processes of cellular metabolism (using computers based on silicon chips; hence *in silico*), refers to computational modeling. Approximations through linearization of complex, stochastic systems have proved useful in enhancing the understanding of known pathways. However, modifications of the pathways still require the bench, biology, and a knowledge of enzymes and biochemistry.

The progression of events in biotechnology has led to gene shuffling where randomness is achieved—one gene at a time—using recombinant techniques to randomly cut and recombine genes. By providing special growth conditions, microbial cells harboring the shuffled genes that enabled the cells to survive were identified (selected), further propagated, and then further altered. The cycle was repeated until a superproducer of a desired "something" was obtained. Since the enzyme, and the shuffling of genes under the direction of a bench scientist, resulted in evolution of an outcome with possible economic advantages (such as a more active enzyme or a higher yield of a metabolite), the description of this process became known as *directed evolution*. This brings us back to the concept of enzymes, specifically, polymers of amino acids with binding domains that dock substrates into an active site and carry out catalysis at temperatures between 15 and 90 °C and 100% selectivity.

INDUSTRIAL ENZYMES

Industrial enzymes are isolated from microorganisms. Prior to the advent of molecular biology, microorganisms were treated with radiation or mutagens to alter their DNA in a random manner, with the hope of isolating survivors with a unique (or enhanced) propensity to produce a desired end product. The Rut C30 fungal strain of *Trichoderma reesei*, developed by Bland Montencort, was obtained in this manner to produce cellulases for hydrolysis of cellulose (Montenecourt and Eveleigh 1977). Once molecular biology developed, microorganisms could be genetically altered in a directed manner to produce specific proteins, with genes derived from other species. Alternatively, enzymes with specialized activity could be obtained through site-directed mutagenesis, or a combination of biocatalysis where specific changes in the genes, and the proteins to which they were translated, was determined through knowledge of protein structure and function, namely, by design (Petsko 1988).

ENZYMES: IN VIVO AND IN VITRO

In vivo, enzymes catalyze a cascade of reactions in a predominately aqueous environment. The majority of the reactions exhibit some degree of reversibility. These reactions are directed to a specific product for industrial applications by growing the microorganisms so that pathways and/or environmental conditions direct the flux of intermediates to a desired end product. This may be done by genetically modifying the organism to add a missing enzyme or enzyme pathway, or to enhance the activity or amount of an enzyme that is already present in the microorganism.

In vitro, soluble enzyme catalysis occurs in both aqueous and nonaqueous environments for industrially important reactions. Industrial enzymes can be used in either a soluble or immobilized form. In the case where the reaction is heterogeneous (i.e., either the product and/or substrate is a solid), soluble enzymes are most likely to be used in a batch reactor. Examples are proteases and lipases in laundry detergents (enzymes act on a solid matrix, i.e., laundry) and cheesemaking (proteases modify colloidal proteins to induce gel formation leading to a solid, i.e., cheese curd). Other examples are the modification of fabrics in which cellulases change the structure of cotton to improve handling, or the hydrolysis of cellulose in wood, cornstalks, or sugarcane bagasse, to sugars that are fermentable to ethanol.

In vitro, immobilized enzymes (i.e., enzymes attached to an insoluble matrix) act on homogeneous substrates where both substrate and product are dissolved in a liquid. In this case, an immobilized enzyme enables ease of processing and the ability to reuse the enzyme. Major processes that use immobilized enzymes are for isomerization of glucose to fructose (HFCS production) and the modification of 6APA (i.e., aminopenicillic acid) (Antrim et al. 1979; Harrison and Gibson 1984). An immobilized enzyme is initially more expensive than soluble enzyme because of the cost of the immobilization process and the need to clean substrate streams so that they do not foul immobilized enzyme in a packed-bed reactor. Since the enzyme is trapped within, or attached to solid particles, the particles and enzyme may be filtered from the reactant solution, or utilized in a packed bed so that the enzyme may be reused. This reduces the net cost of enzyme.

Enzyme immobilization may be achieved by covalently linking the enzyme to a solid material such as porous ceramics or glass, cellulosic materials, or adsorbing enzymes to an ion exchange cellulose through electrostatic or ionic interactions (Fig. 7.1). Alternately, if the enzyme is a high-molecular-weight protein, it can be trapped within the microbial cells in which it was generated, when these cells are heat-treated in order to reduce cell wall permeability. The cells are then extruded into pellets. Enzymes may also be immobilized by entrapment in a membrane reactor. The membrane has pores that are large enough to enable reactant and

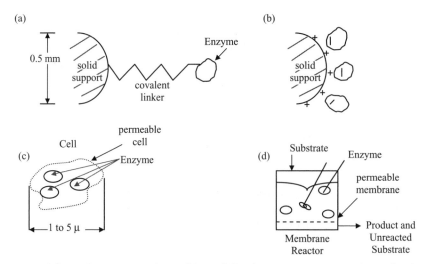

Figure 7.1. Schematic representations of immobilized enzymes: (a) enzyme covalently immobilized on a solid support; (b) enzyme adsorbed on surface of solid particle through ionic exchange; (c) enzyme entrapped within microbial cell that has been killed through heat treatment, while maintaining enzymatic activity as well as permeability of the cell's membrane to low-molecular-weight substrates and products; (d) enzyme retained within membrane reactor.

product molecules to pass, but small enough to retain the higher-molecular-weight enzymes (Rios et al. 2004).

A monograph by Zaborsky (1974) reviews the different types of chemistries and methods used to immobilize enzymes. The IUPCAC classification of immobilized enzymes is based on properties of the free enzyme, and enzyme attachment (Worsfold 1995). The particulate forms of the enzyme are packed in a bed, and the substrate dissolved in buffer is flowed through the packed bed. The benefit of this approach is that the enzyme is reused, thus spreading the cost of the enzyme over a larger volume of substrate being converted. In a batch reactor, the enzyme is added to the solution, used once, and then deactivated after the reaction is completed.

Recombinant technology is applied to enzyme manufacture with the benefits of lower costs, higher activity, and longer life under operational conditions. For example, recombinant chymosin (Chymax® developed by Pfizer) is microbially produced in yeast. It is identical to calves' chymosin (extracted from calves' stomachs), and costs about 50% less and reduces the risk of introduction of an animalborne disease into a milk product (also preprocessed by pasteurization). Chymosin hydrolyzes the κ-casein on the surface of casein micelles to *para*-κ-casein. This induces the micelles to coagulate and milk to clot during the first step of some cheesemaking processes.

Other recombinant enzymes include amylases (by Novo Nordisk, and Nikon Shokuhin Koko/Fujitsu); lipases for detergents and food oil processing (from Novo); a variety of hydrolytic enzymes, isomerases, and oxidases (by Biofood, Germany) (Taylor 1993); and cellulases for bioethanol production from cellulosic materials (Genencor, Novo, AB). The industrial application of enzymes occurs because their use either saves or makes money, according to a practical approach for analyzing enzyme costs presented in a monograph edited by Boyce (1986). Enzymes operate at mild conditions. Their specificity of action facilitates increased product yields, and improved and/or unique products.

FUNDAMENTAL PROPERTIES OF ENZYMES

Enzymes act as stereospecific catalysts since substrates contact enzymes at multiple points (Rodwell 1990). Most substrates form at least three bonds at the enzyme's active site. This confers asymmetry to an otherwise symmetric molecule, as clearly explained by Rodwell (1990). The substrate–enzyme interaction is specific for a particular reaction and selectivity in forming a single type of product. A substrate molecule that is about to attach at three points at an enzyme site (Fig. 7.2) may bind in only one way if the catalytic site of the enzyme is accessible from only one side. Only complementary atoms and sites interact. Rodwell (1990, p. 60) states:

> By mentally turning the substrate molecule in space, note that it can attach at 3 points to one side of the planar site with only one orientation. Consequently, atoms 1 and 3, although identical, become distinct when the substrate is attached to the enzyme. A chemical change thus can involve atom 1 but not atom 3 or vice versa. ... Optically inactive pyruvate, for example, forms L- and not D,L-lactate in an enzyme catalyzed reaction. Only a small number of structurally related compounds are acted upon by a particular enzyme.

An enzyme has optical specificity for at least a portion of the substrate, with most mammalian enzymes acting on L-isomers. Optical specificity may, however, extend to only a portion of the molecule. For example, glycosidases that hydrolyze glycosidic bonds between sugars and alcohols are specific for the ether bond (α or β), but are relatively nonspecific for the aglycone (alcohol) portion of the sugar. β-Glucosidase cleaves the β-1,4 bond of the glucose dimer, cellobiose, and the β-1,4 bond of p-nitrophenyl glucoside but not the α-1,4 bond of maltose (Fig. 7.3). Proteolytic enzymes chymotrypsin and trypsin cleave peptide bonds in which the

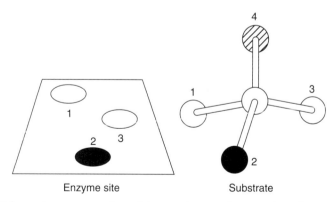

Enzyme site   Substrate

Figure 7.2. Schematic representation of three-point attachment of a substrate to a planar active site of an enzyme [reprinted with permission from Rodwell (1990), Fig. 8.3, p. 60, Appleton & Lange].

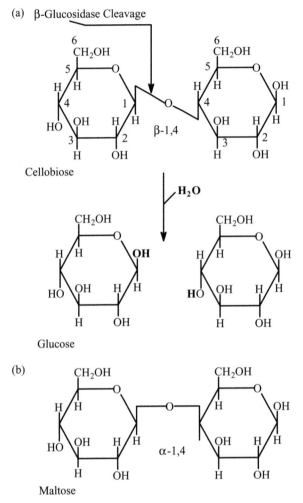

Figure 7.3. Bond specificity of β-glucosidase. Hydrolysis occurs with (a) cellobiose but not with (b) maltose.

carboxyl group is contributed by Phe, Trp, or Tyr, or Lys or Arg, respectively (Fig. 7.4).

Lytic enzymes hydrolyze peptide bonds in which the carboxyl group is contributed by specific amino acids. For biosynthetic processes, oxidoreductases in mammalian cells use NADPH as the reductant, while those in degradative processes use NAD^+ as the oxidant.

CLASSIFICATION OF ENZYMES

Enzymes are classified according to the chemical reaction type and reaction mechanism by the International Union of Biochemistry (IUB) Nomenclature System (Rodwell 1990). This system divides reactions and enzymes that catalyze these reac-

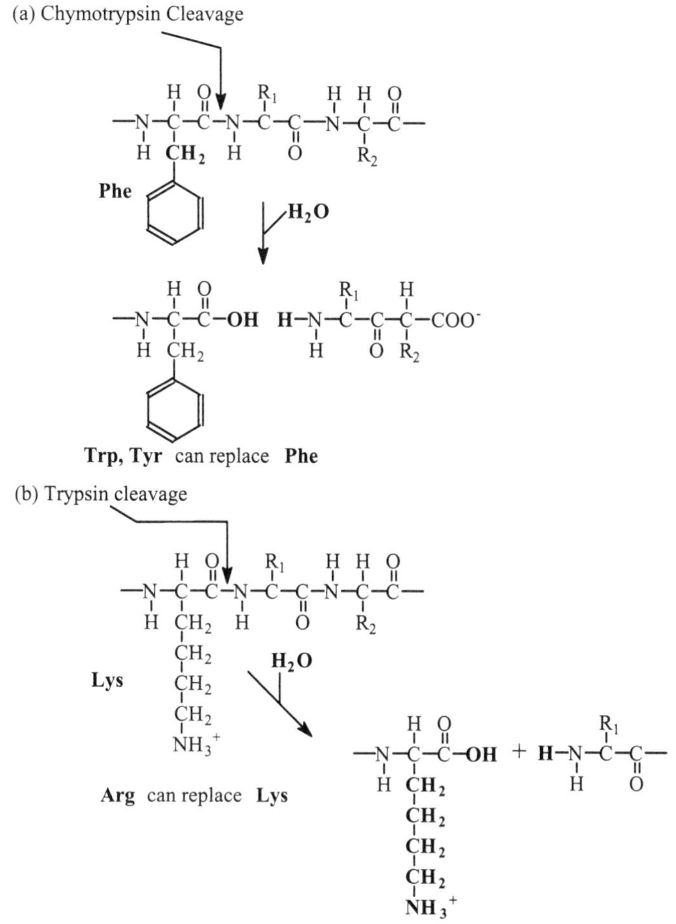

Figure 7.4. Illustration of peptide bond cleavage sites for (a) chymotrypsin and (b) trypsin [adapted from Rodwell (1990)].

tions into six classes. Each class consists of 4–13 subclasses. The enzyme name has two parts. The first part indicates the substrate and the second the type of reaction, with further information given in parentheses, if needed. An enzyme that hydrolyzes cellulose is a cellulase. One that isomerizes glucose to fructose is glucose isomerase.

A code numbering system has been widely adopted because of the often confusing and contradictory IUB names (e.g., for an enzyme that catalyzes a reversible reaction, it might be difficult to determine which compound is the substrate). This code number system for each enzyme characterizes the reaction type:

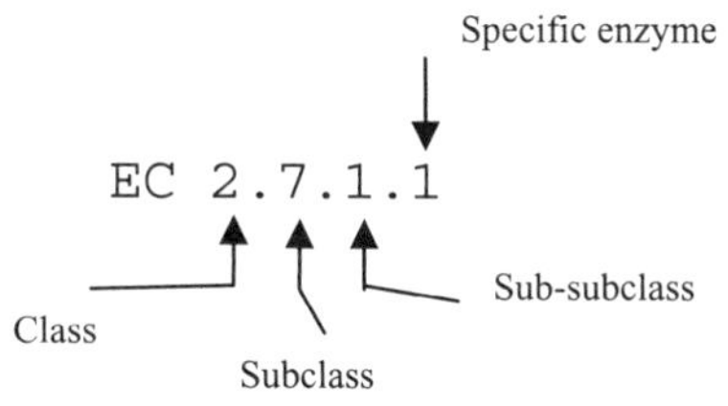

Table 7.1.

International Classification of Enzymes[a]

1. Oxidoreductases
   (oxidation–reduction reactions)
   1.1. Acting on $-\overset{|}{C}H-OH$
   1.2. Acting on $-C=O$
   1.3. Acting on $-CH=CH-$
   1.4. Acting on $-\overset{|}{C}H-NH_2$
   1.5. Acting on
   1.6. Acting on NADH; NADPH

2. Transferases
   (transfer of functional groups)
   2.1. One-carbon groups
   2.2. Aldehydic or ketonic groups
   2.3. Acyl groups
   2.4. Glycosyl groups
   2.7. Phosphate groups
   2.8. S-containing groups

3. Hydrolases
   (Hydrolysis reactions)
   3.1. Esters
   3.2. Glycosidic bonds
   3.3. Peptide bonds
   3.4. Other C–N bonds
   3.5. Acid anhydrides

4. Lyases
   (addition to double bonds
   4.1. $-\overset{|}{C}=\overset{|}{C}-$
   4.2. $-\overset{|}{C}=O$
   4.3. $-\overset{|}{C}=N-$

5. Isomerases
   (isomerization reactions)
   5.1. Racemases

6. Ligases
   (Formation of bonds with ATP cleavage)
   6.1. C–O
   6.2. C–S
   6.3. C–N
   6.4. C–C

[a]Including class names, code numbers, and types of reactions catalyzed.
Source: Reproduced with permission from Lehninger (1970), p. 148, Worth Publishers, Inc., all rights reserved.

where EC is an abbreviation for Enzyme Code. EC 2.7.2.2 denotes an enzyme of class 2 (transferase), subclass 7 (transfer of phosphate), sub-subclass 1 (alcohol is the phosphate acceptor), and enzyme 1, hexokinase or ATP: D-hexose 6-phosphotransferase (Rodwell 1990). The class names, code numbers, and types of reactions catalyzed by the six classes of enzymes were defined a long time ago and are summarized in Table 7.1.

SALES AND APPICATIONS OF IMMOBILIZED ENZYMES

Sales of immobilized enzymes accounted for approximately 20% of the total enzyme market in 1990, and were expected to be one-third of the market in 2000. The single major use of immobilized enzyme in 1990 was glucose isomerase with worldwide sales of $40 million. Most of the remaining commercial applications for immobilized enzyme consisting of penicillin acylase, fumarase, β-galactosidase, and amino acid acylases (Taylor 1993). Since 2000 there has been further growth in enzyme markets, although "firms involved in bioprocessing—the enzymatic conversion of natural raw materials into chemicals—are currently having a hard time making money." Slow market development and the need to develop cheaper production routes are impediments to profits (Scott and Wood 2005).

The belief in the potential for biocatalyst (enzyme markets) seems to continue to grow, however, with Diversa, Dyadic, Danisco (which bought Genencor in 2005), Cognis, DSM, Novo Nordisk, and Chr Hansen, as some of the companies in the field. It is difficult to obtain an estimate of the size of the market. Forward-looking estimates have predicted that global demand for enzymes "is expected to rise 6.5% annually to nearly $5.1 billion in 2009" (Food navigator.com/Europe 2005). Past sales figures would suggest a more modest market size. Regardless of the actual monetary sales volume, the use of enzymes in broad and growing sectors of consumer products is clear. These include dairy (cheese), wine, vegetables, fruits, oils, brewing, fuel ethanol, sweeteners, pulp, paper, leather, cotton, cleaning solutions, and laundry and dishwashing detergents. Other consumer uses of enzymes are in the form of medical diagnostic test kits for humans and for veterinary use, marketed by companies such as Roche and Abbott.

Examples of industrial enzyme reactions are numerous, although the major volume of enzymes used in industry are for hydrolysis processes (Table 7.2). Degradation of pectin and cellulose to improve juice extraction from apples, chymosin in cheesemaking, decolorizing of slaughterhouse blood, modification of soy protein for production of fat-free dairy substitutes, retarding the staling of white bread, processing stiff denim into a supple material (biostoning), and the formulation of laundry detergents are a few examples. The major industrial application of a non-hydrolytic enzyme is the isomerization of glucose to fructose by glucose isomerase (EC 5.3.1.5) to form high-fructose corn syrup (HFCS) (Ladisch 2001). Numerous applications employ other classes of enzymes, including those used in pharmaceutical manufacturing processes, and diagnostic and other medical procedures, listed in Table 7.2.

ASSAYING ENZYMATIC ACTIVITY

Activity is a relative measure of an enzyme's ability to catalyze a reaction under a given set of conditions. Different conditions (pH, temperature, substrate concentration) result in different activities. Enzymatic activity is based on a measured rate or reaction velocity v, which increases with increasing substrate concentration until a maximum velocity V_{max} is approached. This is illustrated in Fig. 7.5 for the following simple reaction:

$$E + S \rightleftharpoons ES \rightarrow E + P \tag{7.1}$$

The reaction velocity v of the reaction sequence of Eq. (7.1) is represented by the Michaelis–Menten equation.

Table 7.2.

Industrial Uses of Carbohydrate-Hydrolyzing Enzymes, Proteolytic Enzymes, Other Types of Hydrolytic Enzymes, Oxidoreductases, Isomerase, and Other Enzymes, and Selected Research, Medical, and Diagnostic Use of Enzymes[a]

Common Name	Enzyme EC Number	IUB Name	Range of Optimal pH	Reaction Catalyzed	Applications
α-Amylase	3.2.1.1	α-1,4-Glycanhydrolase	4.8–6.3	Starch → glucose + maltose + oligosaccharides	Breadmaking Mashing in alcohol production Desizing of fabrics Wallpaper removal
β-Amylase	3.2.1.2	α-1,4-Glycanhydrolase	5.0–5.5	Starch → maltose + dextrin	Maltose production
Amyloglucosidase	3.2.1.3	α-1,4-Glycanhydrolase	3.8–5.5	Starch → glucose + oligosaccharides	Glucose from corn syrup Low-carbohydrate beers
Cellulase	3.2.1.4	System of enzymes β-1,4-Glucanglucanohydrolase β-1,4-Cellobiohydrolase β-Glucosidase	4.0–6.0	Cellulose → glucose + polysaccharides	Treatment of cotton fabrics Laundry detergents Glucose production from cellulose
Dextranase	3.2.1.11	Dextranase	4.0–6.5	Dextran → glucose + oligosaccharides	Reduce viscosity of sugar syrup
Diastase (mixture of α- and β-amylase)	NA	NA	Slightly acid pH[1,2] 6.0–7.0[3]	Starch → glucose	Glucose from corn syrup

Table 7.2. (Continued)

Common Name	Enzyme EC Number	IUB Name	Range of Optimal pH	Reaction Catalyzed	Applications
Chymosin[b]	3.4.23.4	NA[c]	2.5–5.10[4-6]	κ-Casein → p-κ-casein	Cheese production
Hemicellulases	NA	NA	4.5 to 7.5	Xylans, Arabans, Glucans } → Xylose arabinose glucose oligosaccharides	Viscosity reduction, antispoiling agent
Invertase	3.2.1.26	β-Fructofuranosidase	3.0–8.0[7]	Sucrose → glucose + fructose	Confectionery products
Lactase	3.2.1.23	β-Galactosidase	2.4–9.0[8-10]	Lactose → glucose + galactose	Prevent are lactose crystallization in ice cream
Lysozyme	3.2.1.17	Muramidase N-acetylmuramide glycanhydrolase	3.5–8.0[11,12] PH: 6.0–9.0 (*Staphylococcal* and egg white lysozyme)	Bacterial cell walls → acetylglucouronides	Ophthalmic preparations
Naringinase	Naringinase is an enzyme complex consisting of α-rhamnosidase (EC 3.2.1.40) and flavonoid β-glucosidase (EC 3.2.1.21)	Hesperidinase	4.0–6.5	Naringin → Prunin + Rhamnose; Prunin → naringenin + glucose	Removal of bitter taste from grapefruit juice and peel

Table 7-2. (Continued)

Common Name	Enzyme EC Number	IUB Name	Range of Optimal pH	Reaction Catalyzed	Applications
Pectinase	NA	Three major types of pectinase[d]	3.0–10.5[13,14]	Pectin → galacturonic acid	Clarification of fruit juices; Coffee bean fermentation; Citrus oil recovery
Pullulanase	3.2.1.41	Amylopectin 6-glucanohydrolase[1,2]	5.0–8.5[15-17]	Starch → maltose + maltotriose	Treatment of wort (beer brewing)
Bacterial proteases	3.4.24.4	Mixture of enzymes	7.0–11.0	Protein → amino acids (broad specificity)	Laundry detergents
Fungal proteases	3.4.24.3	Mixture of enzymes	6.8–8.5	Protein → amino acids (broad specificity)	Food processing (dough softening, sake production)
Papain	3.4.22.2	Papain	6.0–9.5[18,19]	Hydrolyzes bonds adjacent to basic L-amino acids, of Leu or Gly	Meat tenderizer; Shrinkproofing of wool
Rennin (*see* Chymosin above)	3.4.23.4	Rennet		κ-Casein → p-κ-casein	Cheese manufacture
Streptokinase	3.4.21.7	Fibrinolysin, plasminokinase	8.5 (*Homo sapiens*)[20,21]	Hydrolyzes proteins at Arg and Lys	Anticoagulant (medical use)
Lipase	3.1.1.3	Lipase	4.9–7.5	Fats → fatty acids + glycerol	Flavor improvements in refrigerated dairy products, chocolates

Table 7.2. (Continued)

Common Name	Enzyme EC Number	IUB Name	Range of Optimal pH	Reaction Catalyzed	Applications
Penicillin acylase	3.5.1.11 PS: 3.5.1.14 = aminoacylase (urea cycle and metabolism of amino groups)	Penicillin amydase (penicillin amidohydrolase)	$5.0–9.5^{22,23}$	Hydrolysis of penicillin side chains	Semisynthetic penicillin production
Ribonuclease	3.1.27.1 or 5	Ribonuclease	5.5–8.2	Yeast RNA $\rightarrow$ 5'-UMP	Flavor enhancer production
Aspartase	4.3.1.1	L-Aspartate ammonia-lyase	$6.0–10.5^{24,25}$	Fumarate + NH3 $\leftrightarrow$ aspartate	Amino acid production
Catalase	1.11.1.6	Hydrogen peroxide: hydrogen peroxide oxidoreductase	$4.0–10.5^{26,27}$	$H_2O_2 \rightarrow O_2 + H_2O$	Removal of H_2O_2 after cold sterilization of milk, bleaching
Glucose isomerase	5.3.1.5	D-Xylose ketoisomerase	6.0 to 7.0	D-Glucose $\leftrightarrow$ D-fructose	High-fructose corn syrup (HFCS) production
Glucose oxidase	1.1.3.4	β-D-Glucose: oxygen oxidoreductase	3.4–5.6	D-Glucose + O_2 + H_2O $\rightarrow$ D-gluconic acid + H_2O_2	Stabilizes foods by removing oxygen
Laccase	1.10.3.2	D-Diphenol oxidase	4.0–6.0	2p-Diphenol + O_2 m, p-quinone + H_2O_2	Treatment of cotton fabrics Laundry detergents Glucose production from cellulose

Table 7.2. (Continued)

Common Name	Enzyme EC Number	IUB Name	Range of Optimal pH	Reaction Catalyzed	Applications
Histinase	4.3.1.3	L-Histidine ammonia lyase	4.0–6.5	Histidine $\rightarrow$ m-urocanoate + NH_3	Production of urocanoic acid (used in sunscreen)
Lipoxygenase	1.13.11.12	Lidoxidase	4.5–9.5[28,29]	Caretenoids + linoleic acid + O_2 $\rightarrow$ m-peroxidized linoleic acid	Bread whitening Peroxidized oils used in flavorings
Tryptophanase	4.1.99.1	Tryptophanase	7.0–9.0[30,31]	Indole + pyruvate + NH_3 $\rightarrow$ m-L-tryptophan + H_2O	Amino acid production
Asparaginase	3.5.1.1	L-Asparagine amidohydrolase	5.0–9.5[32–34]	—	Treatment of acute lymphatic leukemia
Cholesterol oxidase	1.1.3.6	Cholesterol oxidase	4.0–8.0[35–37]	—	Assay of cholesterol
DNA polymerase	2.7.7.7	DNA nucleotidyl transferase	6.5–9.2[38–40]	$(DNA)_n$ + dNTP $\rightarrow$ $(DNA_{n+1}$ + $Pp_i)$	DNA synthesis
Glucose oxidase	1.1.3.4	Glucose oxidase	4.0–7.0[41–43]	See Table 4.3	Assay of glucose (used together with catalase)
Lactate dehydrogenase	1.1.1.27	L-Lactate: NAD oxidoreductase	4.5–9.8[44–46]	L-Lactate + NAD^+ $\rightarrow$ pyruvate + NADH	Diagnosis of myocardial infarction and leukemia

Table 7.2. (Continued)

Common Name	Enzyme EC Number	IUB Name	Range of Optimal pH	Reaction Catalyzed	Applications
Acetylcholine esterase	3.1.1.7	Acetylcholine esterase	4.8–9.0[47-49]	Acetylcholine + H_2O → choline + acetic acid	Assay of neuroactive peptides
Uricase	1.7.3.3	Uric acid: oxygen oxidoreductase, urate oxidase	7.0–9.5[50-52]	Uric acid + H_2O + O_2 → allantoin + H_2O	Diagnosis of gout

[a]References cited in this table:

1. Okada (1916)
2. Babacan et al. (2002)
3. Clyde (1959)
4. Foltmann (1969)
5. Mohanty et al. (1999)
6. BRENDA (2008)
7. Bergamasco et al. (2000)
8. De Bales and Castillo (1979)
9. Wierzbickp and Kosikowski (1972)
10. BRENDA (2008)
11. Hawiger (1968)
12. BRENDA (2008)
13. Puri and Banerjee (2000)
14. Kashyap et al. (2001)
15. Hoondal et al. (2002)
16. Saha et al. (1988)
17. Costanzo and Antranikian (2002)
18. BRENDA (2008)
19. BRENDA (2008)
20. Sangeetha and Abraham (2006)
21. Greig and Cornelius (1963)
22. BRENDA (2008)
23. Giordano and Ribeiro (2006)
24. BRENDA (2008)
25. BRENDA (2008)
26. Wang et al. (2000)
27. BRENDA (2008)
28. Shi et al. (2008)
29. BRENDA (2008)
30. Gordon (2001)
31. BRENDA (2008)
32. Fukui et al. (1975)
33. Sukuzi et al. (1991)
34. BRENDA (2008)
35. Mori et al. (1974)
36. Dejong (1975)
37. BRENDA (2008)
38. Yazdi et al. (2001)
39. Doukyu and Aono (1998)
40. BRENDA (2008)
41. Wilson and Kuff (1972)
42. Invitrogen (2008)
43. BRENDA (2008)
44. Szajáni et al. (1987)
45. Shin et al. (1993)
46. BRENDA (2008)
47. Gay et al. (1968)
48. Jonas et al. (1972)
49. BRENDA (2008)
50. Reiner and Aldridge (1967)
51. Fluck and Jaffe (1974)
52. BRENDA (2008)
53. Abdel-Fattah et al. (2005)
54. Zhao et al. (2006)
55. BRENDA (2008)

[b]Other names for chymosin are prorennin, rennin (old name—the term rennin appears rarely today in some fungal milk-clotting aspartic proteinases such as *Mucor* rennin), and prochymosins A, B, and C (for bovine isozymogens).

[c]See I. Gritti, G. Banfi, and G. S. Roi, "Pepsinogens: Physiology, Pharmacology, Pathophysiology, and Exercise," *Pharmacol. Res.* 41(3), 265–281 (2000). Per IUB's Enzyme Nomenclature rules, proteases are designated and classified by three code numbers, referring to class, subclass, and sub-subclass.

[d]Pectinases are classified under three headings according to the following criteria: whether pectin, pectic acid or oligo-D-galacturonate is the preferred substrate, whether pectinases act by *trans*-elimination or hydrolysis, and whether the cleavage is random (endo-, liquefying, or depolymerizing enzymes) or endwise (exo- or saccharifying enzymes). The three major types of pectinase are pectinesterases (PE), depolymerizing enzymes, and protopectinase.

Source: Adapted from Atkinson and Mavituna (1983).

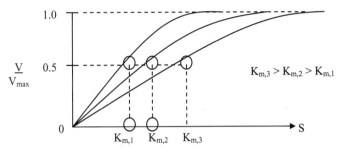

Figure 7.5. Relative velocity (v/V_{max}) as a function of substrate concentration for different values of K_m.

$$v = \frac{V_{max}[S]}{[S] + K_m} \qquad (7.2)$$

and is plotted in dimensionless form as

$$\frac{v}{V_{max}} = \frac{[S]}{[S] + K_m} \qquad (7.3)$$

where [S] represents substrate concentration on a molar basis.

The form of Eq. (7.3) readily leads to the definition of the Michaelis constant K_m, which is the substrate concentration S at which the measured rate is half of the maximum: $v/V_{max} = 0.5$. Figure 7.5 also illustrates the concept that for a given substrate concentration, the reaction rate is below V_{max} inversely proportional to K_m.

An internally consistent comparison of the activity of a specified enzyme requires that the initial substrate concentration, pH, and temperature be the same each time the assay is carried out. The pH is usually selected to lie in a range where the enzyme's activity is maximal, while the temperature is chosen so that thermal denaturation of the enzyme is negligible during the time period over which the enzyme is assayed.

An example assay procedure for an industrial enzyme is given for Alcalase, a proteolytic enzyme manufactured by Novo, Inc., for use in heavy-duty liquid detergents (Boyce 1986). This alkaline protease (Alcalase) has a recommended pH of use of 7.5, and is stable at a temperature of 50°C or lower. It has a relative activity (proportional to v/V_{max}) as a function of temperature and residual activity (proportional to v/v_0, where v_0 is reaction velocity at $t = 0$) as shown in Fig. 7.6. Its activity is measured by a colorimetric assay using denatured hemoglobin (Boyce 1986). The procedure, illustrated in Fig. 7.7, consists of incubating enzyme with hemoglobin for 10 min during which the enzyme hydrolyzes the hemoglobin. Acid is added to precipitate the undigested protein, the supernate recovered, and phenol is added to the supernate. The spectral absorbance of the solution is measured and compared against a tyrosine standard. The enzymatic activity is thus expressed as the milliequivalents (meq) of tyrosine (Tyr) per minute, where 1 meq Tyr/min = 1 Anson unit (AU).

Specific activity is calculated by dividing the activity by the amount of protein in the solution containing the enzyme. Since the enzyme solution will contain other proteins that do not have the specified enzymatic activity, the specific activity will

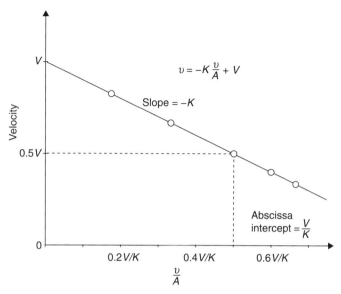

Figure 7.6. Percentages of (a) relative and (b) residual enzymatic activity as a function of temperature and time, respectively [reproduced with permission from Boyce (1986), Fig. 1, p. 24].

usually increase if the enzyme is purified from other proteins. Hence, specific activity is a function of enzyme purity (relative to other proteins) as well as the activity of the enzyme itself. For example, if a protease activity of 12 meq Tyr/10 min is achieved for a solution containing 2 mg protein, the specific activity is 0.6 AU/mg protein.

Industrial enzymes are sold in preparations with a specified activity or strength. After an enzyme is produced, an inert ingredient may be added to give the guaranteed activity on a weight or volumetric basis. Diluents include lactose, sucrose, and salt for solid preparations, or sorbitol, propylene glycol, and glycerol for liquid products. Since the definition of enzyme units on this basis is arbitrary, it is often difficult to directly compare, a priori, the activity of an enzyme in a given weight or volume of one preparation to the same type of enzyme from a different manufacturer. Hence, an enzyme's specific activity based on protein is a useful and internally consistent measure for different enzyme preparations (Boyce 1986).

Enzyme Assays

Enzyme assays define the activity of proteins in solution, and provide the basis on which their purity, function, and value are based. Conversely, known enzymes may be used to probe liquid solutions for the presence of small molecules, and determine their concentration in solution. Coupling of an enzymes action to the reaction of an NAD^+ or $NADP^+$ dependent dehydrogenase gives another type of sensitive enzyme assay. Dehydrogenases depend on the presence of NAD^+ or $NADP^+$ for their activity. Since NADH strongly absorbs light at 340 nm, while NAD^+ does not (Fig. 7.8), the rate of formation of NADH is readily measured from optical density at 340 nm. An enzyme's activity can therefore be measured indirectly if its product

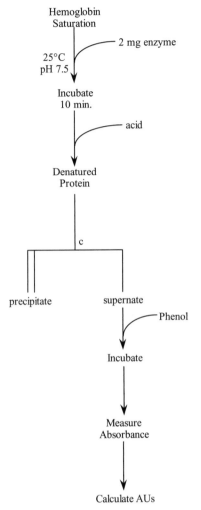

Figure 7.7. Schematic illustration of Anson assay (25 °C, pH 7.5, precipitate with acid after 10 min) for Alcalase [diagram constructed from description of Boyce (1986), pp. 16–17].

is a substrate for the dehydrogenase, which, in turn, reduces NAD^+ or $NADP^+$ to NADH or NADPH as a consequence of its activity (Rodwell 1990).

An example is hexokinase, which phosphorylates glucose to glucose-6-phosphate (Fig. 7.9). The activity of this enzyme is quantitated by measuring the glucose-6-phosphate generated using the glucose-6-phosphate dehydrogenase to reduce the phosphorylated glucose with NADH. A calibration curve is prepared by plotting the rates of NADH disappearance obtained from the slopes of the lines in Fig. 7.9b, against enzyme levels (Fig. 7.10). The quantity of enzyme in an unknown solution may then be calculated from an observed rate of change in optical density at 340 nm (Rodwell 1990). If the objective were to measure the activity of a dehydrogenase itself, the formation (or disappearance) of NADH would, of course, be measured directly.

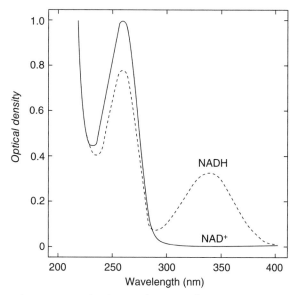

Figure 7.8. Absorption spectra of NAD⁺ and NADH for 44 mg/mL solution for a 1 cm path length; NADP⁺ and NADPH have analogous spectra [reprinted with permission from Rodwell (1990), Fig. 8.4, p. 61, Appleton & Lange].

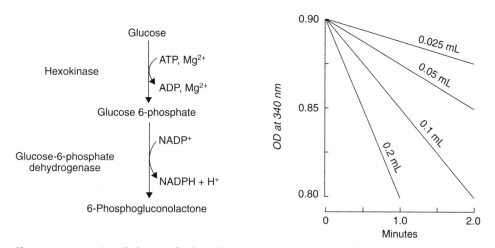

Figure 7.9. (a) Coupled assay for hexokinase activity. The rate of formation of NADPH is monitored spectroscopically [reprinted with permission from Rodwell (1990), Fig. 8.7, p. 62, Appleton & Lange]; (b) assay of an NADH- or NADPH-dependent dehydrogenase—the change in the spectral optical density at 340 nm is due to conversion of reduced to oxidized coenzyme (NADH → NAD⁺) [reprinted with permission from Rodwell (1990), Fig. 8.5, p. 61, Appleton & Lange].

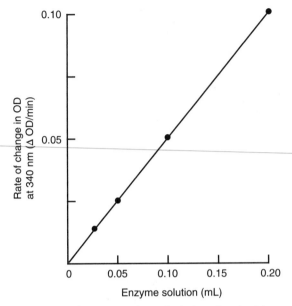

Figure 7.10. Calibration curve for enzymatic analysis [reprinted with permission from Rodwell (1990), Fig. 8.6, p. 62, Appleton & Lange).

The key to quickly isolating, purifying, and characterizing a new enzyme, or developing process conditions for an existing one, is to obtain assay results on the same day, if possible. Enzymes developed for diagnostic purposes may often be useful given their specificity. An example is given by an automated glucose analyzer on which the rate of oxidation of glucose to gluconic acid, catalyzed by glucose oxidase (EC 1.1.3.4), is related to glucose concentration by measuring oxygen disappearance from the solution.

The original instrument consisted of an oxygen electrode probe protruding into the side of a 1.5-mL reaction vessel (as shown in Fig. 7.11). A monitoring circuit in connection with the oxygen electrode measures the initial rate of change of dissolved oxygen. The rate is translated into glucose concentration. Since its development around the mid-1970s, the instrument has become more compact and automated, but the principal reactions have remained the same. Variations of this instrument are now available as handheld devices, sold in pharmacies. These devices use capillary action to deliver microliter volumes of blood to a sensor for regular daily monitoring of blood sugar levels of patients with diabetes.

For a glucose-oxidase-based analyzer, the sequence that commences with the introduction of sample into the enzyme reagent is given by the following reactions (Ladisch 1974):

$$\beta\text{-D-glucose} + O_2 \xrightarrow[\text{(water)}]{\text{glucose oxidase}} \text{gluconic acid} + H_2O_2 \qquad (7.4)$$

$$H_2O_2 + \text{ethanol} \xrightarrow{\text{catalase}} \text{acetaldehyde} + H_2O \qquad (7.5)$$

$$H_2O_2 + 2H^+ + 2I^- \xrightarrow{\text{molybdate}} I_2 + H_2O \qquad (7.6)$$

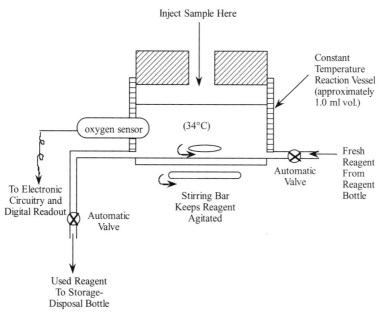

Figure 7.11. Schematic diagram of principal components of the original Beckman glucose analyzer [from Ladisch (1974), Fig. 4].

where the oxygen is consumed at the same rate at which glucose reacts. Reactions (7.4) and (7.5) represent oxygen traps that ensure that the hydrogen peroxide formed in the first reaction does not dissociate to release oxygen back into the enzyme reagent. Enzyme reagent is drained on completion of the analysis, and fresh reagent is filled back in premeasured quantities for each analysis. Each measurement requires approximately 60 s. The instrument is calibrated with a glucose standard. The range of glucose concentrations that such an instrument can handle lies between 0.10 and 5 mg/mL, with a sample volume of 10 μL.

Glucose analysis illustrates the impact of the specificity of the enzyme. Glucose oxidase acts only on β-D-glucose, so only this form of glucose is detected. The instrument is standardized using an equilibrium mixture of α and β forms of D-glucose that are created when β-D-glucose is dissolved in the buffer. Freshly prepared solutions of glucose must be allowed to mutarotate to an equilibrium composition of α and β forms before being injected. Similarly, glucose formed by hydrolysis must also be given sufficient time to form an equilibrium mixture of α,- and β-D-glucose. For example, hydrolysis of cellobiose by cellobiase yields β-D-glucose; hence time must be allowed for mutarotation to occur after the enzyme-catalyzed hydrolysis is stopped.

The effect of mutarotation on measured glucose level is illustrated in Fig. 7.12 (Ladisch 1977). Nonenzyme reactions such as oxidation of iron result in oxygen consumption, and would give a false indication of glucose. The measurement of blood glucose is now commonplace. A consumer product is sold in stores where a drop of blood is introduced to strips containing the necessary enzymes and reagents, and the strip is inserted into a reader.

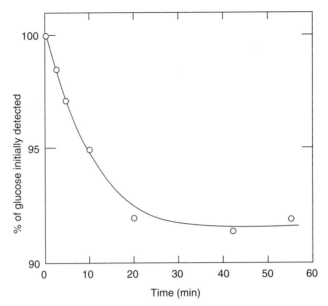

Figure 7.12. Mutarotation timecourse for glucose [reprinted with permission from Ladisch (1977), Fig. 1, p. 160].

Enzymes that catalyze either the appearance of glucose (such as from starch, cellulose, sucrose, or lactose), or the disappearance of glucose (such as glucose isomerization to fructose) can be readily assayed. Two examples are a cellulase complex from *Trichoderma reesei* acting on cellulose, to hydrolyze it to glucose, and glucose isomerase from *Streptomyces* spp. that converts glucose to fructose (Fitzgerald and Vermerris 2005).

Many enzyme assay procedures are based on spectrophotometric measurements or on liquid chromatography. A large number of samples are handled by starting the assay of individual tubes at time intervals of several minutes, and then measuring the samples' absorbence, also at several minute intervals. If the purpose of the assay is to determine enzyme kinetics, rather than just to compare enzymatic activities on an internally consistent basis, the number of assays can easily exceed 1000, given that different enzyme and substrate concentrations, inhibitor cofactor and/or activator levels (where applicable), and temperatures must be examined.

We close this section by emphasizing the importance of taking the time to develop, refine, and validate a quick and reliable enzyme assay. Selecting an enzyme for an industrial application involves defining the enzyme to be used, identifying operational conditions (temperature, pH, time–temperature stability), and integrating upstream and downstream processing steps with the enzyme process. An iterative experimental procedure initiates the process. First the pH and temperature optima must be defined, and then enzyme stability determined. Multiple experiments are required at different pH values for a given temperature, and at different temperatures for a given pH. A practical and easily readable guide to these procedures is given by Boyce (1986).

BATCH REACTIONS

The reactions catalyzed in a living cell occur in an environment where enzymes, substrates, and products all exist in the same mixture, specifically, a batch reactor. The rate at which enzymes catalyze a specified reaction can be characterized using either an initial rate (differential) approach or an integrated rate method combined with experimental data. The initial rate approach is useful when the reaction sequence of enzyme catalyzed steps is already known, and changes in initial reaction rates as a function of initial substrate, product, and/or cofactor concentrations are to be used to postulate an appropriate model. Once developed, such a model can be integrated (analytically or numerically) for the purposes of predicting a reaction timecourse, subject to the constraints that the enzyme is present only at catalytic levels, and that there is no loss of enzymatic activity due to denaturation by thermal or other effects. Initial rate kinetics require that numerous experiments be run at each initial substrate and product concentration, and that product appearance or substrate disappearance be quantitated at conversions of less than 5%. This requires more enzyme and effort compared to an integral kinetics approach that fits the model to timecourse measurements obtained at higher and more easily measured conversions.

THERMAL ENZYME DEACTIVATION

An important criterion for enzyme kinetics is that the enzyme be stable during the time over which its activity is assayed. While the rate of reaction follows the Arnhenius rule with the reaction rate increasing with increasing temperature, enzymatic activity decreases on exposure to higher temperatures for a prolonged period of time (Fig. 7.5). The loss in enzymatic activity reflects a change in enzyme conformation in which disulfide bonds and interactions between hydrophobic domains are disrupted, and the enzyme is denatured.

The study of protein deactivation mechanisms has been facilitated through protein engineering. Although all the rules for protein folding are not yet understood, stable proteins with desired functions can be engineered (Branden and Tooze 1991). For example, the oxidative stability of subtilisin, an alkaline protease obtained from *Bacillus* spp. that is used as a laundry detergent additive to serve for the removal of proteinaceous stains from cloth, was improved by protein engineering. Native *Bacillus amyloliquefaciens* subtilisin is a protease consisting of 275 amino acids, with serine at its active site. Protein engineering was used to replace methionine with leucine at amino acid position 222. The Leu-222 variant was found to be stable in the presence of 0.3% H_2O_2, while the native subtilisin with Met at 222 is extremely sensitive at the same conditions (Fig. 7.13).

There is only a small difference in total energy between a protein's folded structure and a large number of different, rapidly interconverting folded structures (Branden and Tooze 1991). Branden and Tooze present a case study for lysozyme from bacteriophage T4, in which protein engineering studies by Brian Mathews of the University of Oregon, Eugene, revealed some of the factors of importance for protein stability. The entropy of the unfolded structure is favored because of its

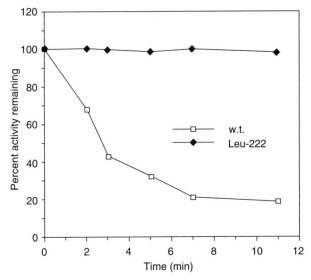

Figure 7.13. Oxidative stability of subtilisins, with comparison of wild-type to Leu-222 variant. Purified wild-type variant subtilisin was mixed with 0.3% H_2O_2 in 0.1 M Tris buffer (pH 8.6). Aliquots were removed at the indicated times and tested for activity. [Reprinted with permission from Crabb (1991), ACS Symp. Ser. 460, p. 89].

large number of structures. Hence, the folded state must compensate for this through various types of chemical and physical interactions, including hydrophobic interactions that bring the hydrophobic side chains from the solvent to the interior of the protein, specific hydrogen bonds, electrostatic interactions, and covalent crosslinks through S–S bridges. As summarized by Branden and Tooze (1991, p. 256): "the observed stability of a native protein structure is the result of a small difference, about 5 to 15 kcal/mol, between the total energies of the folded and unfolded states, each of which is on the order of 10 million kcal/mol." Hence, it is difficult, at best, "to derive a native folded structure by calculations from first principles, which would require not only extremely well-defined energy terms for all interactions, but also applying them to all possible structures of both the folded and the unfolded polypeptide chain." This difference may be measured by calorimetric studies to determine the difference in energy, and circular dichroism to follow changes in protein conformation when the protein is heated or exposed to denaturing agents, and melting points. A difference of 1 °C in the melting point temperature corresponds to a difference in energetic stability of about 0.5 kcal/mol[1].

---

[1] This section of Branden and Tooze's text (1991, pp. 257–260) summarized here, illustrates the increase of protein stability by changing selected amino acids to decrease the number of possible unfolded structures. Lysozyme from bacteriophase T4 contains 164 amino acid residues that folds into two domains (Fig. 7.14). The native protein has no S–S bridges and its cysteine residues, Cys 54 and Cys 97, are far apart in the folded structure. Its melting temperature is 41.9 °C. Novel S–S bridges were found to stabilize the protein. After careful and thorough analysis, five amino acid residues at Ile 3, Ile 9, Thr 21, Thr 142, and Leu 164 were replaced with Cys, and Cys 54 in the wild-type protein was changed to Thr to avoid

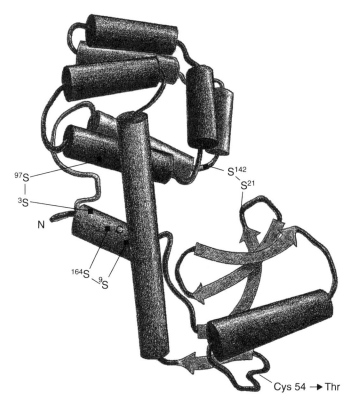

Figure 7.14. The polypeptide chain of lysozyme from bacteriophage T4 folds into two domains [reprinted with permission from Branden and Tooze (1991), Fig. 16.6, p. 257].

The simplest form of an enzyme deactivation equation is given as a first-order process

$$E(t) = E_0 e\left(\frac{-t}{\tau}\right) \tag{7.7}$$

where $E(t)$ = enzymatic activity at time t
$\quad E_0$ = initial enzymatic activity
$\quad \tau$ = a time constant

Enzyme deactivation complicates analysis of kinetic data, and hence experimental conditions are usually chosen at which enzyme deactivation is negligible. These are determined by measuring enzymatic activities at a specified pH, after the enzyme has been exposed to a selected temperature. A typical deactivation curve is illustrated for β-glucosidase in Fig. 7.15. On the basis of this plot, enzyme assays were carried out at 50 °C.

---

formation of incorrect S–S bonds. Protein stability was increased as indicated by increases in melting temperatures. The engineered form with disulfide bonds at Cys 3–Cys 97, Cys 9–Cys 164, and Cys 21–Cys 142 had an increase in T_m of 23 °C over the wild type.

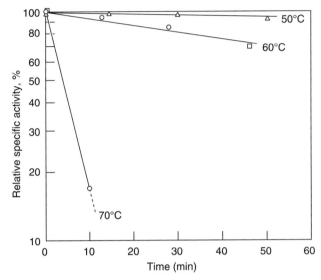

Figure 7.15.   First-order deactivation curve for cellobiase from *Trichoderma viride* [reprinted with permission from Ladisch et al. (1977), Fig. 4, p. 162].

The half-life is the elapsed time corresponding to 50% loss of the initial enzymatic activity. At $t = t_{0.5}$, $E(t_{0.5}) = \frac{1}{2}E_0$, and Eq. (7.6) gives

$$\ln 0.5 = -\frac{t_{0.5}}{\tau} = -0.69315 \quad \text{or} \quad \tau = \frac{t_{0.5}}{0.69315} \qquad (7.8)$$

where τ is the time constant for deactivation. In the case of β-glucosidase, the half-life ranges from 5 min to several hours, or even days, as the temperature decreases (Fig. 7.15).

Enzyme deactivation is a function of its environment, and is slowed or increased by changing the pH, or adding substrate, product, or inhibitor that binds with the enzyme and induces a conformation that is more resistant to deactivation. An example is given by soybean α-galactosidase at 54°C, where deactivation is minimal after 1 h at pH 4.0 in the presence of 100 mM galactose yet is 90% deactivated in 30 min in the absence of galactose. The enzyme deactivates more quickly at pH 7.0 than at pH 4.0, and galactose no longer protects against deactivation at the higher pH. At pH 7.0, the enzyme dissociates into monomers of molecular weights of ~40,000, while at pH 4.0, a tetramer forms that is more resistant to deactivation. In this case, deactivation no longer follows first-order decay (Porter et al. 1992), but instead follows a series mechanism

$$E \xrightarrow{\;k_1\;} E_1^{\alpha_1} \xrightarrow{\;k_2\;} E_2^{\alpha_2} \qquad (7.9)$$

where k_1 and k_2 are deactivation time constants (min^{-1}), $\alpha_1 = E_1/E$ and $\alpha_2 = E_2/E$ (Barclay et al. 1990; Henley and Sandana 1984, 1985, 1986). E, E_1, and E_2 represent three forms of enzymatic activity where the intrinsic activity of these forms follows $E > E_1 > E_2$, and probably reflects the dissociation of the tetramer into monomer as pH increases from pH 4.0 to pH 7.0.

The overall fractional activity remaining, $\alpha = E_+ (t)/E$, results from all three forms of enzyme, which are represented as follows:

$$E_+(t) = E(t) + E_1(t) + E_2(t) \tag{7.10}$$

Henley and Sandana (1986) give the fractional enzymatic activity remaining after a certain time t as

$$\frac{E(t)}{E} = a = \left[1 + \frac{\alpha_1 k_1}{k_2 - k_1} - \frac{\alpha_2 k_2}{k_2 - k_1} \right] e^{(-k_1 t)}$$
$$+ \alpha_2 - \left(\frac{k_1}{k_2 - k_1} \right) (\alpha_1 - \alpha_2) e^{(-k_1 t)} \tag{7.11}$$

If E_2 represents a completely inactive enzyme ($\alpha_2 = 0$), then the fractional activity a is

$$a = \left[1 + \frac{\alpha_1 k_1}{k_2 - k_1} \right] e^{(-k_1 t)} - \left[\frac{\alpha_1 k_1}{k_2 - k_1} \right] e^{(-k_2 t)} \tag{7.12}$$

In the case where both E_1 and E_2 represent completely inactive enzyme, $\alpha_1 = \alpha_2 = 0$, Eq. (7.11) becomes

$$a = \frac{E(t)}{E} = e^{(-k_1 t)} \tag{7.13}$$

which is equivalent to a first-order deactivation process presented previously as Eq. (7.6), where $1/\tau = k_1$.

Henley and Sandana have suggested that the first step in Eq. (7.6) (i.e., a single-step, unimolecular, non-first-order deactivation), where $k_1 \gg k_2$ represents the effect of protecting agents on enzyme stability, where deactivation stops at form E_1. The fractional activity $E_1(t)/E = \alpha_1$ would then represent a coefficient of enzyme stabilization, where

$$\alpha = \left[(1 - \alpha) e^{(-k_1 t)} \right] + \alpha_1 \tag{7.14}$$

Returning to our example of the effect of adding galactose on the heat stability of α-galactosidase at pH 4 and 54 °C (Porter et al. 1992), α_1 and α_2 in Eq. (7.11) are zero in the absence of galactose. α_1 increases from 0.092 (no galactose) to 0.096 (100 mM galactose). At pH 5.5, α_1 increases from 0.085 at 70 °C to 0.533 at 56 °C in the presence of 100 mM galactose since the level of protection increases with decreasing temperature.

Enzyme deactivation may be moderated if the enzyme is immobilized on a solid support. Industrial reactors that utilize immobilized enzymes have half-lives of months to years. This is needed to achieve economical operation. Estimation of reactor volumes for a given throughput requires that expressions for enzyme deactivation be incorporated into the kinetic equations. For determining kinetics and characterizing the enzyme, however, conditions at which loss of enzymatic activity is negligible are usually experimentally determined prior to initiation of kinetic studies. Enzyme stability as a function of time or stream, as well as temperature, pH, and substrate concentration, is further examined after the kinetics are known.

Analysis of various types of initial rate kinetics, and equations for a number of different types of enzyme-catalyzed reaction sequences, are given by Segel (1975), Plowman (1972), and Cornish-Bowden and Wharton (1988). The manner in which these might be integrated into computationally based, hierarchal dynamic models is discussed by Dhurjati and Mahadevan (2008). The next chapter provides an introduction to modeling of single-enzyme reactions for industrial applications and for models in systems biology.

REFERENCES

Abdel-Fattah, Y. R., H. M. Saeed, Y. M. Gohar, and M. A. El-Baz, "Improved Production of Pseudomonas aeruginosa uricase by Optimization of Process Parameters through Statistical Experimental Designs," *Process Biochem.* **40**, 1707–1714 (2005).

Antrim, R. L., W. Colliala, and B. Schnyder, "Glucose isomerase production of high-fructose syrups," in *Applied Biochemistry and Bioengineering*, L. B. Wingard, ed., Academic Press, New York, NY, 1979.

Atkinson, B. and F., Mavituna, *Biochemical Engineering and Biotechnology Handbook*, Nature Press, Macmillan Publishers, Surrey, UK, 1983, pp. 468–513.

Babacan, S., L. F. Pivarnik, and G. G. Rand, "Honey Amylase Activity and Food Starch Degradation," *J. Food Sci.* **67**(5), 1625–1630 (2002).

Barclay, C. D., D. M. Moore, S. R. Lander, and R. L. Legge, "Heat-Denaturation Kinetics of Lignin Peroxidases from Phanerochaete-Chrysosporium," *Enzym. Microbial Technol.* **12**(10), 778–782 (1990).

Bergamasco, R., F. J. Bassetti, F. F. de Moraes, and G. M. Zanin, "Characterization of Free and Immobilized Invertase Regarding Activity and Energy of Activation," *Braz. J. Chem. Eng.* **17**(4–7), 873–880 (2000).

Boyce, C. O. L., ed., *Novo's Handbook of Practical Biotechnology*, 2nd ed., Novo Industri A/S, Denmark, 1986.

Branden, C. and J. Tooze, *Introduction to Protein Structure*, Garland Publishing, New York, 1991, pp. 256–260.

BRENDA, "The Comprehensive Enzyme Information System," http://www.brenda-enzymes.org/ (accessed June 26, 2008).

Clyde, E., "The Optimum pH for Diastase of Malt Activity," *Ohio J. Sci.* **59**(5), 257–262 (1959).

Cornish-Bowden, A. and C. W. Wharton, eds., *Enzyme Kinetics*, IRL Press, Oxford, 1988.

Costanzo, B. and G. Antranikian, "Starch-Hydrolyzing Enzymes from Thermophilic Archaea and Bacteria," *Current Opin. Chem. Biol.* **6**, 151–160 (2002).

Crabb, W. D., "Subtilisin: Commercially Relevant Model for Large-Scale Enzyme Production," in *Enzymes in Biomass Conversion*, G. F. Leatham and M. E. Himmel, eds., American Chemical Society, Washington, DC, 1991.

De Bales, S. A. and F. J. Castillo, "Production of Lactase by *Candida pseudotropicalis* Grown in Whey," *Appl. Environ. Microbiol.* **37**(6), 1201–1205 (1979).

Dejong, P. J., "L-Asparaginase Production by *Streptomyces griseus*," *Appl. Microbiol.* **23**(6), 1163–1164 (1975).

Dhurjati, P. and R. Mahadevan, "Systems Biology: The Synergistic Interplay between Biology and Mathematics," *Can. J. Chem. Eng.* **86**(2), 127–141 (2008).

Doukyu, N. and R. Aono, "Purification of Extracellular Cholesterol Oxidase with High Activity in the Presence of Organic Solvents from Pseudomonas sp. Strain ST-200," *Appl. Environ. Microbiol.* **64**(5), 1929–1932 (1998).

Fitzgerald, J. and W. Vermerris, "The Utility of Blood Glucose Meters in Biotechnological Applications," *Biotechnol. Appl. Biochem.* **41**, 233–239 (2005).

Fluck, R. A. and M. J. Jaffe, "Cholinesterases from Plant Tissues V. Cholinesterase Is Not Pectin Esterase," *Plant Physiol.* **54**, 797–798 (1974).

Foltmann, B., "Prochymosin and Chymosin (Prorennin and Rennin)," *Biochemical J.* **115**(3), 3–4 (1969).

Food Navigator.com/Europe, "New Enzyme Technologies to Fight Downward Price Trend in Saturated Market," *Product Supplier News* (http://www.foodnavigator.com/productnews) (June 9, 2005)

Fukui, S., S.-I. Ikeda, and M. Fujimura, "Comparative Studies on the Properties of Tryptophanase and Tyrosine Phenol-lyase Immobilized Directly on Sepharose or by Use of Sepharose-Bound Pyridoxal 5'-Phosphate," *Eur. J. Biochem.* **5**, 155–164 (1975).

Gay, R. J., R. B. McComb, and G. N. Bowers, Jr., "Optimum Reaction Conditions for Human Lactate Dehydrogenase Isoenzymes as They Affect Total Lactate Dehydrogenase Activity," *Clin. Chem.* **14**(8), 740–753 (1968).

Giordano, R. L. C. and M. P. A. Ribeiro, "Kinetics of β-Lactam Antibiotics Synthesis by Penicillin G Acylase (PGA) from the Viewpoint of the Industrial Enzymatic Reactor Optimization," *Biotechnol. Adv.* **24**, 27–41 (2006).

Gordon, M. H., "The Development of Oxidative Rancidity in Foods," in *Antiooxidants in Food*, M. H. Gordon, N. Yanishlieva, and J. Pokorný, eds., Part 1(2), Woodhead Publishing, 2001, Vol. 7, p. 20.

Greig, H. B. W. and E. M. Cornelius, "Protamine-Heparin Complex as *Substrate for Plasmin*," *Biochim. Biophys. Acta* **67**, 658–668 (1963).

Harrison, F. G. and E. D. Gibson, "Approaches for Reducing the Manufacturing Costs of 6-Aminopenicillanic Acid," *Proc. Biochem.* **19**(1), 33–36 (1984).

Hawiger, J., "Purification and Properties of Lysozyme Produced by *Staphylococcus aureus*," *J. Bacteriol.* **95**(2), 376–384 (1968).

Henley, J. P. and A. Sadana, "A Mathematical-Analysis of Enzyme Stabilization by a Series-Type Mechanism—Influence of Chemical Modifiers," *Biotechnol. Bioeng.* **26**(8), 959–969 (1984).

Henley, J. P. and A. Sadana, "Categorization of Enzyme Deactivations Using a Series-Type Mechanism," *Enzym. Microbial Technol.* **7**(2), 50–60 (1985).

Henley, J. P. and A. Sadana, "Deactivation Theory," *Biotechnol. Bioeng.* **28**(8), 1277–1285 (1986).

Hoondal, G. S., R. P. Tiwari, R. Tewari, N. Dahiya, and Q. K. Beg, "Microbial Alkaline Pectinases and Their Industrial Applications: A Review," *Appl. Microbiol. Biotechnol.* **59**, 409–418 (2002).

Invitrogen, "Taq DNA Polymerase (Native and Recombinant)," www.Invitrogen.com. (accessed June 26, 2008).

Jonas, H. A., R. F. anders, and G. R. Jago, "Factors Affecting the Activity of the Lactate Dehydrogenase of *Streptococcus cremoris*," *J. Bacteriol.* **111**(2), 397–403 (1972).

Kashyap, D. R., P. K. Vohra, S. Chopra, and R. Tewari, "Applications of Pectinases in the Commercial Sector: A Review," *Bioresource Technol.* **77**, 215–227 (2001).

Ladisch, M. R., *Bioseparations Engineering: Principles, Practice, and Economics*, Wiley, New York, 2001.

Ladisch, M. R., *Effect of Glucose Isomerase Purity on Immobilization Economics*, MS, ChE thesis, Purdue Univ., West Lafayette, IN, 1974, pp. 34–38.

Ladisch, M. R., *Enzymatic Hydrolysis of Cellulose: Kinetics and Mechanism of Selected Purified Cellulose Components*, PhD thesis, Purdue Univ., West Lafayette, IN, 1977, pp. 160–161.

Ladisch, M. R., C. S. Gong, and G. T. Tsao, "Corn Crop Residues as a Potential Source of Single Cell Protein: Kinetics of *Trichoderma viride* Cellobiase Action," *Devel. Industr. Microbiol.* **18**, 157–168 (1977).

Lehninger, A. L., *Biochemistry: The Molecular Basis of Cell Structure and Function*, Worth Publishers, 1970.

Mohanty, A. K., U. K. Mukhopadhyay, S. Grover, and V. K. Batish, "Bovine Chymosin: Production of rDNA Technology and Application in Cheese Manufacture," *Biotechnol. Adv.*, **17**, 205–217 (1999).

Montenecourt, B. S. and D. E. Eveleigh, "Preparation of Mutants of *Trichoderma-reesei* with Enhanced Cellulase Production," *Appl. Environ. Microbiol.* **34**(6), 777–782 (1977).

Mori, T., T. Tosa, and I. Chibata, "Preparation and Properties of Asparaginase Entrapped in the Lattice of Polyacrylamide Gel," *Cancer Res.* **34**, 3066–3068 (1974).

Moukheiner, Z., "Biotech—We're Green … ," Forbes Mag. (http://www.Forbes.com80/forbes/00/0221/6504142a.htm) (Feb. 21, 2000).

Okada, S., "On the Optimal Conditions for the Proteoclastic Action of Taka-Diastase," *Biochem. J.* **10**(1), 130–136 (1916).

Petsko, G. A., "Protein Engineering," in *Biotechnology and Materials Science—Chemistry for the Future*, M. L. Good, eds., American Chemical Society, Washington, DC, 1988, pp. 53–60.

Plowman, K. M., *Enzyme Kinetics*, McGraw-Hill, New York, 1972.

Porter, J. E., A. Sarikaya, K. M. Herrmann, and M. R. Ladisch, "Effect of pH on Subunit Association and Heat Protection of Soybean β-Galactosidase," *Enzym. Microbial Technol.* **14**, 609 (1992).

Puri, M. and U. C. Banerjee, "Production, Purification, and Characterization of the Debittering Enzyme Naringinase," *Biotechnol. Adv.* **18**, 207–217 (2000).

Reiner, E., and W. N. Aldridge, "Effect of pH on Inhibition and Spontaneous Reactivation of Acetylcholinesterase Treated with Esters of Phosphorus Acids and of Carbamic Acids," *Biochem. J.* **105**, 171–179 (1967).

Rios, G. M., M. P. Belleville, D. Paolucci, and J. Sanchez, "Progress in Enzymatic Membrane Reactors—a Review," *J. Membrane Sci.* **242**(1–2), 189–196 (2004).

Rodwell, V. W., "Enzymes: General Properties," in *Harper's Biochemistry*, 22nd ed., R. K. Murray, D. K. Granner, P. A. Mayes, and V. W. Rodwell, eds., Appleton & Lange, Norwalk, CT, 1990, pp. 58–88.

Saha, B. C., S. P. Mathupala, and J. G. Zeikus, "Purification and Characterization of a Highly Thermostable Novel Pullanase from *Clostridium thermohydrosulfuricum*," *Biochem. J.* **252**, 343–348 (1988).

Sangeetha, K. and T. E. Abraham, "Chemical Modification of Papain for Use in Alkaline Medium," *J. Molec. Catal. B: Enzym.* **38**, 171–177 (2006).

Scott, A. and A. Wood, "Bioprocessing: Struggling to Grow Profits," *Chem. Week* **167**(5), 15–7 (Feb. 9, 2005).

Segel, I. H., *Enzyme Kinetics: Behavior and Analysis of Rapid Equilibrium and Steady-State Enzyme Systems*, Wiley-Interscience, New York, 1975.

Shi, X., M. Feng, Y. Zhao, X. Guo, and P. Zhou, "Overexpression, Purification and Characterization of a Recombinant Secretary Catalase from *Bacillus subtilis*," *Biotechnol. Lett.* 30, 181–186 (2008).

Shin, K.-S., H.-D. Youn, Y.-H. Han, S.-O. Kang, and Y. C. Hah, "Purification and Characterization of D-Glucose Oxidase from White-Rot Fungus *Pleurotus ostreatus*," *Eur. J. Biochem.* 215, 747–752 (1993).

Sukuzi, S., T. Hirahara, S. Horinouchi, and T. Beppu, "Purification and Properties of Thermostable Tryptophanase from an Obligately Symbiotic Thermophile, *Symbiobacterium thermophilum*," *Agric. Biol. Chem.* 55(12), 3059–3066 (1991).

Szajáni, B., A. Molnár, G. Klámar, and M. Kálmán, "Preparation, Characterization, and Potential Application of an Immobilized Glucose Oxidase," *Appl. Biochem. Biotechnol.* 14(1), 37 (1987).

Taylor, R., "Expanding Applications in the Food Industry for Immobilized Enzymes," *Genetic Eng. News* 13(3), 5 (1993).

Wang, L.-J., X.-D. Kong, H.-Y. Zhang, X.-P. Wang, and J. Zhang, "Enhancement of the Activity of L-Aspartase from Escherichia coli W by Directed Evolution," *Biochem. Biophys. Res. Commun.* 276, 346–349 (2000).

Weil, J., J. Miramonti, and M. R. Ladisch, "Cephalosporin C: Mode of Action and Biosynthetic Pathway," *Enzym. Microbial Technol.* 17, 85–87 (1995a).

Weil, J., J. Miramonti, and M. R. Ladisch, "Biosynthesis of Cephalosporin C: Regulation and Recombinant Technology," *Enzym Microbial Technol.* 17, 88–90 (1995b).

Wierzbickp, L. E. and F. V. Kosikowski, "Lactase Potential of Various Microorganisms Grown in Whey," *J. Dairy Sci.* 5(1), 26–32 (1972).

Wilson, S. H. and E. L. Kuff, "A Novel DNA Polymerase Activity Found in Association with Intracisternal A-Type Particles," *Proc. Natl. Acad. Sci. USA* 69(6), 1531–1536 (1972).

Worsfold, P. J., "Classification and Chemical Characteristics of Immobilized Enzymes" (technical report), *Pure Appl. Chem.* 67(4), 597–600 (1995).

Yazdi, M. T., M. Zahraei, K. Aghaepour, and N. Kamranpour, "Purification and Partial Characterization of a Cholesterol Oxidase from *Streptomyces fradiae*," *Enzym. Microbial Technol.* 28, 410–414 (2001).

Zaborsky, O., *Immobilized Enzymes*, CRC Press, Cleveland, OH, 1974.

Zhao, Y., L. Zhao, G. Yang, J. Tao, Y. Bu, and F. Liao, "Characterization of n Uricase from *Bacillus fastidious* A.T.C.C. 26904 and Its Application to Serum Uric Acid Assay by a Patented Kinetic Uricase Method," *Biotechnol. Appl. Biochem.* 45, 75–80 (2006).

CHAPTER SEVEN
HOMEWORK PROBLEMS

7.1. Total global sales of enzymes were estimated at $580 million in 1990 and $1 billion in 1999, 50% of which were used in the food industry. Genencor's revenues grew to an estimated $309 billion in 1999, and Novo Nordisk had sales of $630 million. A report by Moukheiner (2000) in *Forbes Magazine*, suggested the application of molecular biology resulting in designer enzymes suggests a potential that may be larger. We define designer enzymes as enzymes whose activity has been specifically modified in a directed manner using molecular biology techniques. In 1991, the market sizes for the major enzymes were estimated at $150 million for alkaline proteases from *Bacillus* spp., $70 million for neutral proteases from *Bacillus* spp. and *Aspergillus oryzae*, $100 million for amylases from *Bacillus* spp. and *A. oryzae*, and $60 million for rennin from calf stomach and a *Mucor* species. (Crabb 1991). The value of glucose isomerase from *Streptomyces* spp. is estimated at $45 million by one source (Crabb 1991) and $70 million by another (Petsko, 1988).

(a) Attempt to estimate values of market segments in 2005, for a growth rate 4%/year since 1990.

(b) On the basis of these limited data, estimate the increase in the market values between 1990 and 2010. Show your methods and explain your rationale. Look up current market values of enzyme sales and compare to your calculated estimate. Are these different? Why? Has technology (or decreasing availability) of oil changed the assumptions? What is your projection of the nature of the enzyme market in 2020?

7.2. Give the common names and IUB numbers for enzymes having the following activities. Also give the optimal pH range.

(a) Cellulose → oligosaccharides

(b) Starch → glucose + maltose + oligosaccharides

(c) Dextran → glucose + oligosaccharides

(d) Sucrose → glucose + fructose

(e) Starch → maltose + maltotriose

(f) Hydrolyzes protein at Arg, Lys

(g) $H_2O_2 \rightarrow O_2 + H_2O$

(h) Indole + pyruvate + $NH_3 \rightarrow$ L-trp + H_2O

(i) L-Lactate + $NAD^+ \rightarrow$ pyruvate + NADH

7.3. The enzyme hexokinase is to be used for the determination of glucose in enzyme assay samples for the hydrolysis of cellobiose. After the reaction is completed, the cellobiose-hydrolyzing enzyme (β-glucosidase) is deactivated by placing the test tube in a beaker of boiling water for 3 min, and then transferring the capped tube to an ice bath. An aliquot of the solution is then combined with hexokinase and other reagents.

(a) Briefly describe the basis of the hexokinase assay, and the other molecules that are needed to quantitate the glucose present. Why is spectral wavelength of 340 nm used to determine the amount of glucose present?

(b) The assay is repeated for the same samples using glucose oxidase. The glucose concentration determined by this method is significantly higher than that for the hexokinase assay. The next day, the same samples are again assayed using glucose oxidase, and the glucose readings coincide with those obtained for the hexokinase assay. Explain.

7.4. An enzyme transforms substrate S to product P, as represented by the sequence

$$E + S \rightleftharpoons ES \rightarrow E + P$$

where E = enzyme, S = substrate, P = product. The initial substrate concentration is 100 mmol/L, while P = 0. The rate of deactivation of the enzyme is given by the expression:

$$E(t) = E_0 \exp\left(-\frac{t}{\tau}\right)$$

The reaction is carried out at 60 °C, where the half-life of the enzyme is 2.7 h. V_{max} and enzymatic activity are related by the expression $V_{max} = k_p E(t)$. Assume that $v = dP/dt$, with values of $V_{max,0}$ and K_m given by

$$\begin{aligned} V_{max,0} &= 10\,\text{mmol/min} \\ K_m &= 2\,\text{mmol/L} \end{aligned} \quad \text{(initial enzymatic activity).}$$

Calculate the reaction timecourse, and plot the appearance of product as a function of time over a 4-h period.

CHAPTER EIGHT

ENZYME KINETICS

INTRODUCTION

Enzyme kinetics represent the mathematical formulation of rates at which enzymes transform one molecule to another. Kinetics are an essential part of the biochemist's toolbox for establishing the basis of rate equations, facilitating their derivation, and using them as a tool to analyze probable mechanisms of enzymes [paraphrased from Plowman (1972)]. However, it is also true that enzyme kinetics have become less important as a tool for probing enzyme mechanisms. Molecular biology and site-directed changes to protein structures have made it possible to directly examine the structure and function of proteins at a molecular level.

The application of metabolic pathway engineering, emergence of systems biology, and rediscovery of biocatalysis as a way to carry out industrially relevant biochemical reactions have reinvigorated interest on the part of biochemical engineers in enzyme kinetics, and the engineering of enzyme-based bioreactors. This is particularly true for research and scaleup on enzyme-catalyzed hydrolysis of cellulose to fermentable sugars. Cellulose hydrolysis has been reinvigorated by high oil prices and concern of the environmental footprint of nonrenewable fuels (Houghton et al. 2006; Lynd et al. 2008; Wyman et al. 2005). In a biomedical context the formulation of the binding of substrates and/or products to enzymes sets the stage for modeling receptor interactions at a cellular level. Kinetics facilitate modeling the molecular basis at certain diseases and the rational design of targeted drug candidates (Lauffenberger and Linderman 1993).

This chapter, on enzyme kinetics, provides an introduction to analyzing enzyme reactions in the context of basic mechanisms that enzymes have in common, and the analysis of bioreactors that use enzymes as the catalytic component. Knowledge of the processes involved will be useful to scientists and engineers in the biotechnology field as the discipline of enzyme-based biocatalysis becomes reinvigorated (Schoemaker et al. 2003).

Modern Biotechnology, by Nathan S. Mosier and Michael R. Ladisch
Copyright © 2009 John Wiley & Sons, Inc.

INITIAL RATE VERSUS INTEGRATED RATE EQUATIONS

An integral approach to enzyme kinetics is practical when probable sequences of enzyme action are known or can be postulated on the basis of past experience, and the enzyme and substrate itself is available in sufficient quantity to carry out measurements of product formation. An example is given by 1,4-β-glucanglucanohydrolase, whose action on cellobiose[1] follows the sequence (Ladisch 1980):

$$E + G_2 \underset{k_2}{\overset{k_1}{\rightleftharpoons}} EG_2 \xrightarrow{k_P} E + 2G \tag{8.1}$$

$$E + G \underset{k_4}{\overset{k_3}{\rightleftharpoons}} EG \tag{8.2}$$

where E represents free enzyme; G_2, cellobiose (a dimer of 1,4-β-D-glucose); EG_2, an enzyme–substrate complex; G, glucose; and EG, an enzyme–inhibitor complex. The enzyme is inhibited by the product that it forms (glucose), which is one way in which a living cell regulates a metabolic pathway. When the product of enzyme action builds up, the reaction is slowed down by the product, reversibly binding with the enzyme, causing the enzyme to bind the substrate less effectively. This decreases the efficiency of the catalysis that occurs at the active site of the enzyme and slows the rate of reaction. The velocity or rate equation for this sequence is

$$v = \frac{dP}{dt} = \frac{V_{max}[G_2]}{[G_2] + K_m\left(1 + \dfrac{[G]}{K_i}\right)} \tag{8.3}$$

where

$$V_{max} = 2k_P[E_T] \tag{8.4}$$

$$K_m = \frac{k_2 + k_P}{k_1} \qquad K_i = \frac{k_4}{k_3} \tag{8.5}$$

Kinetic constants determined at high substrate concentrations may appear to work well for industrial enzymes since substrate concentrations at industrial conditions far exceed K_m. Values of K_m are often on the order of 1–10 mM (or 0.01–0.1 g/L, depending on the molecular weight of the substrate), and substrate concentrations

---

[1] The hydrolysis of cellulose in materials such as sawdust, crop residues (stalks and leaves), forages (hay and grasses), and forestry residues is achievable by mixtures of enzymes known generically as *cellulases* and *hemicellulases*. These enzymes work together to break down sugar polymers that make up plant materials into monosaccharides. One of the enzymes that make up a cellulase enzyme mixture is glucanohydrolase and is presented here as an example. Its primary function is not hydrolysis of cellobiose, but rather randomly cleaving cellulose to create reactive sites for other enzyme components that form soluble disaccharides from the cellulose polymer. This enzyme is discussed here to illustrate an example of product inhibition.

may exceed K_m by 10–100-fold, thereby resulting in reaction rates of v that are nearly V_{max}. Thus the reaction rate is essentially constant. In the case of systems biology, this assumption is less likely to be applicable, since intracellular concentrations of substrates and products are at much lower levels. Indeed, intracellular concentrations are often very close to the K_m value as that is the point where the enzyme is most closely regulated by small changes in substrate concentration. Hence, modeling of cellular reaction pathways and regulation requires a more detailed knowledge of the enzymes' kinetic constants than for single- or multistep industrial reactions.

In Eqs. (8.3)–(8.5), V_{max} represents the maximum reaction velocity; k_p is the turnover number; E_T; total enzyme; K_m, the Michaelis constant; and K_i, the dissociation constant for the enzyme inhibitor complex, where the inhibitor is the product of the reaction. The inverted and linear form of Eq. (8.3) is

$$\frac{1}{v} = \frac{K_m}{V_{max}}\left(1 + \frac{[G]}{K_i}\right)\frac{1}{[G_2]} + \frac{1}{V_{max}} \tag{8.6}$$

The line resulting from Eq. (8.6) coincides with the data plotted on a graph of inverse velocity versus inverse substrate concentration in Fig. 8.1, when the values of the constants are:

$$K_m = 1.59\,\text{mM}$$

$$K_i = 0.98\,\text{mM} \tag{8.7}$$

$$\frac{V_{max}}{E_T} = 0.585 \frac{\mu\text{mol G}}{\text{min}\cdot\text{mg protein}}$$

The integration of a rearranged form of Eq. (8.3)

$$V_{max}\int_0^t dt = \int_p^G \left\{1 + K_m\left(1 + \frac{[G]}{K_i}\right)\frac{1}{[G_2]}\right\}dG \tag{8.8}$$

is carried out for

$$[G_2] = [G_{2,0}] - \frac{[G]}{2} \tag{8.9}$$

Since each mole of cellobiose gives 2 mol glucose, $G_{2,0}$ represents the initial cellobiose concentration.

The result is an implicit equation for the glucose concentration as a function of time:

$$V_{max}t = \left(1 - \frac{2K_m}{K_i}\right)[G] + 2K_m\left(1 + \frac{2[G_{2,0}]}{K_i}\right)\ln\left(\frac{[G_{2,0}]}{[G_{2,0}] - \frac{[G]}{2}}\right) \tag{8.10}$$

The values of the kinetic parameters (8.7) together with Eq. (8.10) give the timecourse curves in Fig. 8.2. These are obtained by solving for t on substituting

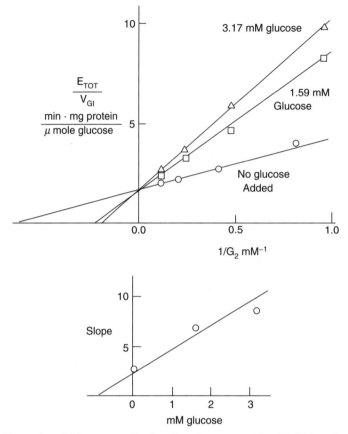

Figure 8.1   Examples of Lineweaver–Burke plots for competitive inhibition [from Ladisch, PhD thesis (1977), Fig. 36, p. 98]. Substrate is cellobiose (G_2).

values of [G] into the equation up to maximum of $G = 2G_{2,0}$. In the absence of product inhibition, the rate equation, and its linear and integrated forms are, respectively

$$v = \frac{V_{\max}[G_2]}{K_m + [G_2]} \tag{8.11}$$

and

$$\frac{1}{v} = \frac{K_m}{V_{\max}} \cdot \frac{1}{[G_2]} + \frac{1}{V_{\max}} \tag{8.12}$$

and

$$V_{\max}t = [G] - 2K_m \ln\left(1 - \frac{[G]}{2[G_{2,0}]}\right) \tag{8.13}$$

The linear form, Eq. (8.12), coincides with the line for no glucose added in Fig. 8.1, and the dashed lines in Figs. 8.2a and 8.2b.

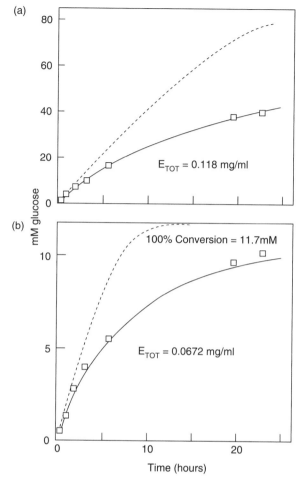

Figure 8.2. Timecourse of cellobiose hydrolysis by endoglucanase. Integrated form of model [Eq. (8.10)] gives solid line. [From Ladisch, PhD thesis (1977), Fig. 37, p. 104].

Both figures show that the accumulation of product results in significant inhibition. The initial rates decrease with increasing levels of added glucose, hence resulting in increasing inverse rates and increasing slopes of the lines in Fig. 8.1. The timecourse curves in Fig. 8.2 show the same effect, where the generation of glucose from the substrate itself, rather than addition of glucose, causes a marked decrease in product accumulation over a noninhibited reaction (denoted by the dashed line).

The constants (8.7) resulted from fitting of data from initial rate experiments to Eq. (8.6) as plotted in Fig. 8.1. These constants were then used to confirm that the integrated rate equation followed timecourse data for initial cellobiose concentrations of 38.3 and 5.85 mM (Fig. 8.2), which extend significantly away from the initial rate region. The results for the integrated equation prove internal consistency and also illustrate the potential difference between determining enzyme kinetics from initial rate data compared to using an integrated equation. In the case of initial rate experiments, each data point represents a separate experiment, while the timecourse (Fig. 8.2) is carried out by taking multiple samples from the same reaction solution from a single experiment. Fitting of the integrated equation [Eq. (8.10)] may also yield the values of K_m, K_i, and V_{max}.

Obtaining Constants from Initial Rate Data Is an Iterative Process

Graphical analysis is useful in obtaining a first estimate of the character of the enzyme's activity, including the type of inhibition, presence of contaminants in the substrate, or presence of multiple enzymatic activities acting on the same substrate. A number of approaches are possible based on rearrangement of the Henri–Michaelis–Menten [Eq. (8.11)]. As discussed by Segel (1975), these include direct linear (Dixon, Eisenthal, and Cornish-Bowden) plots and reciprocal (Lineweaver–Burke) plots in addition to various forms of logarithmic plots. While these plots are helpful in visualizing patterns of enzyme kinetics, statistical fitting of the data is needed to obtain accurate values of kinetic constants.

The commonly used Lineweaver–Burke plot is illustrated by Eq. (8.12) and Fig. 8.1. Best results are obtained when concentrations of the substrates are chosen to be in the same range of values for K_m. Since the K_m for a new enzyme is not initially known, determination of its value entails a trial-and-error approach where a range of substrate concentrations are selected. If [S] and K_m diverge by a factor of ≥10, a second range of substrate concentrations should be selected that are in the neighborhood of the value of K_m, and a second set of kinetic measurements carried out.

If the substrate concentrations are extremely high relative to K_m, the curve may be visualized as being essentially horizontal, making it difficult to accurately determine K_m (Fig. 8.3a) while V_{max} is readily obtained. If the substrate concentrations are extremely low, neither V_{max} nor K_m can be accurately determined, since the reaction is essentially first-order, and V_{max} and K_m appear to be infinite (Fig. 8.3b).

For purposes of measuring kinetic constants, the substrate concentrations should be chosen to be in the range of $0.33-2.0K_m$ (Fig. 8.3) (Segel 1975). The data for the glucanohydrolase in Fig. 8.1 approximate the recommended range (cellobiose concentrations ~$0.16-1.6K_m$). For an enzyme with high substrate affinity (low K_m), the analytical method must be sufficiently sensitive to detect the initial rates of product formation (or substrate disappearance) that correspond to a conversion of less than 5% of the initial substrate.

The kinetic data may also be represented by the Eadie–Hofstee or Hanes plots illustrated in Fig. 8.4. The forms of equations that are represented here are

$$v = -K_m\left(\frac{v}{A}\right) + V_{max} \quad \text{for the Eadie–Hofstee plot} \tag{8.14}$$

and

$$\frac{S}{v} = \frac{1}{V_{max}}S + \frac{K_m}{V_{max}} \quad \text{for the Hanes plot} \tag{8.15}$$

The constants K_m and V_{max} are the same as before, and S is the designation for the substrate in the reaction

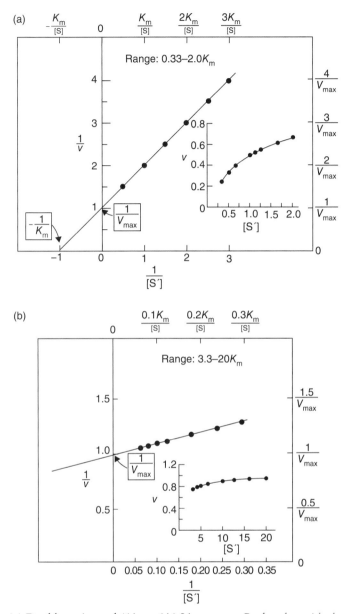

Figure 8.3. (a) Double-reciprocal ($1/v$ vs $1/s$) Lineweaver–Burke plot with the range of substrate concentrations chosen to be optimal for determination of K_m and V_{max}; (b) double-reciprocal plot where the range of substrate concentration S is higher than optimal and the reaction velocity V is relatively insensitive to changes in S [reprinted with permission from Segel (1975), Figs. 11–9 and 11–10, p. 47, Wiley-Interscience].

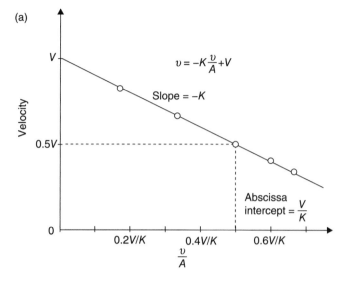

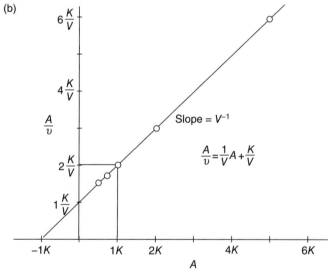

Figure 8.4. (a) Illustration of Hofstee or Eadie plot of rectangular hyperbola ($v = -K_m$ $(v/A) + V_{max}$; (b) Hanes plot of rectangular hyperbola ($S/v = (1/V_{max}) S + K_m/V_{max}$) [reproduced from Plowman (1972), Figs. 1.6 and 1.7, pp. 17–18, McGraw-Hill, New York].

$$S \rightarrow P \qquad (8.16)$$

The Hanes equation [Eq. (8.15)] is obtained by multiplying the general form used for the Lineweaver–Burke plot by S:

$$S\left[\frac{1}{v} = \frac{K_m}{V_{max}}\left(\frac{1}{S}\right) + \frac{1}{V_{max}}\right] \qquad (8.17)$$

which results in Eq. (8.15). As pointed out by Bailey and Ollis (1977), the Line-weaver–Burke equation clearly separates dependent and independent variables but clusters the most easily measured rates (those at higher substrate concentration) near the origin where $1/S \rightarrow 0$. In this case, they recommend that the linear least-squares method not be used to fit the data and extract the constants [unless the data fall in the range suggested by Segel (1975), Fig. 8.3). Equation (8.14) enables a more accurate estimate of V_{max}, since this form spreads the data. The Hanes plot [from Eq. (8.15)] contains $1/v$, which may give larger errors than the Lineweaver–Burke or Eadie–Hofstee (plots). The method used and the accuracy of the constants is often a function of the sensitivity of the analytical method used to quantify small changes of dilute concentrations of substrate. The more accurate the method, the better these changes may be measured against a relatively large background of substrate.

At the other extreme is an enzyme that has an unusually high K_m where detection of small changes against a high substrate concentration background must be achieved. An example is glucose (xylose) isomerase, which has K_m values in the range of 90–920 mM depending on the source of the enzyme (Antrim et al. 1979). Substrate concentrations equivalent to 0.33–$20K_m$ would then fall in a range of 30 mM (0.5 wt/vol%) to 1800 mM (32 wt/vol%), or higher. The viscosities at these substrate concentrations pose practical challenges in appropriately pipetting and diluting the samples, in addition to choosing an assay where background interference from the substrate is low. Since glucose isomerase catalyzes a reversible reaction, other considerations apply in developing the appropriate kinetic expressions, and are discussed later in this chapter.

If K_m is large, either a series of initial rate experiments or a single reaction timecourse could be carried out. The latter approach uses less enzyme (the same test tube is separately sampled), and may facilitate use of less sensitive assay techniques since most of the data are obtained at higher conversion. An integral equation, derived from a sequence such as Eqs. (8.1) and (8.2), is fitted to the data using either graphical or computational methods to determine the constants.

The biochemical literature reports primarily data and results based on initial rate enzyme analysis at low substrate concentrations. The information that an engineer or scientist seeks for designing a process is the performance of an enzyme at high substrate concentrations, and conversions that approach theoretical conversion. Consequently, either analytically or numerically integrated rate equations are needed. In the case of systems biology, initial rate kinetics are useful since concentrations of intermediates in a pathway are low, and "crosstalk" between metabolic enzymes and those involved in genetic regulation occur at low levels of substrate and products.

BATCH ENZYME REACTIONS: IRREVERSIBLE PRODUCT FORMATION (NO INHIBITION)

Rapid Equilibrium Approach Enables Rapid Formulation of an Enzyme Kinetic Equation

The most direct approach to analysis of initial rate enzyme kinetics is the rapid equilibrium treatment originally developed by Henri (in 1903) and Michaelis and

Menten (in 1913) and summarized by Plowman (1972). Segel (1975) gives a clear comparison of rapid equilibrium and pseudo-steady-state approaches to enzyme kinetics. Segel's comparison is repeated here.

The general form of a reaction in which product P forms irreversibly, is given by

$$E + S \underset{k_{-1}}{\overset{k_1}{\rightleftharpoons}} ES \xrightarrow{k_P} E + P \tag{8.18}$$

where E represents enzyme; S, substrate; and P, product, where S and P are expressed in molar concentrations. Rapid equilibrium treatment assumes that

1. E + S forms ES rapidly.
2. E, S, and ES are at equilibrium at all times.
3. S >> E.
4. $ES \xrightarrow{k_P} E + P$ is the rate-limiting step.

The initial rate equation for sequence (8.18) is obtained by writing an activity balance for the enzyme. The total amount of active enzyme is the sum of enzyme [E] that has no substrate or product associated with it, and enzyme to which substrate is bound, ES:

$$[E_t] = [E] + [ES] \tag{8.19}$$

The rate equation for the appearance of product is

$$v = \frac{dP}{dt} = k_P [ES] \tag{8.20}$$

Dividing (8.20) by (8.19), we obtain

$$\frac{v}{(E_t)} = \frac{k_P [ES]}{[E] + [ES]} \tag{8.21}$$

which, when multiplied through by [S]/[S] and rearranged, gives

$$v = \frac{k_P [E_t][S]}{\dfrac{[E][S]}{[ES]} + [S]} \tag{8.22}$$

The ratio [E] [S]/[ES] is a dissociation ratio (or constant) for the enzyme–substrate complex [ES] and is represented by K_s. If all of the enzyme is complexed with substrate, a maximum rate of reaction (i.e., the maximum reaction velocity or V_{max}) results and is given by the definition

$$V_{max} = k_P [E_t] \tag{8.23}$$

where k_p is the turnover number, since it represents the intrinsic rate at which the enzyme processes substrate molecule. Equation (8.22) becomes

$$v = \frac{V_{max}[S]}{K_s + [S]} \tag{8.24}$$

The rapid equilibrium approach is a special case of the pseudo-steady-state approach for reactions where product formation constitutes a clearly identifiable, irreversible, rate-limiting step.

The Pseudo-Steady-State Method Requires More Effort to Obtain the Hart Equation but Is Necessary for Reversible Reactions

The difference between a rapid equilibrium and pseudo-steady-state approximation is sometimes difficult to discern since the resulting equation for reaction sequence (8.18) is identical in form for both approaches. However, the pseudo-steady-state approach is able to handle reversible product-forming reactions. Reversible reactions more reasonably represent many reactions in cellular systems that form the basis of systems biology. For the reaction sequence (8.18), the rate and activity balances are the same. A rate equation is written for the enzyme–substrate complex [ES];

$$\frac{d[ES]}{dt} = k_1[E][S] - (k_{-1} + k_P)[ES] \tag{8.25}$$

A pseudosteady state is attained when the reaction intermediate ES builds up to a small concentration, which does not change during most of the timecourse of reaction (illustrated in Fig. 8.5). Hence

$$\frac{d[ES]}{dt} = 0 \tag{8.26}$$

and Eq. (8.25) becomes

$$[ES] = \frac{k_1}{k_{-1} + k_P}[E][S] \tag{8.27}$$

$d[ES]/dt$ is assumed to be zero because the rate of product formation is no longer an unambiguous rate-limiting step, and the accumulation of the enzyme–substrate complex approaches a constant; that is, the enzyme–substrate complex attains a pseudosteady state.

The rapid equilibrium approach, when applicable, is preferred since it provides a direct way of deriving a rate equation. In comparison, the rate equation obtained from the pseudo-steady-state approach is

$$v = \frac{k_P E_t[S]}{\dfrac{k_{-1} + k_P}{k_1} + [S]} \tag{8.28}$$

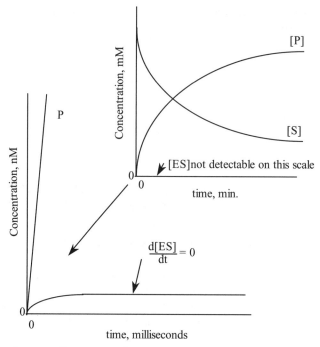

Figure 8.5.   A schematic illustration of pseudo-steady-state assumption.

and requires some thought in order to obtain the constants. In this case, where the reaction sequence is relatively straightforward, the identification of K_m is easily done:

$$K_m = \text{Michaelis constant} = \frac{k_{-1} + k_P}{k_1} \qquad (8.29)$$

In the special case where $k_P \ll k_{-1}$, that is, where the product formation step is limiting, $K_s = K_m = k_{-1}/k_1$. If it is not known that $k_P \ll k_{-1}$, K_m takes the place of K_s and Eq. (8.28) becomes

$$\upsilon = \frac{V_{max}[S]}{K_m + [S]} \qquad (8.30)$$

IRREVERSIBLE PRODUCT FORMATION IN THE PRESENCE OF INHIBITORS AND ACTIVATORS

Segel (1975) gives a compelling example of the utility of the rapid equilibrium approach for an enzyme whose activity is a complex function of both activators and inhibitors (8.31).

$$
\begin{array}{ccc}
EA + S \underset{\beta K_A}{\overset{\beta K_S}{\rightleftharpoons}} & ESA \underset{\beta K_A}{\overset{\beta k_P}{\longrightarrow}} EA + P & \\
K_A \Big\updownarrow & \Big\updownarrow & \\
\begin{array}{c} A \\ + \end{array} & \begin{array}{c} A \\ + \end{array} & \\
E \;+\; S \overset{K_S}{\rightleftharpoons} & ES \overset{k_P}{\longrightarrow} E + P & \quad (8.31)\\
\begin{array}{c} + \\ I \end{array} & \begin{array}{c} + \\ I \end{array} & \\
K_I \Big\updownarrow & \Big\updownarrow \alpha K_I & \\
EI \;+\; S \overset{\alpha K_S}{\rightleftharpoons} & ESI \underset{\alpha k_P}{\longrightarrow} EI + P &
\end{array}
$$

where S is substrate, I is an inhibitor, and A is a nonessential activator. The inhibitor is a molecule that binds with the enzyme such that the enzyme's effectiveness in binding with substrate is decreased but not blocked. The activator is a molecule that binds with the enzyme and enhances its catalytic activity. Since the enzyme displays activity even in the absence of the activator, A is not essential for enzymatic activity and is denoted as a nonessential activator.

The dissociation constants for the complexes [EA] and [ESA], [EI] and [ESI], and the association constants for [EA] and [E] are assumed to be related by the proportionality constants α and β respectively, for purpose of this illustration. The final product, P, is formed from all three complexes (ESA, ES, and ESI). Product formation occurs at a rate proportional to k_P, moderated by the coefficient α and β, as shown.

Derivation of the rate equation for the sequence in (8.41) is readily and rapidly obtained by following the procedure given by Segel (1975). A step-by-step procdure is

1. Write the activity balance:

$$[E_t] = [E] + [EA] + [ESA] + [ES] + [EI] + [ESI] \qquad (8.32)$$

   where E_t is the total enzyme in the system, either as free enzyme or as a complex.

2. Write the rate equation that represents all the pathways from which product is formed:

$$v = \frac{dP}{dt} = \beta k_P[ESA] + k_P[ES] + \alpha k_P[ESI] \qquad (8.33)$$

3. Divide the rate equation Eq. (8.33) by the activity balance (8.42):

$$v = \frac{(\beta[ESA] + [ES] + \alpha[ESI])k_P E_t}{[E] + [ES] + [EI] + [ESI] + [EA] + [ESA]} \qquad (8.34)$$

4. Express the concentration of each species in terms of [E] for the complexes ES, EA, ESA, EI, and ESI in terms of the following ratio:

$$[Complex] = \frac{(\text{product of concentration of ligands})}{(\text{product of dissociation constants between complex and free } E)}$$

For the reaction sequence of Eq. (8.31), equations for the five enzyme complexes are given by

$$[ES] = \frac{[E][S]}{K_S} \tag{8.35}$$

$$[ESA] = \frac{[ES][A]}{\beta K_A} = \frac{[E][S][A]}{\beta K_A K_S} \tag{8.36}$$

$$[EA] = \frac{[E][A]}{K_A} \tag{8.37}$$

$$[EI] = \frac{[E][I]}{K_I} \tag{8.38}$$

$$[ESI] = \frac{[ES][I]}{\alpha K_I} = \frac{[E][S][I]}{\alpha K_I K_S} \tag{8.39}$$

5. Substitute Eqs. (8.35)–(8.39) into Eq. (8.34):

$$v = \frac{\left(\beta \dfrac{[S][A]}{\beta K_S K_A} + \dfrac{[S]}{K_S} + \alpha \dfrac{[S][I]}{\alpha K_S K_A}\right) k_p E_t}{1 + \dfrac{[S]}{K_S} + \dfrac{[I]}{K_I} + \dfrac{[S][I]}{\alpha K_I K_S} + \dfrac{[A]}{K_A} + \dfrac{[S][A]}{\beta K_S K_A}} \tag{8.40}$$

6. Express v in terms V_{max}, K_S, and S, with $V_{max} = k_p E_t$:

$$v = \frac{V_{max}[S]}{K_S \dfrac{\left(1 + \dfrac{[I]}{K_I} + \dfrac{[A]}{K_A}\right)}{\left(1 + \dfrac{\beta[A]}{\beta K_A} + \dfrac{\alpha[I]}{\alpha K_I}\right)} + [S] \dfrac{\left(1 + \dfrac{[I]}{\alpha K_I} + \dfrac{[A]}{\beta K_A}\right)}{\left(1 + \dfrac{\beta[A]}{\beta K_A} + \dfrac{\alpha[I]}{\alpha K_I}\right)}} \tag{8.41}$$

Equation (8.41) is the Henri–Michaelis–Menten form of the rate equation.

INHIBITION

An *inhibitor* is a substance that reduces the velocity of an enzyme reaction. Inhibition occurs through competition of the inhibitor with the substrate for the enzyme's active site, or when the inhibitor binds at a site other than the enzyme's active site in a manner that causes a change in the enzyme's conformation so that its activity is reduced.

Competitive Inhibition

Competitive inhibition is caused by an inhibitor that combines with free enzyme in a manner that prevents substrate binding. Thus, the inhibitor is a compound that *competes* with the substrate for the same binding site in the enzyme. A reaction sequence describing competitive inhibition is given by

$$E + S \underset{k_1}{\overset{k_1}{\rightleftharpoons}} ES \overset{k_P}{\longrightarrow} E + P \qquad (8.42)$$

$$E + I \underset{k_{-2}}{\overset{k_2}{\rightleftharpoons}} EI \qquad (8.43)$$

where *I* represents the inhibitor. The enzyme–inhibitor complex (EI), is unable to bind with the enzyme-substrate complex (ES). Derivation of the rate equation is achieved by writing an activity balance

$$[E_t] = [E] + [ES] + [EI] \qquad (8.44)$$

and rate equation in terms of reaction velocity v

$$v = \frac{dP}{dt} = k_P[ES] \qquad (8.45)$$

In the case where k_P is not clearly the rate–limiting step, a pseudo-steady-state assumption is used for both the enzyme–substrate [ES] and enzyme–inhibitor [EI] complexes:

$$-\frac{d[ES]}{dt} = 0 = -k_1[E][S] + (k_{-1} + k_P)[ES] \qquad (8.46)$$

$$\frac{d[EI]}{dt} = 0 = -k_2[E][I] + k_{-2}[EI] \qquad (8.47)$$

Equations (8.44)–(8.47), when combined, give

$$v = \frac{V_{max}[S]}{K_m\left(1 + \dfrac{I}{K_I}\right) + [S]} \qquad (8.48)$$

where the Michaelis constants are defined by

$$K_m = \frac{k_{-1} + k_P}{k_1} \qquad (8.49)$$

$$K_I = \frac{k_{-2}}{k_2} \qquad (8.50)$$

V_{max} is defined by $k_P[E_t]$ as was the case previously. A linear form of Eq. (8.48) is obtained by inverting it:

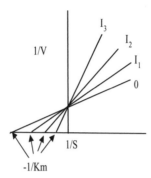

Figure 8.6.   Schematic diagram of competitive inhibition where $I_3 > I_2 > I_1$.

$$\frac{1}{v} = \frac{K_m\left(1 + \dfrac{I}{K_I}\right)}{V_{max}} \cdot \frac{1}{[S]} + \frac{1}{V_{max}} \tag{8.51}$$

Equation (8.51) corresponds to a Lineweaver–Burke form and gives the plot of the type shown in Fig. 8.1 and in the schematic representation of Fig. 8.6.

The slope in Eq. (8.51) is given by m:

$$m = \frac{K_m\left(1 + \dfrac{[I]}{K_I}\right)}{V_{max}} \tag{8.52}$$

On rearrangement, Eq. (8.52) gives

$$m = \frac{K_m}{(V_{max})(K_I)} \cdot [I] + \frac{K_m}{V_{max}} \tag{8.53}$$

Hence, there should be a linear relationship between the slope and the inhibitor concentration for an enzyme subject to competitive inhibition. Because a competitive inhibitor interferes with the substrate binding to the active enzyme, a higher concentration of substrate is required in order to displace the inhibitor at the binding site, thus the apparent K_m for the enzyme–substrate reaction is higher (x-axis intercept in Lineweaver–Burke plot, Fig. 8.6). Since a sufficiently high concentration of substrate will completely outcompete the inhibitor, the apparent maximum enzyme velocity (V_{max}) is unchanged and the y-axis intercept in the Lineweaver–Burke plot ($1/V_{max}$) is constant. Hence, there should be a linear relationship between the slope and the inhibitor concentration for an enzyme subject to competitive inhibition as shown in Fig. 8.7.

Uncompetitive Inhibition

Uncompetitive inhibition reflects an inhibitor that binds to an enzyme–substrate complex away from the active site of the enzyme. This is given by the following reaction sequence:

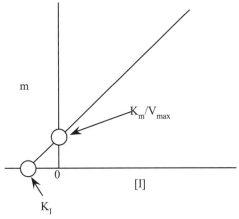

Figure 8.7. Schematic representation of the replot of the slope as a function of inhibitor concentration.

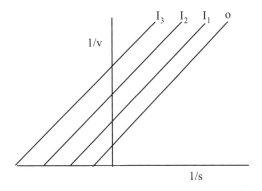

Figure 8.8. Schematic representation of uncompetitive inhibition for $I_3 > I_2 > I_1$.

$$E + \underset{k_{-1}}{\overset{k_1}{\rightleftharpoons}} ES \rightarrow E + P$$
$$ES + I \underset{k_{-2}}{\overset{k_2}{\rightleftharpoons}} ESI$$

$$(8.54)$$

The corresponding rate equation is

$$v = \frac{V_{max}[S]}{K_m + \left(1 + \dfrac{I}{K_I}\right)[S]}$$

$$(8.55)$$

Inverting Eq. (8.55) gives

$$\frac{1}{v} = \frac{K_m}{V_{max}} \cdot \frac{1}{[S]} + \frac{\left(1 + \dfrac{I}{K_I}\right)}{V_{max}}$$

$$(8.56)$$

The resulting Lineweaver–Burke plot is shown in Fig. 8.8.

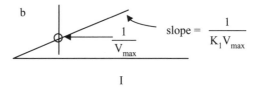

Figure 8.9. Schematic diagram of replot of inhibitor effect.

The affect of uncompetitive inhibition is to change both the apparent K_m and the apparent V_{max} for the enzyme. The intercept term of Eq. (8.56) is therefore a function of the inhibitor concentration

$$b = \frac{\left(1 + \dfrac{I}{K_I}\right)}{V_{max}} \tag{8.57}$$

which is

$$b = \frac{1}{K_I V_{max}} \cdot [I] + \frac{1}{V_{max}} \tag{8.58}$$

Figure 8.9 shows Eq. (8.58) is again linear, with a constant value of b.

(Classical) Noncompetitive Inhibition

The inhibitor and substrate bind randomly, reversibly, and independently at different sites in classical (simple) noncompetitive inhibition. The binding of the inhibitor has no effect on the binding of the substrate, or vice versa, but the resulting complex of enzyme–substrate–inhibitor [ESI] is inactive. This can be represented by the following reaction sequence:

$$
\begin{array}{ccccc}
E & + & S & \underset{}{\overset{K_S}{\rightleftharpoons}} ES & \overset{K_P}{\longrightarrow} E + P \\
+ & & & + & \\
I & & & I & \tag{8.59} \\
\parallel & & & \Big\updownarrow K_I & \\
EI & + & S & \overset{K_S}{\rightleftharpoons} ESI &
\end{array}
$$

Rapid equilibrium kinetics is chosen here since product formation is rate-limiting, and because the development of the rate equation is easier to carry out. The resulting expression is

$$v = \frac{V_{max}[S]}{K_S\left(1 + \dfrac{[I]}{K_I}\right) + \left(1 + \dfrac{[I]}{K_I}\right)[S]} \tag{8.60}$$

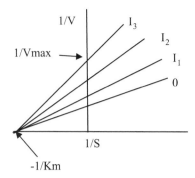

Figure 8.10. Schematic diagram showing pattern for noncompetitive inhibition where inhibitor concentrations follow the order $I_3 > I_2 > I$.

Inverting this equation, we obtain

$$\frac{1}{v} = \frac{K_S\left(1+\dfrac{[I]}{K_I}\right)}{V_{max}} \cdot \frac{1}{[S]} + \frac{\left(1+\dfrac{[I]}{K_I}\right)}{V_{max}} \tag{8.61}$$

The characteristic plot that would result from this mechanism is given by the pattern shown in Fig. 8.10.

Since the binding of the substrate to the enzyme is unaffected, the x-axis intercept ($-1/K_m$) is unchanged.

The binding of the inhibitor to the enzyme prevents the reaction from being catalyzed; thus the apparent V_{max} is reduced (y-axis intercept). An example might be given by the adsorption of an enzyme onto a solid particle in a manner that leaves the active site available but reduces efficiency of binding of substrate (in solution) with the catalytic site of the enzyme. If the value of K_m is independent of the inhibitor concentration (Plowman 1972; Segel, 1975), this type of inhibition may be formulated in terms of rapid equilibrium. If k_P were large relative to k_{-1}, then K_m would be as given by Eq. (8.62):

$$K_m = \frac{k_{-1} + k_P}{k_1} \cong \frac{k_{-1}}{k_1} \tag{8.62}$$

Otherwise the value of K_m would then be a function of an inhibitor. This is not reasonable. Hence k_p must be very small compared to k_{-1}, and K_m must be equivalent to K_s. "Classical noncompetitive inhibition in unireactant systems is obtained *only* under rapid equilibrium conditions" [from Segel, (1975), p. 130].

Substrate Inhibition

Substrate inhibition in a bioreactant, rapid-equilibrium, random system is represented by the sequence

$$\begin{array}{ccccccc} \text{BE} & \rightleftharpoons & \text{B} & + & \text{E} & + & \text{A} & \rightleftharpoons & \text{EA} \\ + & & & & + & & & & + \\ \text{B} & & & & \text{B} & & & & \text{B} \end{array}$$

$$\beta K_B \Updownarrow \qquad\qquad K_B \Updownarrow \qquad\qquad \alpha K_B \Updownarrow \qquad\qquad (8.63)$$

$$\begin{array}{ccccccc} \text{BEB} & \underset{\beta K_I}{\overset{}{\rightleftharpoons}} & \text{B} & + & \text{EB} & + & \text{A} & \underset{\alpha K_A}{\overset{}{\rightleftharpoons}} & \text{EAB} \dashrightarrow \text{E} + \text{P} + \text{Q} \end{array}$$

The balance on enzymatic activity is given by

$$[E_t] = [E] + [EA] + [EB] + [EAB] + [BE] + [BEB] \qquad (8.64)$$

with the rate equation of

$$v = k_P[\text{EAB}] \qquad (8.65)$$

The concentration of each species expressed in terms of [E] is given by

$$[\text{EAB}] = \frac{[E][A][B]}{\alpha K_B K_A} \qquad (8.66)$$

$$[\text{BEB}] = \frac{[E][B]^2}{\beta K_B K_I}$$

with the other quantities defined by the procedure described previously in this chapter. The velocity equation is

$$v = \frac{\dfrac{[A][B]}{\alpha K_B K_A} V_{max}}{1 + \dfrac{[A]}{K_A} + \dfrac{[B]}{K_B} + \dfrac{[A][B]}{\alpha K_A K_B} + \dfrac{[B]}{K_I} + \dfrac{[B]^2}{\beta K_I K_B}} \qquad (8.67)$$

If [A] $\gg$ [B], then [A] can be treated as a constant in an initial rate study where the concentration of B are varied. For this special situation rearrangement of Eq. (8.67) gives

$$v = \frac{[B]V_{max}}{\left(\alpha K_B \left(1 + \dfrac{K_A}{[A]}\right) + [B]\left(1 + \dfrac{\alpha K_A}{[A]} + \dfrac{\alpha K_A K_B}{K_I[A]} + \dfrac{\alpha K_A [B]}{\beta K_I[A]}\right) \right)} \qquad (8.68)$$

and when Eq. (8.68) is inverted, the resulting form is

$$\frac{1}{v} = \frac{\alpha K_B \left(1 + \dfrac{K_A}{[A]}\right)}{V_{max}} \times \frac{1}{[B]} + \frac{\left(1 + \dfrac{\alpha K_A}{[A]} + \dfrac{\alpha K_A K_B}{K_I[A]} + \dfrac{\alpha K_A [B]}{\beta K_I[A]}\right)}{V_{max}} \qquad (8.69)$$

This inverted form differs from the others since both the slope and intercept are a function of the substrate concentrations of [A] and [B]. Consequently the curve has a parabolic shape (Fig. 8.11).

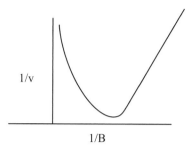

Figure 8.11. Schematic diagram of curve for substrate inhibition with respect to slope B.

where the substrate concentration that gives the highest rate is $[B_{min}]$:

$$[B_{min}] = \sqrt{\beta K_I K_B \left(1 + \frac{[A]}{K_A}\right)} \tag{8.70}$$

Now the velocity [Eq. (8.68)] can be rewritten as:

$$v = \frac{[B]V_{max}^*}{K_b^* + [B] + \dfrac{[B]^2}{K_I^*}} \tag{8.71}$$

where

$$V_{max}^* = \frac{V_{max}}{1 + \dfrac{\alpha K_A}{[A]}\left(1 + \dfrac{1}{\gamma}\right)} \tag{8.72}$$

$$K_B^* = \frac{\alpha K_B\left(1 + \dfrac{K_A}{[A]}\right)}{1 + \dfrac{\alpha K_A}{1}[A]\left(1 + \dfrac{1}{\gamma}\right)} \tag{8.73}$$

$$K_I^* = \beta K_I\left(1 + \dfrac{1}{\gamma} + \dfrac{[A]}{\alpha K_A}\right) \tag{8.74}$$

$$\gamma = \frac{K_I}{K_B} \tag{8.75}$$

inversion of (8.71) gives

$$\frac{1}{v} = \frac{K_b^*}{V_{max}^*}\cdot\frac{1}{[B]} + \frac{1}{V_{max}^*} + \frac{[B]}{K_I^* V_{max}^*} \tag{8.76}$$

where

$$\frac{K_B^*}{V_{max}^*} = \text{asymptote slope}$$

$$\frac{1}{V_{max}^*} = y \text{ intercept}$$

$$\frac{[B]}{K_I^* V_{max}^*} = \text{term that dramatically increase as } \frac{1}{[B]} \to 0$$

EXAMPLE OF REVERSIBLE REACTIONS

The previous examples of classical enzyme reaction kinetics with inhibition assume that the enzyme-catalyzed reaction is irreversible. Several industrially useful enzymes as well as a great many enzymes important in cellular metabolism catalyze reversible reactions. One industrially important enzyme, glucose isomerase, catalyzes a reversible reaction that interconverts glucose and fructose for the production of high fructose.

Corn syrup or HFCS (discussed in Chapter 3) contains 55% fructose and 45% glucose and is widely used in soft (carbonated) drinks, confectionary products, and baked goods. HFCS is made from dextrose (glucose) obtained from the hydrolysis of cornstarch.

The isomerization of glucose to fructose to obtain a 40% fructose product is the first step in producing HFCS. Subsequent chromatography steps enrich the fructose/glucose mixture to obtain 55% fructose. The pseudo-steady-state approach, developed by Briggs and Haldane (1925), is based on the steady-state level (of reaction intermediate) concept of Bodenstein (1913), according to Segel (1975), and applies to the formulation of kinetic expressions for reversible reactions:

$$E + S \underset{k_{-1}}{\overset{k_1}{\rightleftharpoons}} ES \underset{k_{-2}}{\overset{k_2}{\rightleftharpoons}} E + P \tag{8.77}$$

The intermediate, ES, is assumed to attain a pseudo-steady-state concentration shortly after the reaction starts. The derivation of the rate equation follows the same steps as outlined earlier in this chapter for pseudo-steady-state kinetics.

$$\text{Mass balance on enzyme:} \quad [E_t] = [E] + [ES] \tag{8.78}$$

$$\text{Rate equation for formation of P:} \quad v = k_2[ES] - k_{-2}[E][P] \tag{8.79}$$

The change of [ES] with time is

$$-\frac{d[ES]}{dt} = (k_{-1} + k_2)[ES] - k_1[E][S] - k_{-2}[E][P] \tag{8.80}$$

At pseudo-steady state ($-d[ES]/dt = 0$), the expression for [ES] is

$$[ES] = \frac{(k_1[S] + k_{-2}[P])[E]}{(k_{-1} + k_2)} \tag{8.81}$$

The subsequent steps for this derivation follow the same sequence as described previously for pseudo-steady-state conditions. Divide Eq. (8.79) by (8.78), substitute Eq. (8.81) for [ES], multiply out the terms, and rearrange:

$$v = \frac{\left(\dfrac{k_2(k_1[S] + k_{-2}[P])[E]}{(k_{-1} + k_2)} - k_{-2}[E][P]\right)E_t}{[E] + \dfrac{(k_1[S] + k_{-2}[P])[E]}{(k_{-1} + k_2)}} \tag{8.82}$$

$$= \frac{\left(\dfrac{(k_2 k_1[S] + k_2 k_{-2}[P] - k_{-2}k_{-1}[P] - k_{-2}k_2[P])}{k_{-1} + k_2}\right)E_t}{1 + \dfrac{[S]}{\dfrac{(k_{-1} + k_2)}{k_1}} + \dfrac{[P]}{\dfrac{(k_{-1} + k_2)}{k_{-2}}}} \tag{8.83}$$

$$= \frac{\dfrac{k_2 E_t\, S}{\left(\dfrac{k_{-1} + k_2}{k_1}\right)} - \dfrac{k_{-1}E_t\, P}{\left(\dfrac{k_{-1} + k_2}{k_{-2}}\right)}}{\left(1 + \dfrac{S}{\dfrac{k_{-1} + k_2}{k_1}} + \dfrac{P}{\dfrac{(k_{-1} + k_2)}{k_{-2}}}\right)} \tag{8.84}$$

Further simplification of Eq. (8.84) is accomplished by recognizing the groups that represent Michaelis constants and V_{max}:

$$K_{m,s} = \frac{k_{-1} + k_2}{k_1} \quad \text{(forward direction),} \quad \text{i.e.,} \quad E + S \underset{k_{-1}}{\overset{k_1}{\rightleftharpoons}} ES \xrightarrow{k_2} E + P \tag{8.85}$$

$$K_{m,p} = \frac{k_{-1} + k_2}{k_{-2}} \quad \text{(reverse direction),} \quad \text{i.e.,} \quad E + P \underset{k_{-2}}{\overset{k_2}{\rightleftharpoons}} ES \xrightarrow{k_{-1}} E + S \tag{8.86}$$

The maximum reaction velocities are given by

$$V_{max,f} = k_2[E_t] \quad \text{(for the forward reaction)} \tag{8.87}$$

and

$$V_{\mathrm{max,b}} = k_{-1}[\mathrm{E_t}] \quad \text{(for the backward reaction)} \tag{8.88}$$

Since $k_{-1}\mathrm{E}_t$ becomes k_{-1} [ES] if all the active sites are bound with substrate, Eq. (8.84) becomes

$$v = \frac{\dfrac{V_{\mathrm{max,f}}}{K_{\mathrm{m,s}}}[\mathrm{S}] - \dfrac{V_{\mathrm{max,b}}}{K_{\mathrm{m,p}}}[\mathrm{P}]}{1 + \dfrac{[\mathrm{S}]}{K_{\mathrm{m,s}}} + \dfrac{[\mathrm{P}]}{K_{\mathrm{m,p}}}} \tag{8.89}$$

The kinetic constants k_1 and k_{-2} are second-order rate constants, with dimensions of $\mathrm{M^{-1}\,min^{-1}}$, while k_{-2} and k_2 are first-order rate constants, with dimensions of inverse time. This is consistent with the definition of $V_{\mathrm{max,f}}$ and $\mathrm{E_t}$:

$$k_{-2} = \frac{V_{\mathrm{max,f}}}{[\mathrm{E_t}]} = \frac{[\mathrm{M/min}]}{\mathrm{M}} = \frac{1}{\mathrm{min}} \tag{8.90}$$

$V_{\mathrm{max,f}}$ is a pseudo-zero-order constant:

$$V_{\mathrm{max,f}} = k_2[\mathrm{E_t}] = \mathrm{min^{-1} \cdot M} \tag{8.91}$$

at fixed enzyme concentration. The constant k_2 represents the maximum reaction velocity per mole of enzyme, that is, the turnover number.

The constant K_{m} is a zero-order constant and has dimensions of concentration:

$$K_{\mathrm{m,s}} = \frac{k_{-1} + k_2}{k_1} = \left(\frac{\mathrm{min^{-1} + min^{-1}}}{|\mathrm{M^{-1}}||\mathrm{min^{-1}}|} = \mathrm{M} \right) \tag{8.92}$$

In an initial rate study, the product [P] concentration can be zero at $t = 0$. Hence, Eq. (8.89) becomes

$$\therefore v = \frac{V_{\mathrm{max,f}}[\mathrm{S}]}{K_{\mathrm{m,s}} + [\mathrm{S}]} \tag{8.93}$$

If the value of $K_{\mathrm{m,s}}$ is the same as the substrate concentration (i.e., $K_{\mathrm{m,s}} = [\mathrm{S}]$), Eq. (8.93) becomes

$$v = \frac{V_{\mathrm{max,f}} K_{\mathrm{m,s}}}{K_{\mathrm{m,s}} + K_{\mathrm{m,c}}} = \frac{V_{\mathrm{max,f}}}{2} \tag{8.94}$$

K_{m} represents the substrate concentration at which the reaction velocity is half-maximal. K_{m} has significance in that it can indicate an approximate concentration of substrate inside a cell for enzymes that catalyze pathways addressed by systems biology. Approximate magnitudes of kinetic constants for this type of enzyme reaction are in the ranges

$$k_1 \quad 10^7\text{--}10^{10}\,\mathrm{M\cdot min^{-1}}$$

$$k_{-1} \quad 10^2\text{--}10^6\,\mathrm{min^{-1}} \tag{8.95}$$

$$k_2 \quad 50\text{--}10^7\,\mathrm{min^{-1}}$$

$$K_m \text{ typically } \quad 10^{-6}\text{--}10^{-2}\,\mathrm{M} \tag{8.96}$$

At $k_1 \cong 10^{11}$, the rate of diffusion of a small molecule in aqueous solution would become the limiting step. The production of an equilibrium mixture of glucose and fructose using glucose isomerase is atypical of many reversible enzyme reactions because the K_m values are so large (500 mM). The concentrations of the substrates and products within a living cell are 100–10,000 times lower. Nonetheless, the method by which a reversible reaction is formulated into a usable kinetic expression is the same as presented here.

Coenzymes and Cofactors Interact in a Reversible Manner

The reversible reaction for formation of fructose from glucose is unique because it represents industrial adaptation of an enzyme with a different physiological function, e.g. isomerization of xylose to xylulose. Xylose–xylulose interconversion is also important since it is the first step in one metabolic pathway for production of ethanol from xylose, a five-carbon sugar derived from plant biomass. The values of the dissociation constants ($K_{m,s}$) for glucose or fructose are large, and affinity of the enzyme for its substrate is small.

Another reversible reaction of practical interest is the reduction or phosphorylation of small molecules in in vivo pathways. In this case, a molecule known as a *coenzyme* is coupled to other reactions in the cell that facilitate its recycle. The dissociation constants are much smaller, although directionality of the reaction is still a function of the concentration of the coenzyme such as NAD^+ and NADH and ADP. Indeed, many enzymes require "a specific, heat stable, low molecular weight organic molecule, a coenzyme" to attain full activity (Rodwell 1990). The coenzyme, when bound with the apoenzyme (the protein part), forms a holoenzyme that represents the complete catalytic entity.

Reactions that require coenzymes coincide with IUPAC classes 1, 2, 5, and 6 (Table 7.2). A coenzyme may be regarded as a second substrate, with cellular metabolism often depending more on the coenzyme than the substrate. For example, the reaction

$$\tag{8.97}$$

requires NAD^+ to proceed. More important is the reverse reaction where pyruvate goes to lactate, and $NADH^+$ is oxidized to NAD^+. NAD^+ is required for glycolysis

Table 8.1.

Examples of Molecules Utilized in Anaerobic Regeneration of NAD$^+$

Oxidant	Reduced Product	Life Form
Pyruvate	Lactate	Muscle, lactic bacteria
Acetaldehyde	E*t*hanol	Yeast
Dehydroxyacetone phosphate	β-Glycerophosphate	*Escherichia coli*
Fructose	Mannitol	Heterolactic bacteria

Source: Reproduced with permission from Rodwell (1990), Appleton & Lange.

and anaerobic ATP synthesis such as occurs in ethanol fermentation. In microorganisms, the reduction of pyruvate results in metabolites that later serve as oxidants for NADH and are themselves reduced. Systems biology begins to address these interrelated pathways and the cascading effect that they may have on cellular function.

Table 8.1 gives examples of oxidants that result in anaerobic regeneration of NAD$^+$. Coenzymes are a part of this process, and serve as group transfer reagents. For example, in a transamination reaction, the coenzyme as an intermediate group carrier for the group G:

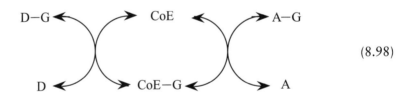

$$(8.98)$$

If the group transferred in reaction (8.95) is hydrogen [denoted by H in Eq. (8.99)], the coenzyme serves to transfer H to reduce molecule A and oxidize molecule D, as is the case for oxidation of acetaldehyde to ethanol:

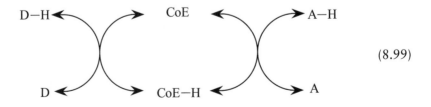

$$(8.99)$$

These coenzymes are required to balance redox reactions or to shift the thermodynamics to favor the reaction (when $\Delta G < 0$), as discussed in Chapter 9.

KING–ALTMAN METHOD

While individual reactions, or possibly sequences of two coupled reactions, represent current industrial processes, more recent developments in in vitro protein synthesis as well as systematic modeling of cellular metabolism for either academic or practical purposes requires a more comprehensive methodology for multistep, multireactant pathways.

If a system of multiple enzymes in a living cell are being modeled, the exact reversible nature of the reaction is less important than a single-step reaction since there is a constant flow of material into and out of a reaction sequence. The reactions are not at equilibrium or at an endpoint, as would be the case for an industrial reaction (Schoemaker et al. 2003; Brooks 2004). As pointed out by Brooks (2004), "all the enzymes of metabolism exist in pathways—enzymes do not function in isolation in the cell." They interact by sharing products and substrates.

Another characteristic of an enzyme reaction in a cell compared to reactions in a test tube or reactor is that they operate at close to steady state, where products and intermediates are at a nearly constant concentration, even though each individual enzyme reaction is not at equilibrium. As clearly stated by Brooks (2004), "there is a constant flow of substrate into, and product out of, a metabolic pathway, and the concentrations of the intermediates of the pathway do not change appreciably." This observation enables mathematical analysis of the flux of intermediates through a specified linear pathway in the cell. The equation that is central to this approach is Eq. (8.89), which describes the net velocity of a reversible reaction. In the case of a multistep reaction, for example

$$\underset{\text{reaction step}}{S} \underset{1}{\rightleftharpoons} P \underset{2}{\rightleftharpoons} Q \underset{3}{\rightleftharpoons} R \underset{4}{\rightleftharpoons} T \tag{8.100}$$

Equation (8.89) would apply to each successive reaction step. The first step would be represented by Eq. (8.89), and the next two steps by Eqs. (8.101) and (8.102):

$$\text{Reaction step 2: } v_2 = \frac{\dfrac{V_{\text{max,f,2}}}{K_{\text{m,P,2}}}[P] - \dfrac{V_{\text{max,b,2}}}{K_{\text{m,Q,2}}}Q}{1 + \dfrac{[P]}{K_{\text{m,P,2}}} + \dfrac{[Q]}{K_{\text{m,Q,2}}}} \tag{8.101}$$

$$\text{Reaction step 3: } v_3 = \frac{\dfrac{V_{\text{max,f,3}}}{K_{\text{m,Q,3}}}[Q] - \dfrac{V_{\text{max,b,3}}}{K_{\text{m,R,3}}}[R]}{1 + \dfrac{Q}{K_{\text{m,Q,3}}} + \dfrac{R}{K_{\text{m,R,3}}}} \tag{8.102}$$

An analogous equation may be written for step 4 [in reaction (8.100)]. The nomenclature is unwieldy, but is needed to distinguish the Michaelis constants and maximum velocities for each reaction step. If the goal were to determine equilibrium concentrations of S and T in vitro, the equilibrium reaction would simply be expressed as

$$E + S \rightleftharpoons E + T \tag{8.103}$$

with the intermediates ignored and the enzyme constants combined into a single equation of the type shown in Eq. (8.89). The parameter V_{max} derived from Eq. (8.103) denotes a combined pseudoenzymatic activity that represents the action of all four enzymes.

If the goal of the kinetic expression is to model metabolic pathway that represents steady-state flux through the system, the steps are accounted for individually. Brooks (2004) uses the nomenclature

$$J_{ss} = v_1 = v_2 = v_3 = v_4 \tag{8.104}$$

where the subscripts correspond to the four reaction steps represented in Eq. (8.100). When the substrate/product pairs are replaced by the nomenclature of Brooks, Eq. (8.103) results, where the concentration of substrate or product that is *not* bound to the enzyme is given by

$$I_{i,free} = \alpha_i [I_{(i+1),free}] + \beta_i \tag{8.105}$$

where

$$\alpha_i = \frac{\bar{k}_i^+ (V_i^- + J_{ss})}{\bar{k}_i^- (V_i^+ + J_{ss})} \tag{8.106}$$

$$\beta_i = \frac{J_{ss}\bar{k}_i^+}{(V_i^+ - J_{ss})} \tag{8.107}$$

Table 8.2 shows how the nomenclature used in Eq. (8.103)–(8.107) is related to an equation obtained for the reaction sequence given by Eq. (8.100). When Eq. (8.105) is used with measured values for multiple enzymes that make up the pathway, a model of steady-state fluxes may be constructed and used to determine responses of linear sequences of enzyme reactions (i.e., sequences that do not have branch

Table 8.2.

Correspondence Between Parameters Used in Eqs. (8.100)–(8.101) and (8.105)–(8.107)

	Molecular Species		Reaction			
Reaction Step	In	Out	$\bar{K}_i^+$	$\bar{K}_i^-$	V_i^+	V_i^-
$i = 1$	S	P	$K_{m,s,1}$	$K_{m,P,1}$	$V_{max,f,1}$	$V_{max,b,1}$
$i = 2$	P	Q	$K_{m,P,2}$	$K_{m,Q,2}$	$V_{max,f,2}$	$V_{max,b,2}$
$i = 3$	Q	R	$K_{m,Q,3}$	$K_{m,R,3}$	$V_{max,f,3}$	$V_{max,b,3}$
$i = 4$	R	T	$K_{m,R,4}$	$K_{m,T,4}$	$V_{max,f,4}$	$V_{max,b,4}$

points) to change in fluxes of entering substrates, or changes in intermediate levels due to change in enzyme properties. This concept may be extended to branched pathways as well.

This brief overview introduces how enzyme pathways relate to higher-level cell responses. This might include the change in metabolism when a cell suddenly experiences a large increase in glucose (important in the study of diabetes) or the control of ATP concentration in an aerated fermentation. The change in ATP was used as the basis of the aerobic bioreactor example in Chapter 6. The introduction of flux analysis would enable a finer level of detail and therefore possibly enable better control or prediction of the cell's response to oxygen limitation. A possible practical result would be the enhancement of the output of the desired end product, such as 2,3-butanediol.

The approach described in this section differs from traditional enzymology and enzyme kinetics in that cell response to perturbations may be used to measure key elements of flux analysis, without the need to isolate, purify, characterize, and determine kinetics for each individual enzyme. The systems approach to enzyme kinetics, now referred to as *systems biology*, differentiates traditional enzymology from the new practice of discerning cellular function in terms of systems of enzymes. This new approach would not be possible without the fundamental scientific basis provided from 100 years of basic enzymology and enzyme kinetics. At the same time, metabolic flux analysis enables a more holistic understanding of living systems based on the biocatalysts (enzymes) from which they are constructed.

The analysis of multisubstrate enzymes that synthesize complex molecules from less complex constituents also involves a kinetic approach, and may require more than flux analysis to understand cell response to a change in reactant levels. This is especially true for cases where a specific enzyme with a complex mechanism has been identified. In this case, a symbolic method known as the King–Altman method can be used to aid in the derivation of the kinetic equation, and is widely used in the biochemical literature. The method determines the number of patterns that can be obtained from a reaction sequence. Allowable interconversion patterns are specified and these patterns are then used and to write kinetic equations for reaction sequences. This method has been described in clear and significant detail by Segel (1975) and is repeated here to introduce the reader to the basic method. For example, consider the following reaction sequence:

$$E + A \underset{k_{-1}}{\overset{k_1}{\rightleftharpoons}} EA \tag{8.108}$$

$$EA + B \underset{k_{-2}}{\overset{k_2}{\rightleftharpoons}} EAB–EPQ \tag{8.109}$$

$$EAB–EPQ \underset{k_{-3}}{\overset{k_3}{\rightleftharpoons}} EQ + P \tag{8.110}$$

$$EQ \underset{k_{-4}}{\overset{k_4}{\rightleftharpoons}} E + Q \tag{8.111}$$

This sequence of reaction steps can also be represented in diagrammatic form as

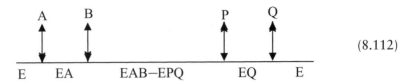

$$(8.112)$$

where the arrows indicate reversible steps. This is a representation for an ordered, sequential B_i, B_i mechanism that is alternatively shown in Eq. (8.111):

$$(8.113)$$

where $E \xrightarrow{k_1[A]} EA$ is an alternate way of expressing $E + [A] \xrightarrow{k_1} EA$, and $EA \xrightarrow{k_2[B]} EAB$ represents $EA + [B] \xrightarrow{k_2} EAB$, and so forth.

The first step in setting up the kinetic expression is to determine the number of patterns P with $(n-1)$ lines, as calculated by

$$P = \frac{m!}{(n-1)!(m-n+1)!} \tag{8.114}$$

where n is the number of enzyme species [the number of corners in the figure represented by Eq. (8.113)] and m is the number of lines. In the case of the reaction sequence of Eqs. (8.108)–(8.111), which resulted in the representation of Eq. (8.113), there are four corners ($n = 4$), and four lines connecting the corners. Therefore

$$P = \frac{4!}{(4-1)!(4-4+1)!} = \frac{4 \cdot 3 \cdot 2 \cdot 1}{(3 \cdot 2 \cdot 1)(1)!} = 4 \tag{8.115}$$

The allowable interconversion patterns are then determined by drawing out the pathways, which, when combined, give the diamond pattern of Eqn. (8.113). The patterns are

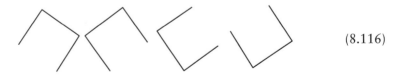

$$(8.116)$$

An activity balance for all of the enzyme species and enzyme–substrate/product complexes is given by the expression for E_T.

$$E_T = E + EA + [EAB–EPQ] + EQ \tag{8.117}$$

The corresponding expressions that describe the pathways, and therefore kinetic sequences for the enzyme E, are shown visually by sequence (8.118).

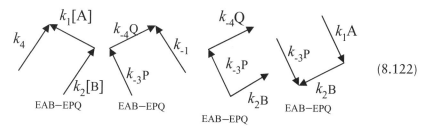

(8.118)

where

$$\frac{E}{E_T} = \frac{k_4 k_{-2} k_{-1} + k_{-1} k_3 k_4 + k_2 k_3 k_4 [B] + k_{-1} k_{-2} k_{-3} [P]}{E + EA + [EAB\text{–}EPQ] + EQ}$$

(8.119)

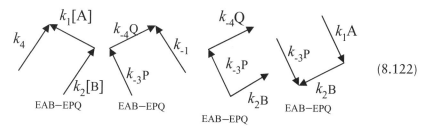

(8.120)

$$\frac{EA}{E_T} = \frac{k_1 k_{-2} k_4 [A] + k_1 k_3 k_4 [A] + k_{-2} k_{-3} k_{-4} [PQ] + k_1 k_{-2} k_{-3} [A][P]}{E + EA + [EAB\text{–}EPQ] + EQ}$$

(8.121)

For intermediate EAB–EPQ, the patterns and associated expression is given by

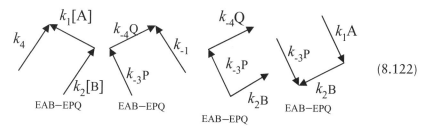

(8.122)

$$\frac{EAB + EPQ}{E_t} = \frac{k_1 k_2 k_4 [A] + k_1 k_2 k_3 [A][B][P] + k_{-4} k_{-3} k_2 [B][P][Q] + k_{-1} k_{-3} k_{-4} [P][Q]}{E + EA + [EAB\text{–}EPQ] + EQ}$$

(8.123)

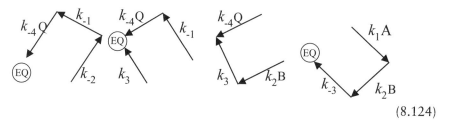

(8.124)

Finally, for intermediate EQ, the expression is

$$\frac{EQ}{E_T} = \frac{(k_{-1}k_{-2}k_{-4}[Q] + k_{-1}k_3k_{-4}[Q] + k_2k_3k_{-4}[B][Q] + k_1k_2k_3[A][B])}{E + EA + [EAB\text{--}EPQ] + EQ} \quad (8.125)$$

The denominator is given by the combination of all the numerator terms, as represented for E_T in Eq. (8.117) or by substituting in the expressions for E, EA, EAB–EPQ, and EQ. Equation (8.126) results:

$$
\begin{aligned}
E_T = \; & k_{-1}k_4(k_{-2}+k_3) \\
& + k_1k_4(k_{-2}+k_3)[A] \\
& + (k_2k_3k_4)[B] \\
& + (k_{-1}k_{-2}k_{-3})[P] \\
& + k_{-1}k_{-4}(k_{-2}+k_3)[Q] \\
& + k_1k_2(k_3+k_4)[A][B] \\
& + k_{-3}k_{-4}(k_{-1}+k_{-2})[P][Q] \\
& + k_2k_3k_{-4}[B][Q] \\
& + k_1k_{-2}k_{-3}[A][P] \\
& + k_1k_2k_{-3}[A][B][P] \\
& + k_2k_{-3}k_{-4}[B][P][Q]
\end{aligned}
\quad (8.126)
$$

The activity balance for enzyme is now expressed in terms of the substrates, products, and kinetic constants. The rate equation for formation of EA is

$$v = k_1[E][A] - k_{-1}[EA] \quad (8.127)$$

Divide the rate equation [Eq. (8.127)] by the activity balance Eq. [(8.126)] to obtain

$$\frac{v}{E_T} = k_1\left(\frac{E}{E_T}\right)[A] + k_{-1}\left(\frac{EA}{E_T}\right) \quad (8.128)$$

Substitute the complete expressions (8.119) and (8.121) into (8.128), and manipulate to obtain the rate expression:

$$
v = \frac{\text{num}_1[A][B] - \text{num}_2[P][Q]}{
\begin{aligned}
&(\text{const} \qquad\qquad + (\text{Coef AB})[A][B] \\
&+ (\text{Coef A})[A] \quad + (\text{Coef PQ})[P][Q] \\
&+ (\text{Coef B})[B] \quad + (\text{Coef BQ})[B][Q] \\
&+ (\text{Coef P})[P] \quad + (\text{Coef AP})[A][P] \\
&+ (\text{Coef Q})[Q] \quad + (\text{Coef ABP})[A][B][P] \\
&\quad + (\text{Coef} \qquad \text{BPQ})[B][P][Q]
\end{aligned}}
\quad (8.129)
$$

The next step is to convert the velocity equation from the coefficient form to one that contains appropriately defined kinetic constants. Cleland's rules, summarized by Segel (1975), are used. Equation (8.129) is multiplied by the factor

$$\text{Factor} = \frac{\text{num}_2}{(\text{Coef AB})(\text{Coef PQ})} \tag{8.130}$$

where the factor is composed of "num₂ divided by the product of the two denominator coefficients which contain as subscripts all the substrates for the reaction in each direction" (i.e., the two denominator coefficients used to define $V_{\text{max,f}}$ and $V_{\text{max,f}}$). The subscripts f and r denote the forward and reverse reactions. Once the factor is defined, Eq. (8.130) is multiplied by the common term:

$$\frac{\dfrac{\text{num}_2}{(\text{Coef AB})(\text{Coef PQ})}}{\dfrac{\text{num}_2}{(\text{Coef AB})(\text{Coef PQ})}} \tag{8.131}$$

This results in the following expression:

$$v = \frac{\dfrac{(\text{num}_1)(\text{num}_2)[A][B]}{(\text{Coef AB})(\text{Coef PQ})} - \dfrac{(\text{num}_2)(\text{num}_2)[P][Q]}{(\text{Coef AB})(\text{Coef PQ})}}{\begin{pmatrix}\dfrac{(\text{const})(\text{num}_2)}{(\text{Coef AB})(\text{Coef PQ})} + \dfrac{\text{num}_2}{(\text{Coef AB})}[A][B]\end{pmatrix}} \tag{8.132}$$
$$+ \frac{(\text{num}_2)(\text{Coef A})}{(\text{Coef AB})(\text{Coef PQ})}[A] + \frac{\text{num}_2}{(\text{Coef AB})}[P][Q]$$
$$+ \frac{(\text{Coef B})(\text{num}_2)[B]}{(\text{Coef AB})(\text{Coef PQ})} + \frac{(\text{Coef BQ})(\text{num}_2)[B][Q]}{(\text{Coef AB})(\text{Coef PQ})}$$
$$+ \frac{(\text{Coef P})(\text{num}_2)[P]}{(\text{Coef AB})(\text{Coef PQ})} + \frac{(\text{Coef AB})(\text{num}_2)}{(\text{Coef AB})(\text{Coef PQ})}[A][P]$$
$$+ \frac{(\text{Coef Q})(\text{num}_2)[Q]}{(\text{Coef AB})(\text{Coef PQ})} + \frac{(\text{Coef ABP})(\text{num}_2)}{(\text{Coef AB})(\text{Coef PQ})}[A][B][P]$$
$$+ \frac{(\text{Coef BPQ})(\text{num}_2)}{(\text{Coef AB})(\text{Coef PQ})}[B][P][Q]$$

The V_{max} in either direction will equal "the numerator coefficient for that direction [either num₁ or num₂] divided by the coefficient of the denominator term which contains the product of concentrations of all substrates for the reaction in that direction":

$$\frac{(\text{num}_2)}{(\text{Coef PQ})} = V_{\text{max,r}} \tag{8.133}$$

$$\frac{(\text{num}_1)(\text{num}_2)}{(\text{Coef AB})(\text{Coef PQ})} = \frac{\text{num}_1}{\text{Coef AB}} \times \frac{\text{num}_2}{\text{Coef PQ}} = V_{\text{max,f}} V_{\text{max,r}} \tag{8.134}$$

By these rules the Michaelis constants are obtained by finding ratios of two denominator coefficients where the denominator of the ratio is the same as that used in defining the V_{max} for that direction, while the numerator of the ratio is chosen so that subscript letters of the numerator and denominator terms cancel to leave only the letter that corresponds to the substrate or product in question. The equilibrium constant is defined so that it is always the ratio of num_1/num_2. Examples of the applications of these rules are given below:

$$\frac{(num_2)(Coef\ A)}{(Coef\ AB)(Coef\ PQ)} = \frac{Coef\ A}{Coef\ AB} \times \frac{num_2}{Coef\ PQ} = K_{m,B}V_{max,r} \tag{8.135}$$

$$\frac{(num_2)}{(Coef\ AB)} = \frac{(num_1)}{(Coef\ AB)} \times \frac{(num_2)}{(num_1)} = V_{max,f} \times \frac{1}{K_{eq}} \tag{8.136}$$

$$\frac{(Coef\ B)}{(Coef\ AB)} \times \frac{num_2}{(Coef\ PQ)} = K_{m,A}V_{max,r} \tag{8.137}$$

$$\frac{(Coef\ BQ)}{(Coef\ AB)} \times \frac{num_2}{(Coef\ PQ)} = \frac{(Coef\ BQ)}{(Coef\ B)/K_{m,A}} \times V_{max,r}$$
$$= \frac{K_{m,A}}{K_{i,Q}} \times V_{max,r} \tag{8.138}$$

Since

$$K_{m,A} = \frac{(Coef\ B)}{(Coef\ AB)} \tag{8.139}$$

$$\frac{(Coef\ P)(num_2)}{(Coef\ AB)(Coef\ PQ)} = \frac{(num_1)}{(Coef\ AB)} \times \frac{(Coef\ P)}{(Coef\ PQ)} \times \frac{(num_2)}{(num_1)} \tag{8.140}$$
$$= V_{max,f}K_{i,Q} \times \frac{1}{K_{eq}}$$

$$\frac{(Coef\ AP)(num_2)}{(Coef\ AB)(Coef\ PQ)} = \frac{(Coef\ AP)}{(Coef\ A)/K_{m,B}} \times V_{max,r} \tag{8.141}$$
$$= \frac{K_{m,B}}{K_{i,Q}} \times V_{max,r}$$

Since

$$K_{m,B} = \frac{(Coef\ A)}{(Coef\ AB)}$$

$$\frac{(Coef\ Q)(num_2)}{(Coef\ AB)(Coef\ PQ)} = \frac{(num_1)}{(Coef\ AB)} \times \frac{(Coef\ Q)}{(Coef\ PQ)} \times \frac{(num_2)}{(num_1)} \tag{8.142}$$
$$= V_{max,f}K_{i,P} \times \frac{1}{K_{eq}}$$

Isoinhibition constants are defined for cases in which a stable enzyme form isomerizes in a step that is partially rate-limiting. These are identified by finding ratios of coefficients containing subscripts from V_{max} to that containing the given ligand as well as that associated with V_{max}:

$$\frac{(\text{Coef } ABP)(\text{num}_2)}{(\text{Coef } AB)(\text{Coef } PQ)} = \frac{(\text{Coef } ABP)}{(\text{Coef } AB)} \times \frac{(\text{num}_2)}{(\text{Coef } PQ)} \tag{8.143}$$

$$= \frac{1}{K_{ii,P} \times V_{max,r}}$$

$$\frac{(\text{Coef } BPQ)}{(\text{Coef } PQ)} \times \frac{(\text{num}_2)}{(\text{num}_1)} \times \frac{(\text{num}_1)}{(\text{Coef } AB)} = \frac{1}{K_{ii,B}} \times \frac{1}{K_{eq}} \times V_{max,f} \tag{8.144}$$

Other terms are

$$\frac{(\text{num}_2)(\text{num}_2)}{(\text{Coef } AB)(\text{Coef } PQ)} = \frac{(\text{num}_1)}{(\text{Coef } AB)} \times \frac{(\text{num}_2)}{(\text{Coef } PQ)} \times \frac{(\text{num}_2)}{(\text{num}_1)} \tag{8.145}$$

$$= V_{max,f} \times V_{max,r} \times \frac{1}{K_{eq}}$$

$$\frac{(\text{const})(\text{num}_2)}{(\text{Coef } AB)(\text{Coef } PQ)} = \frac{(\text{const})}{(\text{Coef } A)/K_{m,B}} \times \frac{(\text{num}_2)}{(\text{Coef } PQ)} \tag{8.146}$$

$$= K_{i,A} \times K_{m,B} \times V_{max,r}$$

where there may be two suitable definitions for some K_i values. For bioreactant mechanisms there are two terms. But for mechanisms that are trireactant or greater, only one definition is workable (Segel 1975). Putting all this together into the rate equation gives

$$v = \frac{V_{max,f} V_{max,r} [A][B] - \dfrac{V_{max,f} V_{max,r}}{K_{eq}} [P][Q]}{\begin{aligned} & K_{i,A} K_{m,B} V_{max,r} + V_{max,r} [A][B] \\ & + K_{m,B} V_{max,r} [A] + \frac{V_{max,f}}{K_{eq}} [P][Q] \\ & + K_{m,A} V_{max,r} [B] + \frac{K_{m,A} V_{max,r}}{K_{i,Q}} [B][Q] \\ & + \frac{K_{i,Q} V_{max,f}}{K_{eq}} [P] + \frac{K_{m,B} V_{max,r}}{K_{i,Q}} [A][P] \\ & + \frac{K_{i,P} V_{max,f}}{K_{eq}} [Q] + \frac{V_{max,r}}{K_{ii,P}} [A][B][P] \\ & + \frac{V_{max,f}}{K_{ii,B} K_{eq}} [B][P][Q] \end{aligned}} \tag{8.147}$$

Determination of the constants can be achieved through initial rate kinetics. For example, for $[P] = [Q] = 0$, Eq. (8.147) becomes

$$\frac{v}{V_{max,f}} = \frac{[A][B]}{K_{i,D,r}K_{m,B} + K_{m,B}[A] + K_{m,A}[B] + [A][B]} \qquad (8.148)$$

If the concentration of B is kept constant, inverting this equation gives a linear form in terms of substrate [A]:

$$\frac{1}{v} = \frac{K_{m,A}}{V_{max}}\left(1 + \frac{K_{i,a}K_{m,B}}{K_{m,A}[B]}\right)\frac{1}{[A]} + \frac{1}{V_{max,f}}\left(1 + \frac{K_{m,B}}{[B]}\right) \qquad (8.149)$$

This example, which was presented by Segel[2] (1975) and repeated here, shows the extraordinary complexity of developing a kinetic expression for a single enzyme with multiple substrates and products. Since a cell has hundreds to thousands of enzymes, the complexity of analytically solving for or developing kinetic equations is readily apparent. Alternately, the need for and utility of metabolic flux analysis, as well as methods for numerical solutions of enzyme kinetic equations are also apparent. This is particularly relevant as structures and functions of numerous enzymes become more readily available as a consequence of genomics and proteomics.

IMMOBILIZED ENZYME

Chapter 7 described how immobilized enzymes are formed by adsorbing, covalently attaching, covalently crosslinking, or entrapping the protein in such a manner that substrate can be passed over the enzyme, while the enzyme itself is retained in a reaction vessel, and solid particles or beads consisting of porous glass, collagen, ceramic, zirconia, cellulose, or ion exchange resin may be used. The enzyme is immobilized by complexing with ion exchange groups on the surface of solid, or by covalently binding it to the bead using a difunctional reagent such as glutaraldehyde (Worsfold 1995; Zaborsky 1974).

A direct form of immobilization is entrapment, where enzyme is retained within the cell walls of the microorganism in which it is generated. An example is glucose isomerase from *Streptomyces* spp, in which the cells are heat-treated to selectively destroy lytic enzymes, therefore avoiding destruction of the cell wall (and the enzyme), while retaining the enzyme within the cell wall. The cells are then agglomerated as hard granules for use in deep-bed reactors. The cells may be cross-linked using glutaraldehyde or otherwise treated to form pellets that have a particle size on the order of millimeters. These give a relatively low-pressure drop when packed in a fixed bed.

---

[2] Segel (1975) presents analytical solutions for numerous types of enzyme mechanisms, and his book is an excellent resource for anyone carrying out enzyme analysis or pathway modeling.

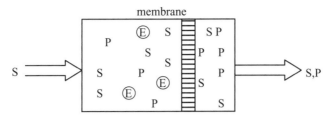

Figure 8.12. Schematic representation of membrane reactor with Ⓔ denoting enzyme, S, substrate; P, product, all dissolved in aqueous buffer.

An enzyme may also be retained in a reactor, if the outlet is covered with a membrane that is impervious to enzyme, but that still allows product and unreacted substrate to pass. Such a configuration is shown schematically in Fig. 8.12. This type of reactor has been used to produce R-mandelic acid, a precursor for semisynthesis of β-lactam antibiotics. This system, discussed in Chapter 10 (on metabolism), is an example of an enzyme bioreactor that uses a regenerable coenzyme (Vasic-Racki et al. 1989).

(R)-Mandelic acid is useful as a raw material for semisynthetic β-lactam antibiotic production, and can be produced from phenylglyoxylic acid. The reaction catalyzed is

Phenylglyoxylic acid   Mandelic acid

$$\text{(8.150)}$$

where MaDH denotes mandelic acid dehydrogenase. In this particular reaction, the NAD^+ was immobilized to polyethylene glycol (abbreviated PEG) so that the molecular weight of the PEG–NAD^+ complex is large enough to prevent it from permeating the membrane. Therefore it is retained in the reactor. The PEG–NAD is recycled within the same reaction vessel by coupling mandelic acid and PEG–NAD^+ formation with the oxidation of NH_4COOH. Hence, the coenzyme NAD is simultaneously reduced to form NADH using ammonium formate. The reaction is catalyzed by formate dehydrogenase:

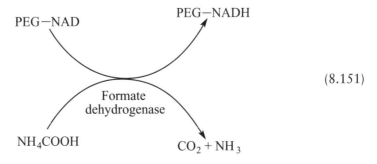

$$\text{(8.151)}$$

This reaction configuration represents an example of an in vitro reaction that utilizes a cofactor to generate a precursor for an antibiotic molecule, and an enzyme. Pyridine-linked dehydrogenases generally exhibit a mechanism of the type given in Eq. (8.112) (Plowman 1972); hence both the oxidation of phenylglyoxylic acid and the reduction of PEG–NAD$^+$ could be assumed to be ordered B_iB_i reactions, with the derivation as given in the previous section. The enzyme kinetic model of Vasic-Racki et al. (1989) appears to be consistent with this scheme.

The pH for the optimum activity of the two enzymes are different for which the type of kinetic expression derived for sequence (8.112) may be applicable. It also illustrates the practical need to select an appropriate pH for the reaction since the pH optima for mandelic acid dehydrogenase and formate dehydrogenase are different. Mandelic acid dehydrogenase activity drops as the pH increases from 5 to 8, while the activity of formate dehydrogenase has a broad optimum between pH 6.5 and 9. A pH of 8 was chosen even though the mandelate dehydrogenase exhibited only 20% of its initial activity at pH 8. The coupled reaction was carried out in a 10-mL reactor with a membrane having a 5000 molecular weight cutoff.

At steady-state operation, the material balance for such a reactor is

$$\nu_{MA} = F([S_0] - [S]) \tag{8.152}$$

where F is the volumetric flow rate, $[S_0]$ is the phenylglyoxylic acid concentration at the reactor inlet, and $[S]$ is its concentration at the outlet. In addition to phenylglyoxylic acid, ammonium formate must also be fed to facilitate regeneration of the immobilized NAD$^+$ back to NADH. The equation is similarly

$$\nu_{NAD} = F([A_0] - [A]) \tag{8.153}$$

where A denotes ammonium formate. For the purposes of this example, $[A_0] \gg A$ such that ν is small in this reactor. Hence, the backward reaction (mandelic acid → phenylglyoxylic acid) can be assumed to be small, and was in fact experimentally confirmed to be less than 1% at the forward reaction.

Online Databases of Enzyme Kinetic Constants

With the explosion in information regarding biological and biochemical systems (bioinformatics), databases for storing, sorting, searching, and retrieving these data have been established by international governments and professional societies to facilitate access to this information anywhere with a connection to the Internet. One of these online databases (BRENDA) is hosted by Cologne University Bioinformatics Center (CUBIC) at the University of Cologne (Germany). BRENDA (http://www.brenda-enzymes.info/) links to proteomic information such as amino acid sequences, enzyme structures, and gene sequences in other online databases. As of the beginning of 2006, BRENDA contained information regarding more than 83,000 enzymes from more than 9800 organisms. One unique feature of BRENDA is the collation in one database of published data regarding kinetic properties of enzymes, including K_m, turnover rates, and K_I for known inhibitors.

REFERENCES

Antrim, R. L., W. Colilla, and B. J. Schnyder, "Glucose Isomerase Production of High Fructose Syrups," *Appl. Biochem. Bioeng.* **2**, 97–155 (1979).

Bailey, J. and D. O. Ollis, *Biochemical Engineering*, McGraw Hill, New York, 1977, pp. 98–100.

Bodenstein, M., "Eine theorie de photochemischen reactions-geschwindig keiten," *Z. Phys. Chem.* **85**, 329–397, (1913).

Briggs, G. E. and J. B. S. Haldane, "A Note on the Kinetics of Enzyme Action," *Biochem. J.* **19**, 338–339, (1925).

Brooks, S. P. J., "Enzymes in the Cell: What's Really Going On?" in *Functional Metabolism Regulation and Adaptation*, K. B. Storey, ed., Wiley-Liss, Hobocken, NJ, 2004, pp. 55–86.

Houghton, J., S. Weatherwax, and J. Ferrell, *Breaking the Biological Barriers to Cellulosic Ethanol: A Joint Research Agenda*, DOE SC-0095, Department of Energy, Washington, DC, 2006.

Ladisch, M. R., *Bioseparations Engineering: Principles, Practice, and Economics*, Wiley, New York, 2001.

Ladisch, M. R., C. S. Gong, and G. T. Tsao, "Cellobiose Hydrolysis by Endoglucanase (Glucan-glucano hydrolase) from *Trichoderma reesei*: Kinetics and mechanism," *Biotechnol. Bioeng.* **22**, 1107–1126 (1980).

Ladisch, M. R., *Enzymatic Hydrolysis of Cellulose: Kinetics and Mechanism of Selected Purified Cellulase Components*, PhD thesis, Purdue Univ., Lafayette, IN, 1977.

Lauffenberger, D. A. and J. J. Linderman, *Receptors: Models for Binding, Trafficking and Signaling*, Oxford Univ. Press, New York, 1993, pp. 9–62.

Lynd, L. R., M. S. Laser, D. Brandsby, B. E. Dale, B. Davison, R. Hamilton, M. Himmel, et al., "How Biotech Can Transform Biofuels," *Nature Biotechnol.*, **26**(2), 169–172 (2008).

Plowman, K. M., *Enzyme Kinetics*, McGraw-Hill, New York, 1972, pp. xi–xiv, 1–39, 41–43.

Rodwell, V. W., "Enzymes: General Properties," in *Harper's Biochemistry*, 22nd ed., R. K. Murray, D. K. Granner, P. A. Mayes, and V. W. Rodwell, eds., Appleton & Lange, Norwalk, CT, 1990, pp. 58–88.

Schoemaker, H. E., D. Mink, and M. G. Wubbolts, "Dispelling the Myths—Biocatalysis in Industrial Synthesis," *Science* **299**, 1694–1697 (2003).

Segel, I. H., *Enzyme Kinetics: Behavior and Analysis of Rapid Equilibrium and Steady-State Enzyme Systems*, Wiley-Interscience, New York, 1975.

Vasic-Racki, D. J., M. Jonas, C. Wandrey, W. Hummel, and M.-R. Kula, "Continuous (R)-Mandelic Acid Production in an Enzyme Membrane Reactor," *Appl. Microbiol. Biotechnol.* **31**, 215–222 (1989).

Worsfold, P. J., "Classification and Chemical Characteristics of Immobilized Enzymes" (technical report), *Pure Appl. Chem.* **67**(4), 597–600 (1995).

Wyman, C. E., B. E. Dale, R. T. Elander, M. Holtzapple, M. R. Ladisch, and Y. Y. Lee, "Coordinated Development of Leading Biomass Pretreatment Technologies," *Bioresource Technol.* **96**, 1959–1966 (2005).

Zaborsky, O., *Immobilized Enzymes*, CRC Press, Cleveland, OH, 1974.

8.1. The enzymatic modification of a polypeptide (S) results in a macropeptide (MP) and a smaller glycopeptide (P). Kinetic studies are carried out by monitoring the appearance of (P). Product formation is found to be the rate-limiting step. Some preliminary rate studies are then carried out with the following observations:

1. For $S_0 = S_{0,1}$, $MP_0 = MP_{0,1}$, and $P_0 = 0$ (i.e., a fixed amount of MP is added to the initial reaction mixture), the rate (dP/dt) is the same as when $MP_0 = P_0 = 0$ at time $t = 0$.

2. For $S = S_{0,1}$, $MP_0 = 0$, and $P = P_{0,1}$ (i.e., a fixed amount of P is added to the initial reaction mixture) the rate (dP/dt) is significantly lower than when $MP = P = 0$ at $t = 0$.

3. For $S = S_{0,2}$, where $S_{0,2}$ is very large and $S_{0,2} \gg S_{0,1}$, the observed rate (dP/dt) is higher than that observed for $S_{0,1}$. However, the addition of $P = P_0$ at $t = 0$ in this case has virtually no effect on the observed initial rate for formation of (P) when, $S = S_{0,2}$.

On the basis of these observations, develop the simplest possible kinetic model that is consistent with the observations given above. Do this by answering the questions below:

(a) What type of inhibition pattern is consistent with the observations given in 1–3, above? (Check one.)

___ Uncompetitive product inhibition

___ Noncompetitive product inhibition

___ Competitive product inhibition

___ Substrate inhibition

(b) Give the reaction sequences (equations) in terms of enzyme, polypeptide (S), macropeptide (MP), and glycopeptide (P).

(c) Derive the rate equation then give its linear form (40 points)

(d) Show the general Lineweaver–Burke plot that would result from this equation when initial rates are measured at five different substrate concentrations for $P_0 = P_{0,1}$, $P_0 = P_{0,2}$, and $P_0 = P_{0,3}$, where $P_{0,3} > P_{0,2} > P_{0,1}$.

(e) How is V_{max} obtained from the resulting Lineweaver–Burke plot?

(f) How is K_S determined?

(g) How is (are) the inhibition constant(s) determined using values of other parameters obtained from the Lineweaver–Burke plot?

8.2. The formation of xylulose from xylose is a reversible reaction catalyzed by the enzyme, xylose isomerase. This reaction is of practical interest since ordinary yeast can ferment xylulose to ethanol, but not xylose. Major amounts of xylose are formed when biomass is hydrolyzed to fermentable sugars. The rate expression for the reaction is

$$E + X \underset{k_{-1}}{\overset{k_1}{\rightleftharpoons}} EX \underset{k_{-2}}{\overset{k_2}{\rightleftharpoons}} E + Xu \qquad \text{(HP8.2a)}$$

where E is enzyme, X is xylose, and Xu is xylulose. The rate expression for this equation, developed from the pseudo-steady-state assumption, is

$$v = \frac{\dfrac{v_{\text{max,f}}}{K_{\text{m,X}}}[\text{X}] - \dfrac{v_{\text{max,b}}}{K_{\text{m,Xu}}}[\text{Xu}]}{1 + \dfrac{[\text{X}]}{K_{\text{m,X}}} + \dfrac{[\text{Xu}]}{K_{\text{m,Xu}}}} \qquad \text{(HP8.2b)}$$

where $K_{\text{m,X}}$ and $K_{\text{m,Xu}}$ are Michaelis constants for xylose and xylulose, respectively, and $V_{\text{max,f}}$ and $V_{\text{max,b}}$ are the maximum rates for formation of Xu and X, respectively. An important parameter for designing a xylose-to-xylulose process is the maximum conversion that can be achieved in a batch enzyme reactor. Use Eq. (HP8.2a) to derive an expression for maximum conversion from initial rate kinetic studies. Show how initial rate kinetics may give values of the appropriate parameters. Be specific; use Lineweaver–Burke plots as part of your explanation.

8.3. Xylose isomerase is commercially available in an immobilized form (also used to produce HFCS). Hence, it is decided to design a process for xylose-to-ethanol production where immobilized xylose isomerase is to be used as catalyst. The process involves

1. Converting xylose (in aqueous solution) to xylulose so that the conversion is 5% of the maximum possible. This is to be accomplished in a single pass through the immobilized enzyme column.

2. Passing the resulting solution (containing xylulose at molar ratio of xylulose to xylose of 5–95) over an immobilized yeast column to convert the xylulose to ethanol.

3. Pressurizing the effluent from the immobilized cell column, heating, and then spraying into a flash chamber to remove most of the ethanol formed.

4. Cooling the remaining solution and recycling it to the immobilized xylose isomerase column for another pass.

This approach is to be examined in an attempt to minimize enzyme costs. Preliminary experiments indicate that the isomerization reaction in the immobilized enzyme column is controlled by external diffusion effects, where the rate of xylulose formation is related to diffusion by

$v \quad = k_{\text{mt}}a_{\text{m}}\left([\text{S}_{\text{b}}] - [\text{S}_{\text{s}}]\right)$

$k_{\text{mt}} \quad =$ mass transfer coefficient

$a_{\text{m}} \quad =$ surface area/unit volume of support

$[\text{S}_{\text{b}}] \quad =$ substrate concentration in bulk phase

$[\text{S}_{\text{s}}] \quad =$ substrate concentration at surface of immobilized enzyme.

Since the conversion for each pass is small, the reaction equation

$$\text{E} + \text{X} \underset{k_{-1}}{\overset{k_1}{\rightleftharpoons}} \text{EX} \underset{k_{-2}}{\overset{k_2}{\rightleftharpoons}} \text{E} + \text{Xu}$$

is simplified, as a first approximation, to

$$E + X \underset{k_{-1}}{\overset{k_1}{\rightleftharpoons}} EX \xrightarrow{k_2} E + Xu$$

Using these data

(a) Derive the relationship between diffusion and reaction, and express V_{max} in terms of $k_{mt}a_m$, S_b, S_s, and K_m. (50 points)

(b) If the value of V_{max} is known, show how K_m and k_{mt} can be determined from the observed rate of reaction and measurement of the bulk-phase concentration S_b (use a plot to explain your answer). (35 points)

(c) How can V_{max} be determined for this case? (15 points)

8.4. Glucose isomerase catalyzes the following reaction:

$$G + E \rightleftharpoons EG \rightleftharpoons E + F$$

This enzyme is widely used in the wet-milling industry in the immobilized form. An important parameter in designing and evaluating such a process is the maximum rates of conversion that are possible. Lloyd and Khaleeluddin [*Cereal Chem.* 53(2), 270–282 (1976)] discuss this in some detail. They state that "glucose isomerase is a relatively fast acting enzyme since most enzymes have turnover number of 10 to $10^4 \, min^{-1}$" and glucose isomerase exhibits a turnover number of $7300 \, min^{-1}$.

(a) Derive the equation from which the turnover number is calculated. State all assumptions and show your work.

(b) A key equation in the analysis of the immobilized enzyme reactor is equation (3) in the article by Lloyd and Khaleeluddin. Clearly show how this equation is obtained from their equations (1) and (2).

(c) On page 276 the authors give the relation

$$\frac{1}{I} = \frac{1}{I} e^{(-0.693t/\tau)} + \frac{1}{I_e}$$

Derive this equation. What are the limitations for the use of this relation to obtain the equilibrium product concentration?

(d) For this type of immobilized support is mass transfer limitations a factor? Why or why not?

8.5. One of the enzyme components involved in cellulose hydrolysis (a cellulase) is found to follow the mechanism

$$E + S \underset{k_2}{\overset{k_1}{\rightleftharpoons}} ES \xrightarrow{k_3} E + P$$

$$ES + I \underset{k_6}{\overset{k_5}{\rightleftharpoons}} ESI$$

$$E + I \underset{k_8}{\overset{k_7}{\rightleftharpoons}} EI$$

where E represents free enzyme; S, substrate; I, inhibitor, and ES, ESI, and EI represent enzyme–substrate, enzyme–substrate–inhibitor, and enzyme–inhibitor complexes, respectively.

(a) Derive the equation

$$\frac{1}{v} = \frac{K\left(1 + \dfrac{1}{K_{i,2}}\right)}{V} \cdot \frac{1}{S} + \frac{\left(1 + \dfrac{1}{K_{i,1}}\right)}{V}$$

where $K_{i,1} = k_6/k_5$ and $K_{i,2} = k_8/k_7$.

Use the pseudo-steady-state assumption. (30 points)

(b) Data giving the initial rate v as a function of substrate concentration at 0, 5 mM, 10 mM, and 15 mM glucose, is given. Show the form of the plots that would result if the expression in (a) is valid. Clearly indicate the location of the intercept. Can K_m and V_{max} be determined from this plot? Why or why not?

METABOLISM

INTRODUCTION

The building blocks—nucleic acids, proteins, lipids, and polysaccharides—that self-assemble into living cells are generated by metabolic processes. These processes couple the release of energy from sugars, fats, oxygen, methane, and/or sulfur with reactions that synthesize the building blocks. The catalysts that drive these processes are enzymes. Metabolism organizes the reaction pathways into a highly integrated, self-regulating networks where each node is an intermediate molecular species (Fig. 9.1). The description of the combined pathways that make up the working cell may be addressed by a systematic approach: systems biology. However, the effective modeling of metabolism and a comprehensible explanation of the system require that pathways be defined and their effect on cell function be understood. This chapter connects the basics of enzymes and enzyme kinetics to the biochemistry of some of the key pathways that either utilize or generate energy in living cells (via ATP, ADP, NADH, NAD$^+$, NADPH, NADP$^+$, FAD, and cyclic AMP).

We organize metabolism into anaerobic and aerobic pathways and describe ways in which these pathways overlap. This sets the stage for a discussion of the biological energetics of cells in Chapter 10. Enzymes play a catalytic role, but there are other molecules needed for metabolism to occur and for a living cell to function. Adenosine triphosphate (ATP) and adenosine diphosphate (ADP) are chemical compounds that store and transfer energy (structure in Fig. 9.2a); cyclic AMP (cAMP) has regulatory functions (Fig. 9.2b), and nicotinamide adenine dinucleotide (NAD), which accepts hydrogen and electrons to be used to form covalent bonds (Fig. 9.2c).

Metabolism encompasses anabolic, catabolic, and amphibolic pathways. Anabolic pathways synthesize small molecules and macromolecules required for the cell's functioning. Catabolic pathways provide the energy and chemical building blocks required by the anabolic pathways. Catabolic pathways are oxidative processes that release free energy in the form of high-energy phosphate bonds or reducing equivalents. Amphibolic pathways link anabolic and catabolic pathways (Fig. 9.3).

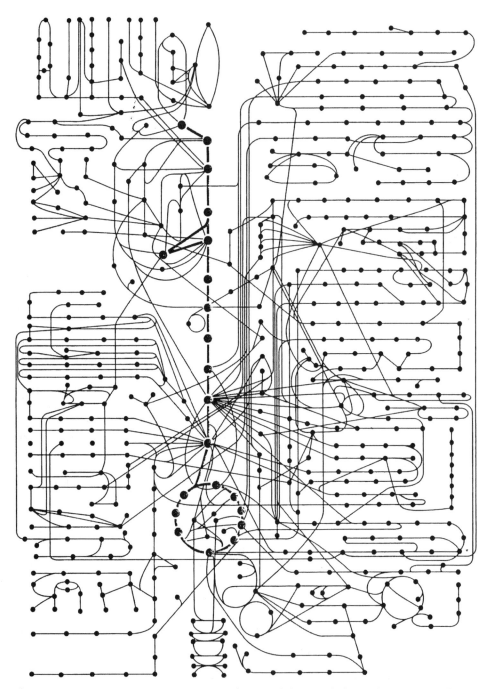

Figure 9.1. Diagrammatic representation of some of the metabolic pathways in a cell. The dots represent reversible reactions that connect these intermediates, and that are catalyzed by enzymes. Some reactions are compartmentalized into organelles of a cell, while most occur in the cytoplasm or cytosol. [Figure reproduced from Alberts et al. (1989), Chapter 2, p. 40, Garland Publishing, New York].

(a) ATP, ADP, AMP
(b) cyclic AMP

(c) NAD$^+$, NADH

(d) NADP$^+$, NADPH

Same as NAD$^+$, NADH; R' = O$-$P$=$O

(e) FAD

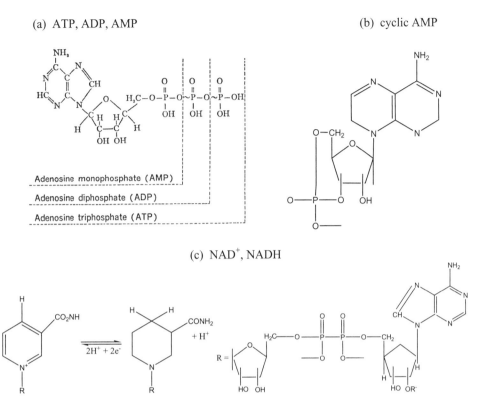

Figure 9.2. Structures of important energy transfer molecules in the cell. [The representation of ATP, ADP, and AMP is from Aiba et al. (1973), p. 57, Academic Press].

AEROBIC AND ANAEROBIC METABOLISM

Glycolysis Is the Oxidation of Glucose in the Absence of Oxygen

Glycolysis occurs in almost every living cell and results in formation of 2 mol of ATP (adenosine triphosphate; structure shown in Fig. 9.2a) per mole of glucose

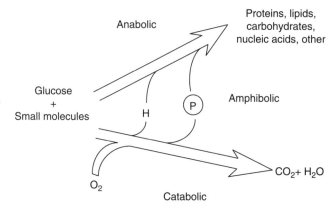

Figure 9.3. Metabolism follows catabolic (energy-generating) and anabolic (synthesizing) pathways connected through amphobilc pathways [adapted from Murray et al. (1990)].

processed (Alberts et al. 1989). ATP stores the energy from glucose metabolism in the form of high-energy phosphate bonds in ATP or ADP. These bonds have standard free energies of hydrolysis of -30.5 kJ/mol (which is equivalent to -7.3 kcal/mol) for ATP $\rightarrow$ ADP + P_i, and -27.6 kJ/mol (equivalent to -6.6 kcal/mol) for ADP $\rightarrow$ AMP + P_i, where P_i denotes inorganic phosphate (Murray et al. 1990).

Oxidation Is Catalyzed by Oxidases in the Presence of O_2, and by Dehydrogenases in the Absence of O_2

Chemical or biochemical oxidation is the removal of electrons from an electron donor. Oxidation is accompanied by reduction of an electron acceptor that gains electrons. Enzymes that catalyze oxidation directly with oxygen as the electron acceptor are referred to as *oxidases* and form either water or hydrogen peroxide (H_2O_2) as the reduced product (Fig. 9.4). R is shorthand for a compound that represents the oxidized form of RH_2. The sequences in Fig. 9.4 represent a respiratory conversion since oxygen is directly involved.

Oxidation of a metabolite in the absence of molecular oxygen (O_2) occurs through the transfer of hydrogen from one substrate to another in a coupled oxidation–reduction (redox) reaction. These reactions are reversible and are catalyzed by dehydrogenases (Fig. 9.4a). Many dehydrogenases require either NAD^+ (Fig. 9.2c) or $NADP^+$ (Fig. 9.2d) as cofactors or coenzymes.[1] The reversible character

---

[1] A *coenzyme* is a heat-stable, low-molecular-weight organic molecule that binds with an inactive form of an enzyme (referred to as the *apoenzyme*). The cofactor may bind covalently or noncovalently to the apoenzyme to yield an active enzyme referred to as the *holoenzyme* (Murray et al. 1990). A coenzyme functions as a group transfer molecule where a functional group, X, is transferred from a donor molecule D to an acceptor molecule A through the participation of the coenzyme/enzyme (CoE). Further explanation is given by Rodwell in Murray et al. (1990):

$$D-X \quad \diagdown\diagup \quad CoE \quad \diagdown\diagup \quad A-X \quad \text{(reduced)}$$
$$D \quad \diagup\diagdown \quad CoE-X \quad \diagup\diagdown \quad A \quad \text{(oxidized)}$$

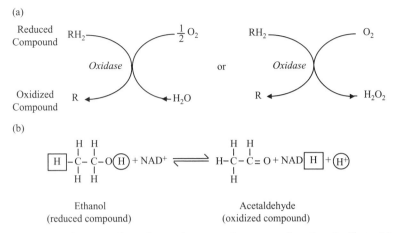

Figure 9.4. (a) Oxidases catalyze the oxidization of compounds using O_2 [from Murray et al. (1990), Fig. 13.1, p. 106]; (b) ethanol dehydrogenase uses NAD^+ to oxidize ethanol to acetaldehyde.

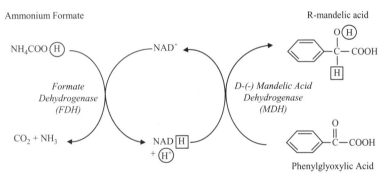

Figure 9.5. NADH acts as a reducing agent in the synthesis of β-lactam for the synthetic production of antibiotics [adapted from Vasic-Racki et al. (1989)].

of the reaction enables reducing equivalents to be readily transferred throughout the cell (Murray et al. 1990). This concept is illustrated by the oxidation of an alcohol in Fig. 9.4b.

A Membrane Bioreactor Couples Reduction and Oxidation Reactions (*R*-Mandelic Acid Example)

The in vitro conversion of ammonium formate to CO_2 and NH_3, accompanied by the reduction of the NAD^+ to NADH, is another example of oxidation of a reactant, $NH_4 COOH$, with NAD^+ as the cofactor (Vasic-Racki et al. 1989). The NADH then reduces phenylglyoxylic acid to *R*-mandolenic acid, as is illustrated in schematically in Fig. 9.5. *R*-Mandolenic acid is a chemical building block for manufacture of β-lactam antibiotics by combining fermentation-derived precursors as starting

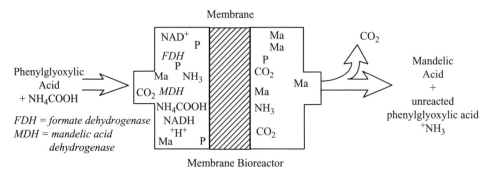

Figure 9.6. An in vitro membrane bioreactor to generate precursors for the synthetic production of antibiotics. The phenylglyoxylic and mandelic acids are denoted by P and Ma, respectively. [Adapted from Vasic-Racki et al. (1989)].

components for chemical synthesis. Since both biocatalytic and chemical synthesis steps are involved, this is referred to as *semisynthesis*.

A stoichiometric amount of NADH is needed, and hence a specially designed reactor was used that retains the enzymes and cofactor behind a membrane (Fig. 9.6) while allowing the unreacted substrate and product to pass through. The reduction of the NAD^+ to NADH was achieved in a cost-effective manner by using ammonium carbonate as the reducing agent and by attaching the NAD^+ and NADH to a water-soluble polymer that is too large to pass through the membrane. The NAD^+ and NADH are retained to the left of the membrane and the reaction occurs in the left compartment of the system at a productivity of $700\,g/(L \cdot day)$ (Vasic-Racki et al. 1989). This type of in vitro reaction system captures the advantages of enzyme-catalyzed transformations: specificity, high yield, and reaction at close to ambient conditions and pH 8.

The membrane reactor partially emulates the reactions that take place in a living cell since reactants, products, and cofactors are all in intimate contact. The membrane bioreactor differs from an in vivo biochemical system in that it couples the activities of only two enzymes in the same compartment rather than hundreds of enzymes, substrates, products, and cofactors interacting/reacting in a living cell (Fig. 9.1).

Three Stages of Catabolism Generate Energy, Intermediate Molecules, and Waste Products

Polysaccharides, fats, and proteins provide nutrients for both prokaryotic and eukaryotic cells. These nutrients are utilized in three stages (Fig. 9.7) (Alberts et al. 1989). First the macromolecules are broken down into their constituent monomeric units (i.e., polysaccharides to monosaccharides, proteins to amino acids, and fats into fatty acids and glycerol). The second step utilizes the appropriate monomers that have been transformed to glucose or fructose. The glucose passes through a sequence of 9 or 10 reactions to produce pyruvate or acetyl coenzyme A (abbreviated acetyl CoA; Fig. 9.8).

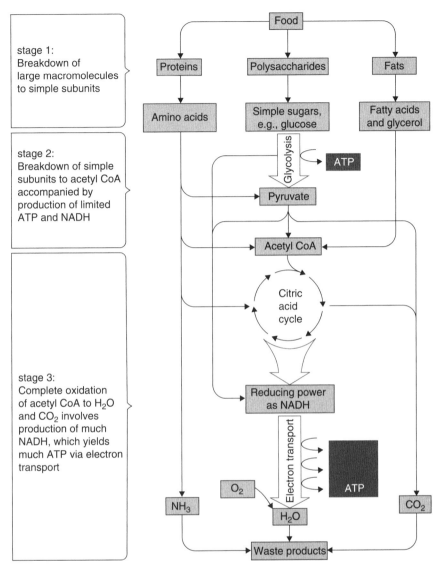

Figure 9.7.   Structure of acetyl-CoA [from Alberts et al. (1989), Fig. 2.19, p. 65, Garland Publishing, New York].

The Glycolysis Pathway Utilizes Glucose in Both Presence (Aerobic) and Absence (Anaerobic) of O_2 to Produce Pyruvate

A molecule of glucose that is derived from starch, glycogen, or cellulose is processed to two molecules of pyruvate and two molecules of ATP through the glycolytic pathway, which is also referred to as the *Embden–Meyerhoff pathway* (illustrated in Figs. 9.9 and 9.10). In the absence of oxygen, only 2 mol of ATP are produced per mole of glucose, and the product of the pathway, pyruvate, is the starting intermediate for other reactions in the cell. In the presence of oxygen, the pyruvate

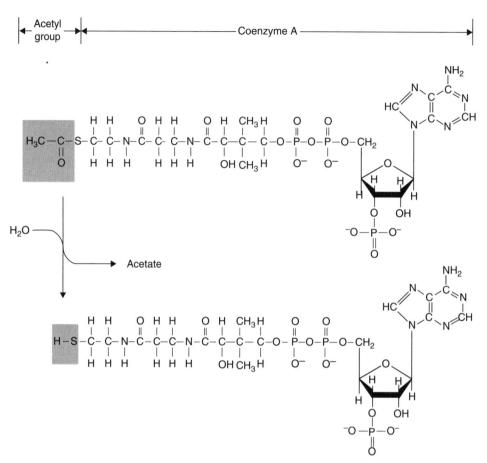

Figure 9.8. Simplified diagram of three stages of catabolism [from Alberts et al. (1989), Fig. 2.18, p. 64, Garland Publishing, New York].

derived from the Embden–Meyerhoff pathway is processed further via a catalytic cycle referred to as the *citric acid cycle*, also called the *Krebs cycle* or *tricarboxylic acid cycle* (abbreviated TCA) as illustrated in Fig. 9.7. The citric acid cycle in prokaryotes (bacteria) is found in the cytosol. In eukaryotes, the citric acid cycle is associated with organelles known as the *mitochondria*.

Glycolysis Is Initiated by Transfer of a High-Energy Phosphate Group to Glucose

The first step in glycolysis transfers a phosphate group from ATP to glucose. Glucose-6-phosphate is transformed to its chemical isomer, D-fructose-6-phosphate, and finally to D-fructose-1,6-diphosphate as shown in Fig. 9.9. The D-fructose-1,6-diphosphate then forms two molecules of glyceraldehyde-3-phosphate. Figure 9.9 is part of a larger representation of the glycolytic sequence by Murray et al. (1990, p. 165). Each molecule of glyceraldehyde-3-phosphate is further phosphorylated with inorganic phosphate (abbreviated P_i) to give 1,3-diphosphoglycerate, which,

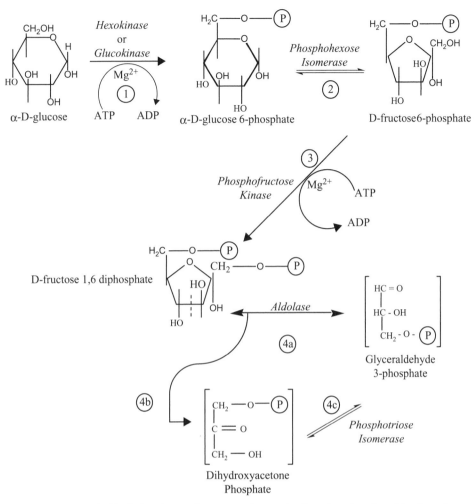

Figure 9.9. First half of glycolysis where α-D-glucose is phosphorylated and broken down into a three-carbon molecule [from Murray et al. (1990), p. 165, Appleton & Lange].

in turn, transfers its phosphate back to ADP to form a molecule of ATP and 3-phosphoglycerate. The 3-phosphoglycerate is isomerized to 2-phosphoglycerate. The 2-phosphoglycerate is then chemically dehydrated to phosphoenolpyruvate, which, in turn, transfers its phosphate to ADP to give a second molecule of ATP as well as pyruvate. The interconversion of the enol to the keto forms of the pyruvate occur spontaneously (Murray et al. 1990). This path gives an example of the prevalence of phosphate transfer reactions, and the enzymes (kinases[2]) that catalyze them.

---

[2] Kinases represent one of the six classes of enzymes that are defined by the International Union of Biochemistry (IUB) Nomenclature System on the basis of chemical reaction type and reaction mechanism. Kinases transfer phosphate groups from ATP to the substrate molecule (Murray et al. 1990).

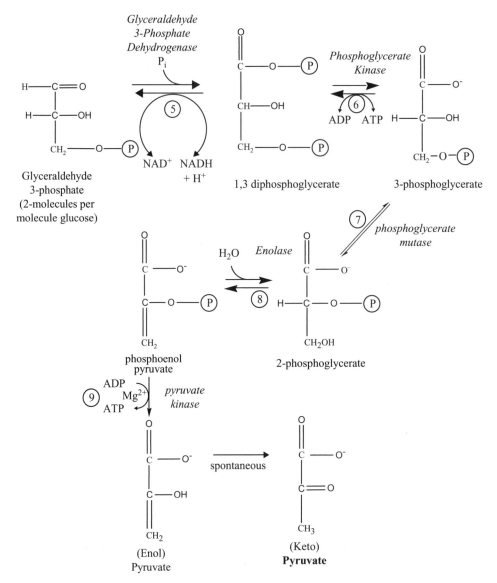

Figure 9.10. Second half of glycolysis [from Murray et al. (1990), Fig. 19.2, p. 165 Appleton & Lange].

Both microbial prokaryotes (bacteria) and eukaryotes (yeast) prefer glucose as a substrate. If more than one monosaccharide is present in the culture medium, the glucose will usually be utilized first. This is particularly true when glucose is present in a mixture of glucose (a six-carbon or C_6 sugar) and pentose (a five-carbon or C_5 sugar) for microorganisms that can utilize both C_6 and C_5 substrates. Glucose is preferred since the cell must otherwise carry out extra steps to transform the pentose into a hexose that can be utilized by the glycolysis pathway. The pyruvate is then transported into the mitochondria in a eukaryotic cell, where it is processed further

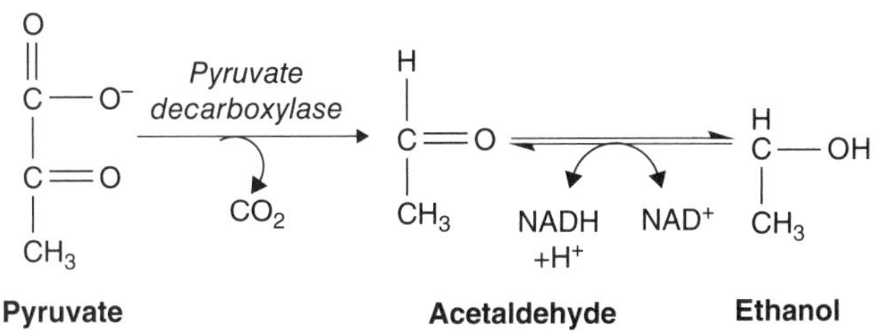

Figure 9.11. The product of glycolysis (pyruvate) is further processed to ethanol in order to recycle NADH to NAD$^+$ to allow glycolysis to continue.

via the citric acid cycle if molecular oxygen is present, or in the case of ethanol fermentation is converted to ethanol (Fig. 9.11).

Products of Anaerobic Metabolism Are Secreted or Processed by Cells to Allow Continuous Metabolism of Glucose by Glycolysis

Three examples where an alcohol or organic acid must be secreted by the cell in order to enable glycolysis to continue are ethanol, lactic acid, and succinic acid fermentations. Ethanol is formed from pyruvate by the reaction shown in Fig. 9.11, resulting in one molecule each of ethanol, CO_2, and NAD$^+$ per molecule of pyruvate. The overall yield per molecule of glucose is two molecules each of ethanol, CO_2, and NAD$^+$. The ethanol is formed and secreted by the cell as a consequence of the cell's energy metabolism and the need to regenerate NAD$^+$ from NADH to enable glycolysis to continue under anaerobic conditions. The reactions given in Fig. 9.11 combined with glycolysis (Embden–Meyerhoff pathway) give the Embden-Meyerhoff–Parnas pathway (abbreviated EMP) (Bailey and Ollis 1977).

Formic, acetic, lactic, propionic, and butyric acids are produced by microorganisms growing at fermentative conditions where oxygen is not present to serve as a hydrogen acceptor for the NADH produced by glycolysis. Fermentation harnesses the energy released by partial oxidation of hydrogen-rich glucose by transferring the hydrogen atom to another organic molecule that becomes more reduced. These reduced compounds (ethanol, lactic acid, formic acid, or other organic acids) are metabolic waste products and must be excreted by the cell in order to avoid their toxic effects.

Hence, lactic acid is produced from pyruvate, in order to regenerate the needed NAD$^+$, as illustrated in Fig. 9.12. The overall stoichiometry for the formation of lactic acid under anaerobic conditions is given in Fig. 9.13 (Alberts et al. 1989).

Regeneration of NADH to NAD$^+$ occurs by transfer of the hydride ions (H:) from NADH to pyruvate to produce lactic acid. The overall reaction, starting with glucose processed through the Embden–Meyerhoff glycolysis pathway is

$$\text{Glucose} + 2P_i + 2\text{ADP} \rightarrow 2 \text{ lactic} + 2\text{ATP} + 2H_2O \qquad (9.1)$$

where $\Delta G° = -32.4\,\text{kcal/mol}$ for the overall reaction. The free-energy change for glucose conversion to lactic acid is $-47\,\text{kcal/mol}$. The difference in the free energies (14.6 kcal) represents the energy that is conserved in the high-energy phosphate

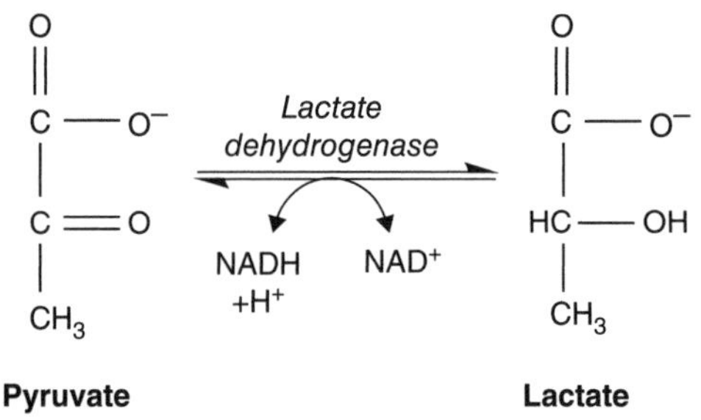

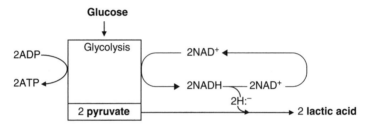

Pyruvate   **Lactate**

Figure 9.12. The product of glycolysis (pyruvate) is further processed to lactate in order to recycle NADH to NAD$^+$ to allow glycolysis to continue.

Glucose

Glycolysis

2ADP

2ATP

2NAD$^+$

2NADH ——— 2NAD$^+$

2H:$^-$

2 **pyruvate**   2 **lactic acid**

Figure 9.13. Overall stoichiometry of lactic acid fermentation from glucose [Alberts et al. (1989), Fig. 7.56, p. 382, Garland Publishing].

bonds in ATP generated during the metabolism of glucose. This is equivalent to 7.3 kcal/mol ATP or 14.6 kcal for two molecules of ATP (Bailey and Ollis 1977).

The reduction of formic acid and fumarate is a function of the energy-conserving electron transport chain in the plasma membrane of *E. coli*. Formic acid at the exterior of the cell membrane is converted to two electrons, two protons (H$^+$), and CO_2. The electrons are used inside the cell to reduce fumarate to succinate. The sequence of reactions shown in Fig. 9.14 illustrates how protons are generated outside the cell and electrons are used inside the cell. This is equivalent to pumping protons into the cell, although the protons do not physically cross the membrane. Rather, a membrane-based electron transport system generates an electrochemical proton gradient across the membrane. The thermodynamic driving force of the reaction is expressed as the redox potential. The redox potential for formic acid/CO_2 is -0.42 V, while that of fumarate/succinate is $+0.030$ V (Alberts et al. 1989).

The production of succinate may utilize the intermediate, fumarate, by an alternate pathway by the regenerating two molecules of NAD$^+$ from NADH as shown in Fig. 9.15 (Alberts et al. 1989). In this case only one molecule of pyruvate is transformed into succinate. The second molecule of pyruvate is utilized in biosynthesis inside the cell for each molecule of succinate that is excreted.

The stoichiometry of reaction shows that the yields per weight of glucose are highest for lactic acid (Fig. 9.13), lower for succinic acid (Fig. 9.15), and lowest for ethanol (Figs. 9.9–9.11). The maximum or theoretical yields follow the moles of oxygen incorporated into the acid or alcohol structure per mole of glucose entering the glycolytic pathway. Lactic acid retains six oxygens (three in each of the two

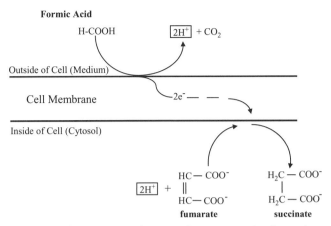

Figure 9.14. Formic acid fermentation showing electron transfer driven by external reduction of formate [with permission from Alberts et al. (1989), Fig. 7.57, p. 383, Garland Publishing].

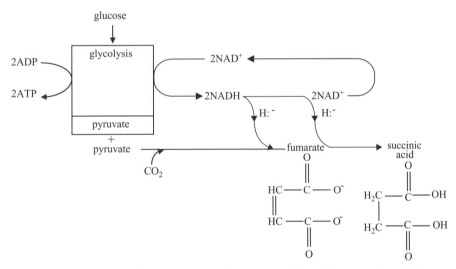

Figure 9.15. Succinic acid fermentation [with permission from Alberts et al. (1989), Fig. 7.56, p. 382, Garland Publishing].

molecules formed), succinic has four, and ethanol has two (one oxygen per molecule × two molecules of ethanol).

Other Metabolic Pathways Utilize Glucose Under Anaerobic Conditions (Pentose Phosphate, Entner–Doudoroff, and Hexose Monophosphate Shunt Pathways)

The dominant glycolytic pathway given in the biochemical engineering literature for prokaryotes and eukaryotes is the Embden–Meyerhoff glycolytic pathway (Figs. 9.9

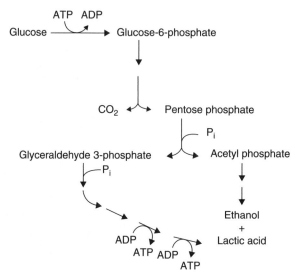

Figure 9.16. Partial diagram for glucose monophosphate pathway [adapted from Bailey and Ollis (1977), McGraw-Hill, New York].

and 9.10). However, there are at least three other pathways that utilize glucose. These are reviewed by Bailey and Ollis (1977) and include the pentose phosphate and Entner–Doudoroff pathways as well as a hexose monophosphate shunt pathway that yields 2.5 mol ATP per mole of glucose, instead of the 2 mol of ATP per mole of glucose of the Embden–Meyerhoff glycolytic pathway.

The pentose phosphate shunt entails phosphorylation of glucose, splitting off of CO_2 to give pentose phosphate, phosphorylation of the pentose phosphate, and lysis of the pentose into glyceraldehyde-3-phosphate and acetyl phosphate. The glyceraldehyde-3-phosphate is oxidized to lactic acid, while the acetyl phosphate is oxidized to ethanol, as shown in Fig. 9.16. Only 1 mol of ATP is generated per mole of glucose that passes through this pathway. Note that Fig. 9.16 does not show all the species in this reaction. A detailed description is given by Bailey and Ollis (1977).

The Entner–Doudoroff pathway proceeds through 3-keto-3-deoxy-6-phosphogluconic acid, which splits into pyruvate and glyceraldehyde-3-phosphate as shown in Fig. 9.17. This pathway also only yields 1 mol ATP per mole of glucose utilized.

The third pathway, occurs in *Bifidobacterium*, and gives the greatest amount of ATP per mole of glucose for any known fermentation route. Two molecules of glucose monophosphate are isomerized to fructose-6-phosphate. These are transformed to three molecules: heptose phosphate, triose phosphate, and acetic acid. The heptose and triose phosphate molecules then pass through the sequence shown in Fig. 9.18 to give two molecules of lactic acid and an additional two molecules of acetic acid. The sequence of steps for this "mixed-acid fermentation" is described in greater detail by Bailey and Ollis (1977, Fig. 5.10, p. 240), from which the pathway in Fig. 9.18 is derived.

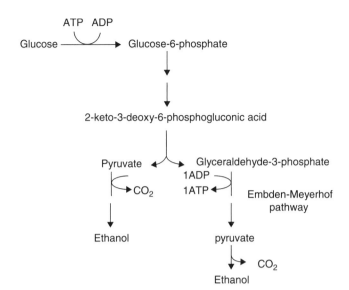

Figure 9.17.   Partial diagram of Entner–Doudoroff pathway.

Knowledge of Anaerobic Metabolism Enables Calculation of Theoretical Yields of Products Derived from Glucose

The production of ethanol from glucose has a maximum possible (i.e., theoretical) yield of 0.511 g/g

$$\frac{2\,\text{mol ethanol} \cdot 46\,[\text{g ethanol}/(\text{g} \cdot \text{mol ethanol})]}{1\,\text{mol glucose} \cdot 180\,[\text{g glucose}/(\text{g} \cdot \text{mol glucose})]}$$

since two moles of pyruvate are obtained from each mole of glucose (MW = 180) and 1 mol of pyruvate would give 1 mol ethanol (MW = 44) if all the pyruvate were converted to ethanol. However, since the glucose (and therefore, the pyruvate) provides not only energy required for the cell's functioning but also the carbon atoms needed to form other molecules utilized in the cell's biological and biochemical systems, not all of the glucose or pyruvate will be oxidized into ethanol.

The stoichiometry of the Embden–Meyerhoff–Parnas pathway enables calculation of a maximum yield of ethanol. Such a calculation is useful for estimating best-case fermentation process yields and generating a first estimate of the contribution of substrate cost to the product cost. Thus, if glucose costs 11 ¢/kg (5 ¢/lb), the contribution of the cost of the glucose to the cost of manufacturing fermentation ethanol would be

$$\left| \frac{1\,\text{kg glucose}}{0.511\,\text{kg ethanol}} \right| \frac{11¢}{1\,\text{kg glucose}} \right| = 21.5 \frac{¢}{\text{kg ethanol}} \qquad (9.2)$$

Similar calculations can be carried out for lactic and succinic acids, which can also be obtained under anaerobic conditions from pyruvate. The a priori calculation of

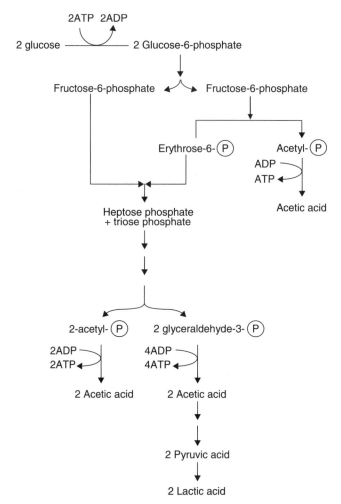

Figure 9.18. Metabolic pathway for the mixed-acid fermentation of *Bifidobacterium* [adapted from Bailey and Ollis (1977), Fig. 5.10, p. 240, McGraw-Hill, New York].

theoretical yields under aerobic conditions is more difficult, and sometimes not possible, because the metabolites may be generated or utilized by more than one pathway, with interactions occurring between pathways.

Economics Favor the Glycolytic Pathway for Obtaining Oxygenated Chemicals from Renewable Resources

Renewable resources are defined as plant materials and organisms that fix CO_2 into organic molecules via photosynthesis using light. These molecules include starch and cellulose. When broken down into their monomer units, these polymers yield glucose. Since the glucose, or molecules derived from it, is ultimately transformed into CO_2 and water and the CO_2 is recycled through the biosphere back into plant or other phytomatter, the materials that yield glucose are referred to as being "renewable."

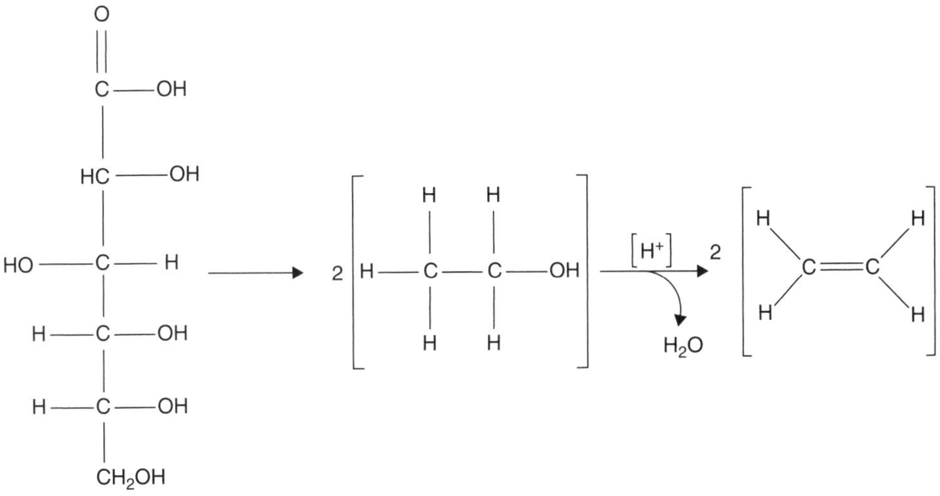

Glucose (MW = 180, \$5.00)

Ethanol (MW = 46, \$8.50)

Ethylene (MW = 28, \$6.20)

Figure 9.19. Minimum economic values of ethanol and ethylene derived by fermentation of glucose to ethanol followed by the catalytic dehydration of ethanol to ethylene.

The comparison of maximum yields leads to the heuristic observation that the most economically attractive molecules to be derived from renewable resources (via glucose) would appear to be those with oxygen in their structures, specifically, oxygenated chemicals. Molecules with the most oxygen are the most attractive since the yield per unit weight of glucose is likely to be higher, and the cost of making these compounds from nonrenewable, oxygen-deficient hydrocarbons (coal, natural gas, or petroleum) is high. Alternately, the reduction of oxygen-containing, fermentation-derived products to hydrocarbons is likely to be economically unattractive. A higher degree of oxidation of the starting substrate corresponds to a lower economic potential of transforming it into a hydrocarbon that contains little or no oxygen. As summarized by Hacking (1986, p. 77), "It is the reason why biotechnology, with its reliance on carbohydrate feedstocks, is predicted to have more impact on the production of oxidized bulk chemicals such as lower aldehydes, ketones, and acids rather than on higher alcohols or most polymers."

Consider the fermentation of 100 lb glucose (assumed value of 5¢/lb) to 51 lb ethanol (the theoretical maximum). Assume that the ethanol would correspond to a value of \$1.10/gal (or 16.6¢/lb). The catalytic dehydration of ethanol would give 1 g of ethylene per 1.64 g of ethanol or a maximum theoretical yield of 61%, based on ethanol, for the reaction shown in Fig. 9.19. The cost of the resulting 31 lb ethylene, at 20¢/lb would be \$6.20. This would be unattractive unless the price of ethylene were to exceed that of ethanol by more than this amount.

CITRIC ACID CYCLE AND AEROBIC METABOLISM

The third stage of catabolism under aerobic conditions that follows (1) glycolysis and (2) formation of acetyl CoA is (3) the citric acid cycle, which is named after the first acid formed in this pathway. The reactions are represented by a cycle because the acetyl group is attached to oxaloacetate and then oxidized, with the oxaloacetate returning to its original state at the end of the sequence of reactions, that is, one turn of the cycle as illustrated in Fig. 9.20. The cycle serves a catalytic function since no input of carbon other than the substrate occurs, and since the reaction sequence that makes up the cycle both begins and ends with citric acid.

The citric acid cycle is a source of a pool of precursors for the synthesis of amino acids (Aiba et al. 1973). The citric acid cycle is also referred to as the *Krebs cycle* or the *tricarboxylic acid* (TCA) cycle since the starting acid, formed from acetyl CoA, is citric acid. The citric acid cycle was discovered in 1937, as a result of efforts to define pathways of aerobic metabolism that would explain how cells produce CO_2 and water in the presence of oxygen, and lactic acid or ethanol in the absence of oxygen.

Respiration Is the Aerobic Oxidation of Glucose and Other Carbon-Based Food Sources (Citric Acid Cycle)

In eukaryotes, aerobic respiration occurs in the mitochondria,[3] where the citric acid cycle can generate as many as 12 molecules of ATP for each molecule of acetyl-CoA from pyruvate that enter it. Hence, if 2 molecules of acetyl-CoA are formed from 1 molecule of glucose, up to 24 molecules ATP would be produced per molecule of glucose. The complete oxidation of glucose would result in 36 molecules of ATP, if the ATP generated in the steps leading to acetyl CoA is accounted for (Alberts et al. 1989).

The citric acid cycle in aerobic prokaryotes is analogous to that of eukaryotes, except that the citric acid cycle pathway in prokaryotes is found in the cytosol since prokaryotes do not have mitochondria. In both prokaryotes and eukaryotes, the carbon atoms in the acetyl groups are converted to CO_2, while the H is transferred to other molecules. Oxygen and hydrogen (in the form of water) enter the cycle in the steps leading to citrate, succinate, and malate formation (Fig. 9.20). There are many parts of the cell's metabolisms that are common to most organisms. These include pathways that lead to formation of amino acids and proteins, lipids, and polysaccharides. The citric acid cycle provides precursors for biosynthesis of these components (Alberts et al. 1989).

The Availability of Oxygen, under Aerobic Conditions, Enables Microorganisms to Utilize Pyruvate via the Citric Acid Cycle

The citric acid cycle oxidizes over half of the carbon compounds in cells, and generates the NADH that passes high-energy electrons through the respiratory chain to form ATP from ADP. The cycle ends when electrons reduce O_2 to H_2O [paraphrased from Alberts et al. (1989), Ref. 1, pp. 347–348]. It is a key metabolic pathway since

---

[3] Mitochondria are small enclosed structures, found in all eukaryotic cells, that have a large internal surface area with which are associated enzymes that convert metabolites of glucose, in the presence of oxygen, to chemical forms of energy that are used to drive cellular reactions (Alberts et al. 1989).

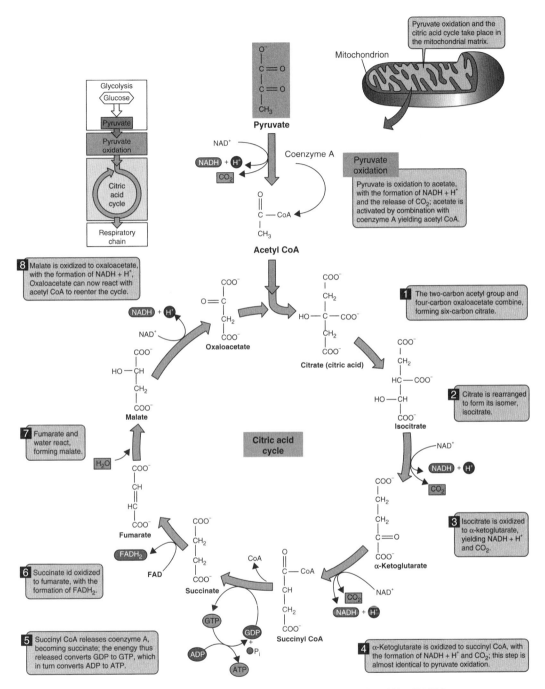

Summary: $CH_3\ COCOOH + 4NAD^+ + FAD \rightleftharpoons CO_2 + 4NADH + 4H^+ + FADH_2$

Figure 9.20. Simplified representation of citric acid cycle. The intermediates that are boxed in denote starting compounds for synthetic reactions to form amino acids, porphyrins, or carry out acetylation reactions. [From Aiba et al. (1973), Fig. 3.8, p. 68, Academic Press, New York, 1973; figure is reproduced from Purves et al. (2003), Fig. 7.8].

it provides both the carbon skeletons and the energy (as ATP) needed for biosynthetic and other reactions elsewhere in the cell (Aiba et al. 1973). The overall balance based on pyruvate entering the citric acid cycle is

$$CH_3 - \overset{O}{\overset{\|}{C}} - COOH + 4NAD^+ + FAD \rightleftharpoons 3CO_2 + 4NADH + 4H^+ + FADH_2 \quad (9.3)$$

Pyruvic acid

where the structures of NAD^+ and FAD^+ are as given in Figs. 9.2c and 9.2e. The carboxylic acids that are diverted from the citric acid cycle for biosynthetic purposes elsewhere in the cell must be replaced. Otherwise, the TCA cycle would stop and the cell would not be able to use oxygen to produce energy to power biochemical reactions. The carboxylic acids are replaced by fixing CO_2 to C_3 intermediates[4] of the glycolytic pathway.

Since the citric acid cycle is catalytic, the yields of molecules synthesized from components of the cycle or precursors generated through the citric acid cycle are not directly proportional to the acids that make up the cycle. The calculation of theoretical yields is further complicated by the reactions that occur elsewhere in the cell that use intermediates of the citric acid cycle as starting reactants.

Another pathway deaminates alanine, serine, cysteine, glycine, and threonine to give pryuvic acid. This is indicated by an arrow pointing from the amino acids to the pyruvic acid. Acetyl-CoA has a reversible interaction with lipids, while other amino acids would be deaminated to give back acetyl-CoA. The citric acid cycle also contributes to, or receives carbon backbones from, various amino acids. Prophyrins, methionine, and lysine are obtained from succinate. The energetics of the cycle, discussed earlier in this chapter, result in NADH, which supplies energy to other reactions.

As two examples, glutamate can be formed from ammonia and α-ketoglutarate (from the TCA cycle), or directly from ammonia and a carbohydrate. Alternately, bacteria combine ammonia with fumarate (also from the TCA cycle) to form aspartate.

Glutamate and aspartate donate amino groups to α-keto acids, including pyruvic, oxalacetic, and α-ketoisovaleric acids to form other amino acids. Pyruvate is a precursor to acetyl-CoA and the other acids that are part of the TCA cycle. In addition, succinyl-CoA from the citric acid cycle combines with glycine to form pyroles, which, in turn, are used in the synthesis of porphyrins. Porphyrins[5]

---

[4] A common nomenclature for describing metabolism of different carbon compounds refers to the number of carbon atoms in the backbone of the molecule that is undergoing transformation. Hence, glucose with six carbons in its structure would be a C_6 compound; xylose with five carbons would be a C_5; phosphoenol pyruvate with three carbons, a C_3.

[5] The *porphyrins* are cyclic compounds formed from linkage between four pyroles, with the resulting structure sometimes being presented as a cross. The shorthand was proposed by Fischer.

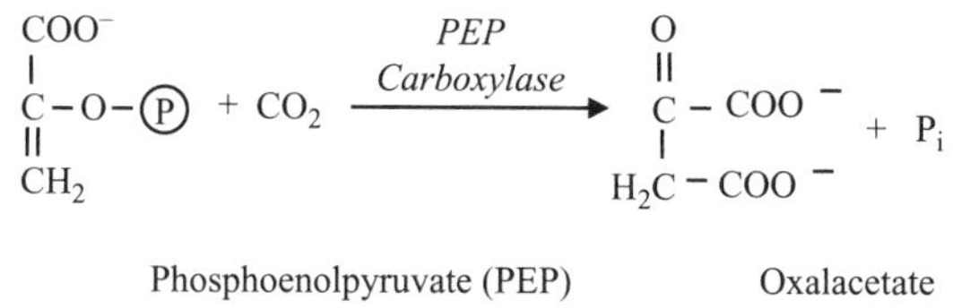

Figure 9.21.   Conversion of phosphoenolpyruvate (PEP) to oxalacetate.

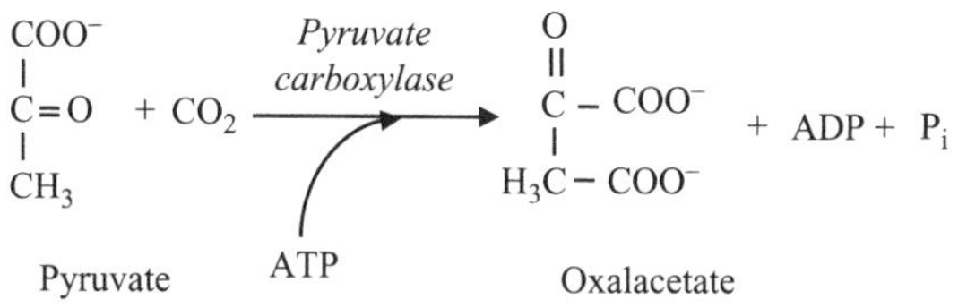

Figure 9.22.   Conversion of pyruvate to oxalacetate.

are incorporated into the structure of cytochromes and chlorophyll[6] and associate with the iron and magnesium, respectively, which are essential to the functioning of these proteins, for photosynthesis, or for oxygen uptake and transport in higher organisms.

Aiba et al. (1973) give the example of phosphoenolpyruvate (PEP) for enteric bacteria, shown in Fig. 9.21, where PEP is converted to oxaloacetate and inorganic phosphate (P_i). ATP provides the energy to drive reaction for formation of oxalo-acetate from pyruvate and CO_2 (Fig. 9.22).

In addition, the pathways that we presented as semiindependent (glycolysis and TCA cycle) are in fact intimately intertwined. For example, the third enzyme in the glycolysis pathway, phosphofructosekinase (Fig. 9.9), is inhibited by high levels of citric acid (TCA cycle). Hence accumulation of this acid would slow down the glycolysis pathway and the production of acetyl-CoA derived from pyruvate, which in turn would slow down citric acid production. Synthesis of some amino acids also derive their carbon skeletons from intermediates of the glycolysis pathway, such as tyrosine, phenylalanine, and tryptophan. Similarly, pyruvic acid can contribute carbon backbones to lysine, isoleucine, valine, and leucine.

---

[6] *Cytochromes* are a family of colored proteins that act as agents for oxidation–reduction reactions. Cytochrome *c*, for example, has a molecular weight of about 13 kD and binds one molecule of iron per molecule of protein, due to the attached heme group consisting of an iron atom held by four nitrogen atoms. The iron atom changes from the ferric [Fe(III)] to the ferrous [Fe(II)] state whenever it accepts an electron. The iron sequestered in porphyrins associated with hemoglobin and myoglobin bind oxygen through the sixth coordination position of the iron. The porphyin associated with a hydrophobic tail region makes up chlorophyll, where the Mg forms coordination bonds with the nitrogen of the pyrrole groups. The electrons in the porphyrin ring are excited by absorbed sunlight, which moves an electron from its resident molecular orbital to a one of higher energy. The energy is transferred to an electron acceptor that eventually generates ATP and NADH to drive the combination of CO_2 and water to form carbohydrates (Alberts et al. 1989).

The Citric Acid Cycle Generates Precursors for Biosynthesis of Amino Acids and Commercially Important Fermentation Products

The anaerobic glycolysis (Embden–Meyerhoff) pathway results in pyruvate and the anaerobic Embden–Meyerhoff–Parnas (EMP) pathway converts pyruvate to ethanol. The aerobic citric acid cycle is responsible for production of commercially important, high-volume fermentation products. These include the carboxylic acids (acetic, fumeric, citric and succinic acids)[7] and amino acids [monosodium glutamate (MSG)], lysine, tryptophan, and phenylalanine). These are derived through aerated fermentations, unlike ethanol fermentation. The citric acid cycle is also of fundamental importance since it represents a catalytic sequence of reactions that results in biosynthesis of the 20 amino acids that are the basis of the protein catalysts, namely, enzymes that enable the cell to function. A summary of the amino acids, their properties, and abbreviated names are given in Figs. 9.23 and 9.24. About half of the amino acids are nonpolar at pH 7, as shown in Fig. 9.23. The remaining amino acids are acidic, basic, or uncharged polar as shown in Fig. 9.24.

Glucose Is Transformed to Commercially Valuable Products via Fermentation Processes: A Summary

Strictly speaking, *fermentative conditions* refer to metabolism that occurs in the absence of oxygen. However, the term *fermentation* is commonly used to describe a microbial culture process under anaerobic, microaerobic, or aerobic conditions for either prokaryotes or eukaryotes. For example, fermentation is used to describe processes for generation of citric acid (under aerobic conditions), ethanol (under microaerobic conditions), and butanol, acetone and lactic and succinic acids (at anaerobic conditions). These products diffuse out of the cells into the surrounding liquid medium known as the *fermentation broth*. The metabolism leading to alcohols is referred to as being *anaerobic* ("in the absence of oxygen"), although ethanol is actually generated in the presence of trace amounts of oxygen, at microaerobic conditions. Yeast grows at both aerobic and microaerobic conditions, although the end products are different at the two conditions.

Clostridia bacteria that produce butanol and acetone are strict anaerobes (Ladisch 1991). Oxygen at the parts per billion (ppb) level is lethal to the living

---

[7] Citric acid, a flavoring agent for soft drinks, has an annual market of at least $1 billion worldwide with major producers consisting of Haarman and Reimer (division of Bayer), ADM, F. Hoffmann-LaRoche Ltd. (of Switzerland), and Jungbunzlauer International AG (of Switzerland) (Kilman 1996a,b). Lysine, an animal feed additive, had sales of at least $600 million worldwide, with ADM's sales estimated at $300 million in 1996 (Kilman 1996). Major producers of lysine include Archer Daniels Midland (ADM) of the United States, Ajinomoto and Kyowa Hakko Kogyo of Japan, and Sewon America of S. Korea (Kilman 1996a, 1998a,b). Fumeric acid, a food acidulant and beverage ingredient, is derived exclusively through chemical synthesis, due to economics (1998a). Monosodium glutamate production exceeds 220,000 tons per year and is widely used as a flavor enhancer and seasoning (Hacking 1986). L-Tryptophan was sold in US healthfood stores as a remedy for insomnia, and as a dietary supplement, until 1990, when it was banned after the supplement was traced to a blood disorder: eosinophilia–myalgia syndrome (EMS) (Roberts 1990; Chandler 1990). The causative agent is still not known. L-Phenylalanine is used in the synthesis of the noncaloric sweetener, aspartame, trade name Nutrasweet (Hamilton et al. 1985).

NONPOLAR (at pH =7)

	Alanine	Cysteine	Glycine	Isoleucine	Leucine	Methionine	Phenylalanine	Proline	Tryptophan	Valine
mRNA Codon (5′ 3′)	GCA GCC GCG GCU	UCG UGU	GGA GGC GGG GGU	AUA AUC AUU	UUA UUG CUA CUC CUG CUU	AUG	UUC UUU	CCA CCC CCG CCU	UGG	GUA GUC GUG GUU
Amino Acid	Ala	Cys	Gly	Ileu	Leu	Met	Phe	Pro	Trp	Val
	A	C	G	I	L	M	F	P	W	V
Side Chain	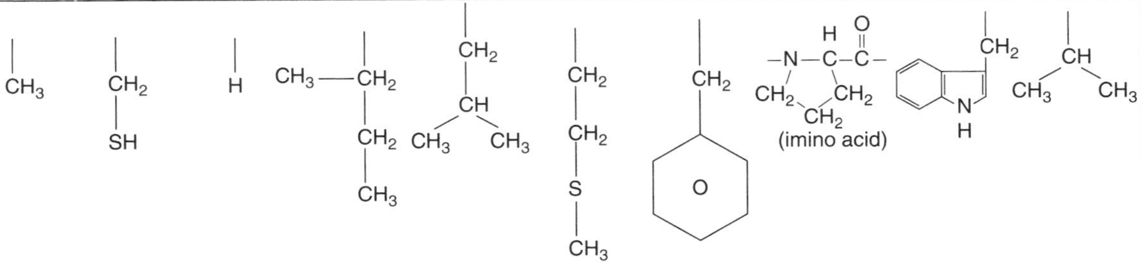 $-CH_3$	$-CH_2-SH$	$-H$	$-CH(CH_3)-CH_2-CH_3$	$-CH_2-CH(CH_3)-CH_3$	$-CH_2-CH_2-S-CH_3$	$-CH_2-$ (phenyl ring)	(imino acid) ring structure	$-CH_2-$ (indole ring)	$-CH(CH_3)-CH_3$
MW	71	103	57	113	113	131	147	97	186	99
pI	6.0	5.07	5.97	6.02	5.98	5.74	5.48	6.30	5.89	5.96
pK_{COOH}	2.34	1.96	2.34	2.36	2.36	2.28	1.83	1.99	2.38	2.32
pK_{NH_3}	9.69	10.28	9.60	9.68	9.60	9.21	9.13	10.60	9.39	9.62
pK_R	9.02	8.18	—	—	—	—	—	—	—	—
Solubility @ 25°C g/100 g water	16.5	NA	25.0	4.11	2.19	3.35	2.97	162.3	1.14	8.85

Figure 9.23. Properties, structures, and nomenclature for uncharged amino acids.

	ACIDIC (at pH =7)		BASIC (at pH =7)			UNCHARGED POLAR (at pH =7)				
mRNA Codon	5' 3' GAC GAU	GAA GAG	5' 3' AAA AAG	AGA AGG CGA CGC CGG CGU	CAC CAU	5' 3' AAC AAU	CAA CAG	AGC AGU UCA UCC UCG UCU	ACA ACC ACG ACU	UAC UAU
Amino Acid	Aspartic	Glutamic	Lysine	Arginine	Histidine	Asparagine	Glutamine	Serine	Threonine	Tyrosine
	Asp	Glu	Lys	Arg	His	Asn	Gln	Ser	Thr	Tyr
	D	E	K	R	H	N	Q	S	T	Y
MW	114	128	129	157	137	114	128	87	101	163
pI	2.77	3.22	9.74	10.76	7.59	5.41	5.65	5.68		5.66
pK_{COOH}	3.65	2.19	2.18	2.17	1.82	2.02	2.17	2.21	2.63	2.20
pK_{NH_3}	9.60	9.67	8.95	9.04	9.17	8.80	9.13	9.15	10.43	9.11
pK_R	3.86	4.25	10.53	12.48	6.00	—	—	—	—	10.07
Solubility @ 25°C g/100 g water	0.50	0.84	NA	NA	4.29	3.11 @ 28°C	3.6 @ 18°C	5.0	20.5	0.045

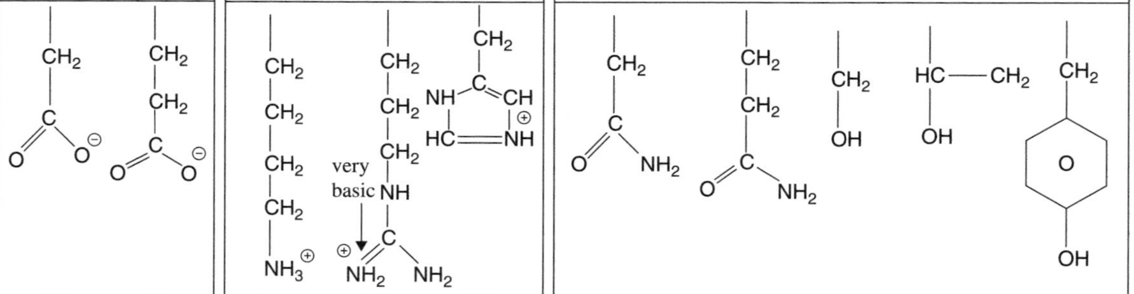

Figure 9.24. Properties, structures, and nomenclature for charged amino acids, and uncharged polar amino acids.

organism but not to spores formed by the bacteria.[8] The carbon sources for fermentations are not restricted to glucose or other sugars but can also include fatty acids and glycerol. These are transformed to acetyl-CoA (structure in Fig. 9.7), while amino acids are transformed to pyruvate or acetyl CoA, as well as forming the waste product ammonia (NH_3). Under aerobic conditions, pyruvate and acetyl CoA enter the citric acid cycle. However, glucose and some pentoses are preferred since they are less expensive than many other carbon sources.

Essential Amino Acids Not Synthesized by Microorganisms Must Be Provided as Nutrients (Auxotrophs)

While all 20 amino acids are required for normal functioning of cells (i.e., life), not all of the 20 amino acids are made by every organism. For example, 9 out of the 20 amino acids in proteins cannot be synthesized by humans, and must therefore be obtained through the diet. The amino acids that are essential to the human diet are threonine, methionine, lysine, valine, leucine, isoleucine, histidine, phenylalanine, and tryptophan. Essential amino acids are made by other organisms, "usually by long and energetically expensive pathways that have been lost in the course of vertebrate evolution" (Alberts et al. 1989, p. 70). The other 11 amino acids are called "nonessential" since they can be generated by human cells. Living organisms synthesize proteins composed of L-amino acids. D-Amino acids occur in nature in bacterial peptides that exhibit antimicrobial, opiate, or prey toxic properties (Kreil 1994; Heck et al. 1994).

The Utilization of Fats in Animals Occurs by a Non–Tricarboxylic Acid (TCA) Cycle Mechanism

Fats consist of fatty acids that are covalently bound to glycerol to form a triglyceride having the generic structure shown in Fig. 9.25 which is insoluble in water. A fatty acid is cleaved from the triglyceride and converted into a molecule of fatty acyl-CoA, which then enters a four step fatty acid oxidation pathway, in which two carbon atoms are removed from the fatty acid to form acetyl-CoA, each time the fatty acid passes through the cycle of reactions. Alberts et al. (1989) give a more detailed discussion of fatty acid metabolism.

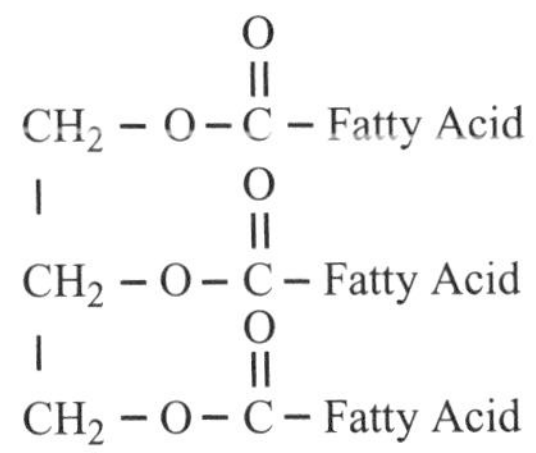

$$
\begin{array}{c}
\qquad\qquad\quad O \\
\qquad\qquad\quad \parallel \\
CH_2-O-C-\text{Fatty Acid} \\
|\qquad\qquad\quad O \\
\qquad\qquad\quad \parallel \\
CH_2-O-C-\text{Fatty Acid} \\
|\qquad\qquad\quad O \\
\qquad\qquad\quad \parallel \\
CH_2-O-C-\text{Fatty Acid}
\end{array}
$$

Figure 9.25. Glycerol forms the backbone for triglyceride fats.

---

[8] *Clostridium tetani* is another type of strict anaerobe. It causes tetanus. If a deep wound introduces spores of this bacterium into the body and the wound then cuts off blood circulation, anaerobic conditions may develop, enabling the spores to germinate and grow.

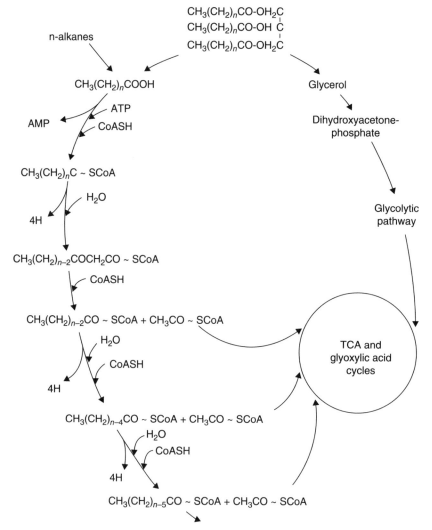

Figure 9.26. Pathways for growth of microorganisms on fat and *n*-alkanes, and oxidation of fat. The fatty acid derived from hydrolysis of the triglyceride is activated and oxidized at the β-carbon atom; the resulting β-keto acid is hydrolyzed to yield acetyl ~ SCoA units, which are further metabolized via the glyoxylic acid and TCA cycles. *n*-Alkanes are first oxidized to the corresponding fatty acid, then activated by ATP and CoA; further oxidation occurs by β-oxidation. [Figure reproduced from Aiba et al. (1973), Ref. 5, Fig. 3.10, p. 74.]

Some bacteria and fungi are able to grow on fat as a sole source of carbon. Organisms cleave the triglyceride into glycerol and its three constituent fatty acids. The glycerol enters the glycolytic pathway via dihydroxyacetone phosphate, and the fatty acids to form an acyl ~ SCoA ester in the presence of coenzyme A. This is oxidized at the β-carbon atom and the resulting β-ketoacyl-SCoA is hydrolyzed to give acetyl ~ SCoA and an acyl ~ SCoA ester with two less carbon atoms, as illustrated in Fig. 9.26 (Aiba et al. 1973).

Some Bacteria and Molds Can Grow on Hydrocarbons or Methanol in Aerated Fermentations (Single-Cell Protein Case Study)

The metabolism of hydrocarbons starts with the oxidation of the *n*-alkane to the corresponding fatty acid, followed by activation with ATP and CoA (Aiba et al. 1973). The production of yeast as a source of single-cell protein (SCP) for use as food or animal feed from petrochemical derived *n*-alkanes, gas oil or methane, was apparently considered by many petroleum companies, including British Petroleum, Chinese Petroleum (Taiwan), Dianippon and Kyowa Hakko (Japan), and Shell (Atkinson and Mavituna 1983), between about 1965 and 1975. Huge demonstration plants were designed that ranged from 1000 to 100,000 tons/year, although interest in these processes disappeared when oil price increases in the early 1970s made use of petrochemical feedstock uneconomical (Hacking 1986). According to Hacking (1986), only one large plant was operational in the United States, Canada, Britain, or continental Europe, by 1980, and this was the Pruteen process by ICI in Billingham, England. This plant was based on petroleum derived methanol as a feedstock, rather than a hydrocarbon.

Methanol was chosen since it had a high solubility in water and therefore avoided phase transfer resistances encountered when water insoluble *n*-alkanes were used. Methanol reduced the explosion hazard that was prevalent in aerated methane fermentations, and reduced the amount of oxygen required as well. A decreased oxygen requirement translated to a reduced cooling load, since the heat generated during aerated yeast fermentation is directly proportional to the oxygen consumed. The use of methanol also avoided carryover of polycylic aromatic compounds (that were possible carcinogens present in petrochemical feedstocks) into the final protein product.

Despite its large scale (70,000 tons/year produced in a 150-m^3 fermentor) and efficient design, the process was not economical because of the high methanol prices (about \$110/ton). Furthermore, competing sources of vegetable protein were less expensive. Soy meal sold at \$200/ton and fish meal at \$350–\$400/ton at the time that this analysis was published by Hacking (1986). The *capital* cost of the ICI Pruteen plant was on the order of \$100 million (1980 prices), and contributed \$150/ton to the cost of manufacturing the single-cell protein. Half of the operational cost and 40% of the total cost was due to the fermentation substrate: methanol. The equipment for dewatering and drying the an SCP product contributed 31% of the overall capital cost of the plant, compared to only 9% for the compression[9] equipment, and 14% for the fermentation system that consisted of a 150-m^3 airlift fermentor. Hacking (1986) estimated that a price in excess of \$500/ton would be necessary to make this type of process economical if the market were for animal feed. A higher-value end use, such as SCP for human food, may have made the product profitable. However, this never materialized because of regulatory concerns

---

[9] The metabolic processes that drive growth of microbial biomass are based on respiration. Therefore significant amounts of oxygen are needed. Oxygen is supplied by bubbling compressed air through the fermentation broth consisting of the carbon source, methanol, and other nutrients. A large energy-consuming air compressor is needed to supply the compressed air, which is passed through the bottom of a specially designed vessel that promotes mixing of the air bubble in the broth as the air rises. The modeling of aerobic fermentations is discussed in Chapter 6.

about carryover of hydrocarbon residues, and negative publicity about this problem encountered in another plant.

Many factors can impact the economic feasibility of a new fermentation process, even if the initial yield estimates based on the organism's metabolism appear to be favorable. The prime objective of ICI, in constructing their large plant, was apparently to gain expertise in large-scale fermentation. The experience of ICI and other manufacturers served to catalyze the development of technology for aerated fermentations for inexpensive, bulk products. Developments that may have benefited plants that followed included advances in designs for production-scale airlift fermentors, industry experience with the benefits and limits of economics of scale for fermentation technology, and improvements in downstream processing equipment that accounted for 31% of the capital costs. This example also illustrates the complexity of scaling up a biotechnology for a commodity product. Large investments in biotechnology, whether then or now, do not by themselves guarantee business success.

Extremophiles: Microorganisms that Do Not Require Glucose, Utilize H$_2$, and Grow at 80–100 °C and 200 atm Have Industrial Uses

An archaebacterium, *Methanococcus jannaschii*, discovered in 1983 by J. A. Leigh in a sediment sample collected from the bottom of the Pacific Ocean at 2600 m depth, produces energy by reducing CO_2 with H_2 to produce methane. Sequencing of this microorganism's genome showed that genes for all the known enzymes and enzyme complexes associated with methanogenesis were present (Morell 1996). While *M. jannaschii*[10] can use H_2 and formate as substrates, it lacks the genes for using methanol or acetate.

The unusual characteristics of this type of microorganism have also been found for lithoautotrophs[11] in 1000-m-deep aquifers in the basalt rocks near the Columbia

---

[10] *M. jannaschii* grows at pressures exceeding 200 atm, has an optimum temperature for growth of 85 °C, is a strict anaerobe, and produces methane (Bult et al. 1996). The complete genome of this extremophile (i.e., loves extreme conditions) was reported in 1996, and 56% of the "archeon's 1783 genes are entirely new to science, unlike any found in the two other branches of life" (eukaryotes and prokaryotes). The metabolism of this unusual microorganism is similar, in some respects, to bacteria, while genes that control translation and transcription of DNA, and genome replication are similar to yeast and other eukaryotes (Morell 1996). It should be noted that the catalytic reduction of CO_2 does not require high pressures or temperatures, as was demonstrated by the efficient synthesis of isocyanate and carbodimide from CO_2 at close to room conditions by using a bisamide that contains geranium or tin as a metal catalyst site (Freemantle 1996).

[11] An autotroph requires only CO_2 as a carbon source, while a heterotroph derives its carbon needs from organic molecules. An example of an autotroph is *Clostridium thermoaceticum*. The acetyl-CoA pathway in this anaerobe (the Wood pathway) combines two molecules of CO_2 with coenzyme A to form acetyl-CoA, which then serves as a substrate for growth. This pathway apparently occurs in some humans where acetate, rather than methane, is produced. Wood estimated that 2.3×10^6 metric tons of acetate/day is formed from CO_2 (Alberts et al. 1989). Chemoautotrophs and chemoheteroautotrophs obtain energy from chemical sources (e.g., glucose), while photoautotrophs or photoheterotrophs derive their energy from light. A lithotroph derives its energy from hydrogen, or other inorganic reductants. Lithotrophs that obtain their carbon from CO_2 and that are also autotrophs are named *lithoautotrophs*.

River in Washington State. These organisms use H_2 that evolves from the basalt and water (Borman 1995; Kaiser 1995). One hypothesis places these microorganisms on Earth about 2.8 billion years ago, before photosynthesis evolved (Borman 1995) with fossils of photosynthetic *Grypania spiralis* from the Empire mine (Marquette, Michigan) placing the origin of organelle-bearing eukaryotic cells at about 2.1 billion years ago. *Grypania* are photosynthetic alga that grew to 0.5 m in length, and were 2 mm wide (Han and Runnegar 1992).

The utility of these organisms, and the enzymes that they produce, include applications for heat-stable enzymes in chemical processing, cleaning up toxic wastes, food processing, wood pulping, beer brewing (News Item 1996; Burton 1997), and other processes that require temperatures in the 80–100 °C range or are carried out at biologically extreme conditions. An example of the commercial utility of an enzyme derived from an extremophile is the DNA polymerase used for the polymerase chain reaction (PCR), which was isolated from the Mushroom Pool, a hotspring in Yellowstone Park (Burton 1997). Other examples include an alkaline cellulase that opens up amorphous regions of cellulose to release dirt (Japan's "Attack" detergent), and an alkaline amylase that joins glucose molecules together to make circular starch molecules, known as *cyclodextrins*, that are used to sequester flavoring agents in foods (Myers and Anderson 1992).

The Terminology for Microbial Culture Is Inexact: "Fermentation" Refers to Both Aerobic and Anaerobic Conditions While "Respiration" Can Denote Anaerobic Metabolism

Bailey and Ollis (1977) succinctly define "the sequence of steps that extract energy under anaerobic conditions" as fermentations. This convention follows that of Pasteur, who called fermentation "life without air." Since most early microbiological processes such as winemaking were anaerobic, the term was also used to describe these processes. Microbial conversion processes are usually called *fermentations* in industry even though many of these processes are aerobic. "The exact meaning of fermentation must often be inferred from context" (Bailey and Ollis 1977, p. 232). This chapter, as well as other sections of the book, attempt to clearly define the nature of the transformations carried out by living organisms, although the term fermentation will be used in an industrial context, that is, to denote microbial cultures and transformations under both aerobic and anaerobic conditions.

The definition of respiration in the context of industrial processes is also sometimes unclear. While respiration is a process where organic or reduced inorganic compounds are oxidized by inorganic compounds to produce energy, oxidants other than molecular oxygen (O_2) can be involved. For example, sulfate (SO_4^{2-}) is the oxidant for *Desulfovibrio*, and nitrate (NO_3^-) is used by denitrifying bacteria. Bacterial sulfate reduction is believed to be possible at temperatures of up to 110 °C, according to measurements made at deep ocean hydrothermal vents (Jorgenson et al. 1992).

Most industrial fermentations are aerobic and have metabolisms that can be characterized by aerobic respiration. *Anaerobic respiration* refers to metabolism in which an oxidant other than oxygen is involved. *Aerobic respiration* refers to processes where O_2 is the oxidizing agent.

METABOLISM AND BIOLOGICAL ENERGETICS

Biochemical processes of energy extraction and chemical reducing power from the environment have been described in this chapter. Energy and redox potential are utilized by all living cells to synthesize the biochemical building blocks for growth and reproduction. Balancing the needs of the living cell against the production of useful chemicals through metabolic engineering requires an understanding of the biochemical pathways as well as the biochemical energetics involved.

The biochemical reactions of all metabolic pathways, like all chemical reactions, must obey the laws of thermodynamics. The presence of enzyme catalysts increases the rate of the reaction but cannot overcome thermodynamics. Thermodynamically unfavorable reactions are coupled with the thermodynamically favorable hydrolysis of ATP to ADP to make the composite reaction favorable. Electrons transferred from the oxidation of NADH to NAD^+ (or NADPH to $NADP^+$) are also required to balance the redox of numerous chemical reactions. Understanding biochemical energetics is as important as understanding the metabolic pathways that make up the biochemical reactions of a cell. The focus of the next chapter is on biological energetics.

REFERENCES

Aiba, S., A. E. Humphrey, and N. F. Millis, *Biochemical Engineering*, 2nd ed., Academic Press, New York, 1973, pp. 57, 68.

Alberts, B., D. Bray, J. Lewis, M. Raff, K. Roberts, and J. D. Watson, *Molecular Biology of the Cell*, 2nd ed., Garland Publishing, New York, 1989, pp. 58, 65, 78, 205–216, 341, 345–348, 353–354, 359, 362–363, 373, 381–383, 601.

Atkinson, B. and F. Mavituna, *Biochemical Engineering Biotechnology Handbook*, the Nature Press, New York, NY (1983).

Bailey, J. E. and D. F. Ollis, *Biochemical Engineering Fundamentals*, McGraw-Hill, New York, 1977, pp. 3, 14–15, 221–254, 338–348, 357, 371–383.

Borman, S., "Newly Found Bacteria Thrive on Rocks and Water," *Chem. Eng. News* 73(43), 8–9 (1995).

Bult, C. J., O. White, G. J. Olsen, L. Zhou, R. D. Fleischmann, G. G. Sutton, J. A. Blake, et al., "Complete Genome Sequence of the Methanogenic Archeon, Methanococcus jannaschii," *Science* 273(5278), 1058–1072 (1996).

Burton, T. H., "Yellowstones Geyers Spout Valuable Microorganisms," *Wall Street J.* B1, B7 (Aug. 11, 1997).

Chandler, C., "U.S. Suits Cloud Showa Denko's Future," *Wall Street J.* (Aug. 17, 1990).

Freemantle, M., "Carbon Dioxide Fixation: Facile Method Converts CO_2 to Industrially Important Compounds," *Chem. Eng. News* 74(46), 8 (1996).

Hacking, A. J., *Economic Aspects of Biotechnology*, Cambridge Studies in Biotechnology 3, Cambridge Univ. Press, 1986, pp. 77–79, 98–102.

Hamilton, B. K., H.-Y. Hsiao, W. E. Swann, D. M. Anderson, and J. J. Delente, "Manufacture of L-Amino Acids with Bioreactors," *Trends Biotechnol.* 3(1), 64–68 (1985).

Han, T.-M. and B. Runnegar, "Megascopic Eukaryotic Algae from 2.1 Billion Year-Old Negaunee Iron-Formation, Michigan," *Science*, 257(5067), 232–235 (1992).

Heck, S. D., C. J. Siok, K. J. Krapcho, P. R. Kelbaugh, P. F. Thadeio, M. J. Welch, R. D. Williams, et al., "Functional Consequences of Post-Translational Isomerization of Ser46 in a Calcium Channel Toxin," *Science* **266**(5187), 1065–1068 (1994).

Jorgenson, B. B., M. F. Isaksen, and H. W. Jannasch, "Bacterial Sulfate Reduction above 100°C in Deep-Sea Hydrothermal Vent Sediments," *Science* **258**(5089), 1756–1757 (1992).

Kaiser, J., "Can Deep Bacteria Live on Nothing but Rocks and Water?" *Science* **270**(5235), 377 (1995).

Kilman, S., "In Archer-Daniels Saga, Now Executives Face Trial: Price-Fixing Case Will Be Biggest Criminal Antitrust Showdown in Decades," *Wall Street J.* B10 (July 9, 1998a).

Kilman, S., "Jury Convicts Ex-Executives in ADM Case," *Wall Street J.*, A3–A4 (Sept. 18, 1998b).

Kilman, S., "Ajinomoto Pleads Guilty to Conspiring with ADM, Others to Fix Lysine Price," *Wall Street J.* B5 (Nov. 15, 1996a).

Kilman, S., "ADM Settles Two Suits for $65 Million, Raising Hopes for Plea in Federal Probe," *Wall Street J.* A3, A6 (Sept. 30, 1996b).

Kinoshita, S., "Glutamic Acid Bacteria," in *Biology of Industrial Microorganisms*, A. L. Demain and N. A. Solomon, eds., Benjamin/Cummings, Menlo Park, CA, 1985, pp. 115–142.

Kreil, G., "Conversion of L to D-Amino Acids, a Post-Translational Reaction," *Science* **266**(5187), 996–997 (1994).

Ladisch, M. R., "Fermentation-Derived Butanol and Scenarios for Uses in Energy Related Applications," *Enzym. Microbial Technol.* **13**, 280–283 (1991).

Morell, V., "Life's Last Domain," *Science* **273**(5278), 1043–1045 (1996).

Murray, R. K., P. A. Mayes, D. K. Granner, and V. W. Rodwell, *Harper's Biochemistry*, 22nd ed., Appleton & Lange, Norwalk, CT, 1990, pp. 10–14, 58, 68–69, 102–106, 146, 160–161, 165, 199–200, 318–322, 336–338, 549–550, 555.

Myers, F. S. and A. Anderson, "Microbes from 20,000 feet under the Sea," *Science* **255**(5040), 255–256 (1992).

Nakao et al. *Agric. Biol. Chem.* **37**, 2399 (1973).

News Item, "Decoding of Microbe Genome Suggests Biotech Applications," *Genet. Engi. News* **16**(15), 4 (1996).

Purves, W. K., D. Sadava, C. Heller, and G. H. Orians, *Life: The Science of Biology*, 7th ed., Sinauer Associates and W. H. Freeman, 2003.

Roberts, L., "L-Tryptophan Puzzle Takes New Twist," *Science* **249**(4972), 988 (1990).

Shuler, M. L. and F. Kargi, *Bioprocess Engineering: Basic Concepts*, Prentice-Hall, Englewood Cliffs, NJ, 1992, pp. 48–56, 59, 158–159.

Vasic-Racki, D. J., M. Jonas, C. Wandrey, W. Hummel, and M.-R. Kula, "Continuous (R)-Mandelic Acid Production in An Enzyme Membrane Reactor," *Appl. Microbiol. Biotechnol.* **31**, 215–222 (1989).

CHAPTER NINE
HOMEWORK PROBLEMS

9.1. *Glutamic Acid Fermentation.* Biotin is added in trace amounts (0.1 mg/L) to a microbial fermentation that transforms glucose into glutamic acid. The productivity, which is related to cell mass, increases. When the biotin concentration is increased further to (1 mg/mL) in a second experiment, cell growth slows and glutamic acid production is also lower. When biotin is not added, the cells do not grow, and glutamic acid is not produced. Substrate depletion is not a factor since $S \gg K_s$ in all three experiments, where

$$\mu = \frac{\mu_{max}[S]}{K_s + [S]}$$

While accumulation of glutamic acid in the fermentation broth does not cause inhibition of growth, intracellular accumulation of glutamic acid does cause inhibition.

Explain the role of biotin in achieving maximal glutamic acid accumulation in the fermentation broth and relate to the pathway for glutamic acid production.

9.2. *Mass Balance from Knowledge of Glycolytic Pathway.* The provisions of the Clean Air Act Amendment require that ground-level ozone concentrations, CO, and volatile organics be controlled, with particular emphasis on non-attainment (geographic) areas. These levels can be achieved by blending oxygenates into gasolines, which are the major source of these emissions. Ethanol is one of the oxygenates being used.

Your company is considering producing fuel ethanol from waste cellulose. The initial objective is to estimate the fermentation volume and supply of municipal solid waste (containing 50% cellulose, dry-weight basis) required to produce 20 million gal/year of fuel-grade alcohol containing 0.6% water. The microorganism to be used is *Saccharomyces cerevisiae* var. anamensis with a 16.7% glucose feed concentration. How many tons per year of municipal solid waste (dry basis) will be needed for a 20-million-gal/year facility? How many liters of fermentation volume will be needed to produce this amount? Assume that the turnaround time between batches is 8 h, with an additional 4 h for the lag phase. The working volume of the fermentor, which is agitated, is 70% of the total volume.

9.3. *Citric Acid Production.* An alternate business strategy is to produce citric acid from dextrose obtained from corn at a plant located in the Decatur, Illinois area. What is the theoretical maximum yield obtainable from a bushel of corn? Assume that there are 56 lb of corn per bushel, at a moisture level of 15.5%, and a starch content of 75% (dry-weight basis). Express your answer in terms of lb citric acid/lb glucose and lb citric acid/bu (bushel) of corn. Is the theoretical yield for citric acid higher (or lower) than that for ethanol?

9.4. *Glutamate Production.* An abbreviated representation of metabolic pathways leading to various amino acids is as shown. A *Corynebacterium* species, which is biotin-deficient and lacks α-ketoglutarate dehydrogenase, utilizes some part of this pathway to produce glutamate (i.e., sold as monosodium glutamate or MSG). In commercial processes using this microbe, the yield is 50% of (molar) theoretical when the biotin content in the medium is maintained at 2 μg/L. Broth concentrations of glutamate reach up to 90 g/L (see Fig. HP9.4).

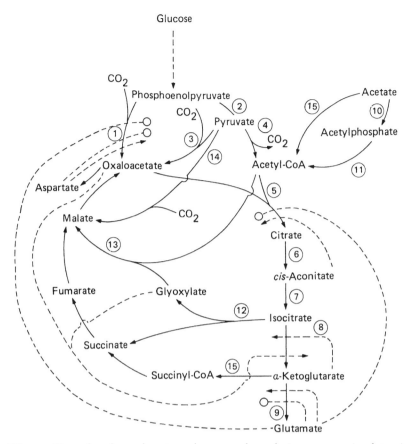

Figure HP9.4. Central and anaplerotic pathways and regulation patterns in glutamic acid bacteria: 1—phosphoenolpyruvate carboxylase; 2—pyruvate kinase; 3—pyruvate carboxylase; 4—pyruvate dehydrogenase; 5—citrate synthase; 6,7—aconitase; 8—isocitrate dehydrogenase; 9—glutamate dehydrogenase; 10—acetate kinase; 11—phosphate acetyltransferase; 12—isocitrate lyase; 13—malate synthetase; 14—malic enzyme; 15—α-ketoglutarate dehydrogenase, (—>, inhibition; - - -○, repression) [reproduced from Kinoshita (1985). Fig. 5.4, p. 127].

(a) What key step is blocked that allows glutamate to be produced? Draw a likely pathway to glutamate formation. Explain (briefly).

(b) Biotin deficiency results in production of cell membranes deficient in phospholipids (Nakao et al. 2399, 1973), and thereby allows glutamate to be excreted. When the biotin concentration is increased from 2 μg/L to 15 μg/L, the glutamate concentration in the broth levels off. In another fermentation, when penicillin is added during log growth phase, excretion of phospholipids is noted, and glutamate production appears to be triggered. Give a possible explanation of why decreased biotin levels or penicillin addition during log-phase growth gives high yields of glutamate.

(c) Using a different microbial strain, it is claimed that glutamate can be produced from ethanol at 66% of theoretical yield (based on alcohol consumed) to give product concentrations of 60 g/L. If ethanol costs \$1.65/gal and dextrose costs \$0.11/lb, which of the two processes would have lower substrate cost:

1 Glucose → glutamate?

2 ETOH → glutamate?

Explain. (Show your work!). Calculate substrate costs in ¢/lb glutamate.

(d) What is the total annual production of MSG (worldwide basis)?

(e) Can Y_{ATP} be calculated for glutamate production? If you have insufficient data, explain what would be needed.

9.5. *2,3 Propanediol Production.* The calculation of theoretical yields of propanediol from a fermentable substrate enables estimation of substrate costs for the best-case scenarios. Assume that glucose costs 5¢/lb (calculated on a dry-weight basis). What is the cost of substrate if

(a) A hypothetical electrochemical route is able to split water and <u>all</u> of the glucose is converted to a mixture of 1,2-propanediol and 1,3-propanediol. The reaction sequence for producing the propanediol is as shown in Eq. (HP9.5), with the [H] requirement obtained from the dissociation of the water:

$$C_6H_{12}O_6 + 8[H] \rightarrow 2C_3H_8O_2 + 2H_2O \qquad (\text{HP9.5})$$

(b) What is the maximum yield of propanediol per lb glucose if the glucose provides both the carbon that is incorporated into the propanediol, and sufficient hydrogen (reducing equivalents) needed for formation of the propanediol, as well as energy requirements of the cell (results in formation of CO_2)? Explain your rationale, and show your work.

(c) What is the best possible yield under anaerobic conditions? Explain your rationale, and show your work.

(d) What is the best possible yield under aerobic conditions (involving the TCA cycle)?

BIOLOGICAL ENERGETICS

INTRODUCTION

The functioning of the cell's metabolism depends on capture of and transfer of energy in the form of electrons or reducing equivalents, and high-energy phosphate bonds. The energy is used for synthesis as well as for moving food sources into the cell and wastes out of the cell. Microbial cells are able to generate and utilize energy under both aerobic and anaerobic conditions as discussed in Chapter 9.

REDOX POTENTIAL AND GIBBS FREE ENERGY IN BIOCHEMICAL REACTIONS

The Gibbs free-energy change of a chemical reaction represents the maximum amount of work that can result from the reaction at a constant temperature. The Gibbs free-energy change provides a numerical measure of the tendency of a reaction to occur spontaneously. Reactions with negative (less than zero) Gibbs free energy will occur spontaneously. Reactions with positive (greater than zero) Gibbs free energy can occur only if energy is supplied to the reaction. The reduction–oxidation potential relates reduction, or the gain of electrons, of one molecule, to the oxidation, or removal of electrons, of a second molecule (Murray et al. 1990). The reduction–oxidation (abbreviated redox)[1] potential provides a numerical measure of the tendency of reactants to accept or donate electrons. For example, the reaction of O_2 with hydrogen, represented in Eq. (10.1), has a redox potential

---

[1] Redox is the abbreviation used to denote reduction–oxidation potential. The nomenclature in the literature pairs the reactant whose bond releases energy with the resulting product. Thus the formation of CO_2 from formic acid is denoted by formic acid/CO_2 and the formation of succinic acid from fumaric acid is a fumaric acid/succinic acid pair.

Modern Biotechnology, by Nathan S. Mosier and Michael R. Ladisch
Copyright © 2009 John Wiley & Sons, Inc.

E_0 of +0.82 V for a mixture containing 0.5 mol O_2 (=1 mol O), and 2 mol H^+. The large positive value of the redox potential reflects the strong tendency of oxygen to accept electrons (Alberts et al. 1989):

$$\frac{1}{2}O_2 + 2e^- + 2H^+ \rightarrow H_2O, \quad E_0 = +0.82 \text{ V at } 25°C \text{ pH } 0 \qquad (10.1)$$

This reaction proceeds in the direction for which the free-energy change at a standard state $\Delta G°$ is negative. For equimolar amounts of oxygen (O), hydrogen (H), and H_2O, the free-energy change is $\Delta G° = -37.8$ kcal/mol, at the standard state (25°C, pH 0). Hence, the formation of water is strongly favored. Even though this reaction is thermodynamically possible, it does not occur unless an event initiates it (e.g., if a flame is applied to gaseous H_2 and oxygen) or appropriate enzymes are present in the liquid phase. Additionally, while a large negative $\Delta G°$ indicates that the formation of water is thermodynamically favorable, this does not give us any insight into the rate of the reaction, once initiated.

The free-energy change at the standard state of 25°C, pH 0, and reactants at 1.0 mol/L (abbreviated M) is insufficient for determining whether a reaction will proceed for any chemical or biochemical reaction. A thermodynamically favorable reaction may require a catalyst if the reaction is to occur at a measurable rate. An unfavorable thermodynamic reaction may occur anyhow, if coupled to another reaction that provides favorable energetics or displaces the equilibrium concentration of the product. For example, consider the conversion of glyceraldehyde-3-phosphate to dihydroxyacetone phosphate (between reaction steps 4 and 5 in Fig. 10.1a). This reaction sequence (Fig. 10.3) shows the glycolytic pathway that occurs under anaerobic conditions in which the cell utilizes 1 mol glucose to generate 2 mol ATP (Fig. 10.2) and 2 mol NADH. The pyruvate that is formed may yield a number of useful products under anaerobic conditions (Fig. 10.1b).

The free-energy change of −1830 cal/mol favors dihydroxyacetone phosphate. The equilibrium reaction sequence can be represented by

$$aA \rightleftharpoons aA^* ... bB^* \rightleftharpoons bB \qquad (10.2)$$

where $a + b = 1$; A^*, B^* represent a transition state; and A and B are represented by Fig. 10.2. However, the concentration of glyceraldehyde-3-phosphate (Fig. 10.3) is less than its equilibrium concentration since it is continually processed into pyruvate (see reaction pathways in Fig. 10.1a), hence pulling the reaction away from the dihydroxyacetone phosphate.

The concentration of cell constituents is likely to be less than 1 M, and hence the condition for the standard state would not be satisfied. Table 10.1, in which Battley et al. (1991) have assigned approximate molecular weights to represent the different types of molecules, shows that the concentration of these macromolecules is likely to be significantly less than 1 M. The concentration of low-molecular-weight precursors from which these macromolecules are synthesized are also likely to be less than 1 M. For example, even if 10% (or 64 mg) of the weight of protein in *E. coli* cells (Table 10.1) were in the form of amino acids (average MW assumed at 130), their concentration would be less than 0.5 M.

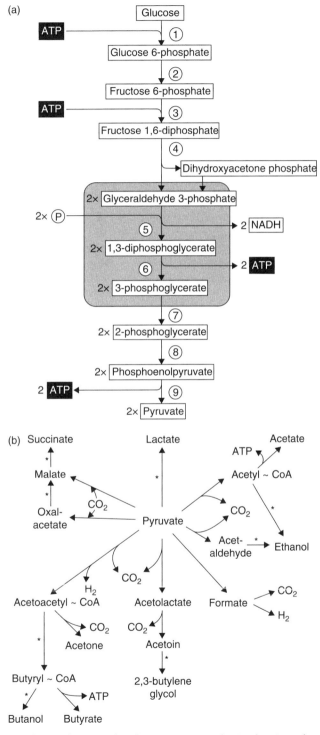

Figure 10.1. (a) Pathway showing glycolysis. Step 4 results in cleaving of a 6-carbon sugar (fructose 1,6-diphosphate) into two 3-carbon sugars. Steps 5 and 6 result in net synthesis of ATP and NADH molecules [From Murray et al. (1990), Fig. 2.20, p. 66]. (b) Products from anaerobic metabolism of pyruvate by the indicated microorganisms do not involve the citric acid cycle [from Aiba et al. (1973), Fig. 3.7, p. 67].

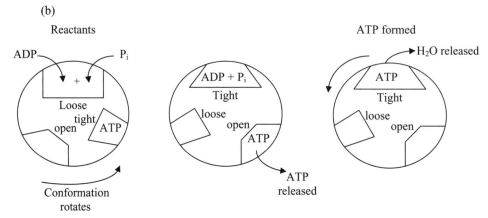

Figure 10.2. (a) Structures representing ATP, ADP, and AMP [from Aiba et al. (1973), p. 57, Academic Press]; (b) partial representation of ATP synthase [adapted from Borman (1997)].

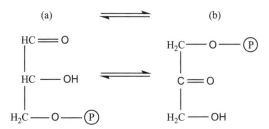

Glyceraldehyde 3-phosphate   **Dihydroxyacetone phosphate**

Figure 10.3. Equilibrium reaction between glyceraldehyde 3-phosphate and dihydroxyacetone phosphate.

   Since the standard condition (1 M concentration) is not appropriate for determining Gibbs free energies for reaction occurring in vivo, the Gibbs free energy must be determined on the basis of the actual, or estimated, concentration. The free-energy change in terms of concentration is (Levenspiel 1972)

Table 10.1.

Concentrations of Macromolecular Cellular Components of E. coli K-12 Based on Analysis of Their Constituent Monomers

Component	Per g Dry Cell		Approximate Average MW[a] (Da)	$\mu mol^b/L$
	μmol of Monomer	Weight (mg)		
Proteins	5081	643	50,000	12.9
Nucleotides				
RNA	630	216	100,000	2.2
DNA	100	33	2×10^6	1.65×10^{-5}
Lipopolysaccharides	218	40	1,000	40.0
Peptidoglycan	166	28.4	10,000	2.8
Polyamines	41	2.2	1,000	2.2
Total	6236	962.6	NA	NA

[a]Authors' estimates (magnitude of order).
[b]Assumes that volume of a single cell is 10^{-15} L, and 1 g dry cells = 5 g wet cells (equivalent to 0.004 L water/g dry cells or 250 g dry cells/L, based on water held *inside* the cell only. This does not include water adhering to or trapped between cells [this estimate is based on Bailey and Ollis (1977)]. *Source*: Battley (1991), Table I, p. 336.

$$\Delta G_b = (\text{free energy of product(s)}) - (\text{free energy of reactant(s)})$$

$$= (bG_B) - (aG_A) = -RT \ln K_{eq} = -RT \ln \frac{[B]_{eq}^b}{[A]_{eq}^a} \qquad (10.3)$$

The equation denotes the equilibrium concentrations of A and B, while a and b are the stoichiometric coefficients. Rodwell [in Murray et al. (1990)] derives the expression of free energy given in equation terms of a half-cell reaction equation (10.4a), where ΔG_F is the free energy associated with the formation of the transition state, [A* ... B*], to form products. The corresponding expression for the formation of the transition state is given by

$$\Delta G_F = \Delta G_F^0 + RT \ln \left[\frac{A^*...B^*}{A} \right] \qquad (10.4a)$$

where A* ... B* represents the transition state, R is the gas law constant and T is the absolute temperature. The expression for the free energy of dissociation ΔG_D is

$$\Delta G_D = \Delta G_D^\circ + RT \ln \left[\frac{B}{A^*...B^*} \right] \qquad (10.4b)$$

The degree (°) superscript represents the standard state (20 °C, pH 0; all concentrations are 1 M). The overall free-energy change ΔG results from the combination of ΔG_F and ΔG_D:

$$\Delta G = \Delta G_F + \Delta G_D$$

$$= \left(\Delta G_F^\circ + \Delta G_D^\circ \right) + RT \left(\ln[A * \ldots B *] - \ln A + \ln B - \ln[A * \ldots B *] \right) \quad (10.4c)$$

$$\Delta G = \Delta G^\circ + RT \ln \left[\frac{B}{A} \right]$$

ΔG indicates how far the reaction is from equilibrium, where $R = 2 \times 10^{-3}$ kcal/(mol.K).

The standard free-energy change ΔG° is normally calculated or measured at 20 °C, pH 0; and concentrations of all species, at 1 M. Biological systems may operate at temperatures of 4–60 °C, although 25 °C is an accepted base temperature. However, the pH will likely be far from 0 (and closer to 7). Since pH influences Gibbs free energy, Gibbs free energy is standardized for biochemical thermodynamic conditions of 25 °C and pH 7.0, and the concentration of all molecular species is still assumed to be 1 M. The symbol $\Delta G^{\circ\prime}$, is used in the biochemistry literature to indicate that the value is based on biochemical standard conditions, indicated by the prime sign next to the degree superscript.

At equilibrium the change in free energy is zero ($\Delta G = 0$), and Eq. (10.4c) gives an expression for the standard free energy ΔG° for equilibrium concentrations of $[A]_{eq}$ and $[B]_{eq}$ as shown in the following equation:

$$\Delta G^\circ = -RT \ln \left(\frac{[B]_{eq}}{[A]_{eq}} \right) = -RT \ln K_{eq} \quad \text{at 20 °C, pH 0} \quad (10.4d)$$

While measured concentrations of metabolic intermediates for prokaryotes are difficult to find, an example from Lehninger for human erythrocytes, cited by Bailey and Ollis (1977), places concentrations of intermediates in the Embden–Meyerhoff glycolysis pathway at between 14 and 138 μmol/L or about 10,000 times less than the concentration at the standard state. Despite this, Gibbs free energies at the standard state are able to explain the energetics of the cell.

Levenspiel (1972) points out that the standard states commonly chosen for thermodynamics of solutes in liquid are either at 1 molar concentrations, or at such dilute concentrations that the solution activity approaches unity. Hence, if the standard free energy of Eq. (10.4d) is defined in terms of fugacities,[2] Eq. (10.5) results

$$\Delta G^\circ = -RT \ln K_{eq} = -RT \ln \frac{(f/f^\circ)_B^b}{(f/f^\circ)_A^a} \quad \text{at 25 °C, pH 0} \quad (10.5)$$

where f = fugacity of the component at equilibrium conditions, f° = fugacity of the component at arbitrarily selected standard state at temperature T, and the superscripts represent the stochiometric coefficient for the reaction $aA \rightarrow bB$

---

[2] *Fugacity* is related to the chemical potential by $\mu = RT \ln f + \Phi$, where Φ is a function of temperature and μ, the chemical potential, is defined for one mole of material. It corrects for deviations of concentrated solutions from the ideal behavior associated with dilute solutions (Smith and van Ness 1959). The fugacity ratio for a solid component is 1 ($f/f^\circ = 1$). For a gas the fugacity coefficient corrects for deviation from ideal gas behavior.

The expression in Eq. (10.5) simplifies to a ratio of concentrations since $(f/f°)_A \rightarrow [A]$, and $(f/f°)_B \rightarrow [B]$ for dilute solutions. This approximation may require reexamination if quantitative modeling of metabolic pathways is the goal. However, for purposes of explaining the energetics of the cell, this appears to be acceptable (Alberts et al. 1989; Lewin 1997; Murray et al. 1990; Aiba et al. 1973; Shuler and Kargi 1992; Bailey and Ollis 1977).

The relation between the standard free-energy change at equilibrium $\Delta G°$ and redox potential is given by Aiba et al. as (1973)

$$\Delta G° = nF \, \Delta E_0 \quad \text{at } 20°C, \text{ pH } 0 \tag{10.6}$$

where n = number of electrons transferred

F = Faraday's constant [23,063 cal/(V.eq)]

ΔE_0 = difference in potential (volts) between product and reactant

These authors point out that for biological systems, n is generally 2 (since two electrons are transferred). A prime sign is used since the system is at standard biochemical conditions (Aiba et al. 1973, p. 58):

$$\Delta G' = -nF \, \Delta E_0' = (-2)\left(23,063 \frac{\text{cal}}{\text{V} \cdot \text{eq}}\right)\Delta E_0' = -46.1, \quad \Delta E_0' \text{ kcal/eq} \tag{10.7}$$

For molecules where equivalents and moles are the same, $\Delta G'$ is expressed as kcal/mol by Eq. (10.7).

The reaction will proceed in the direction in which the free energy is less than zero or the redox potential is positive. The NAD$^+$/NADH equilibrium is shown in Fig. 10.4 (Aiba et al. 1973) has $E_0 = -0.32 \, V$.

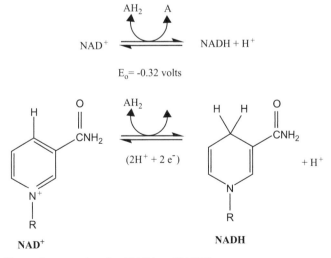

Figure 10.4. The redox reaction for NAD$^+$ to NADH.

(a)

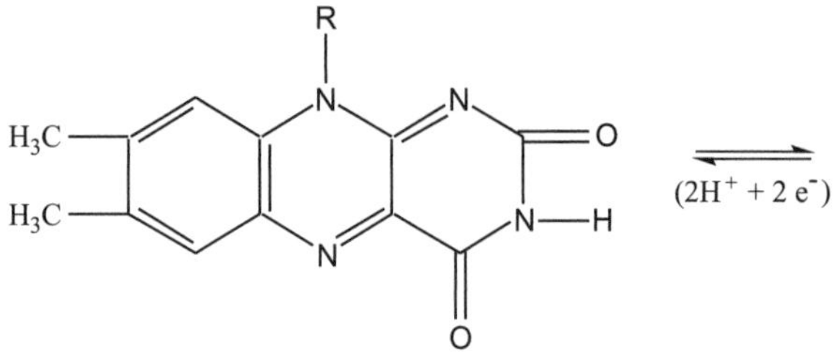

FAD

$$\overrightarrow{}$$

$(2H^+ + 2 e^-)$

(b)

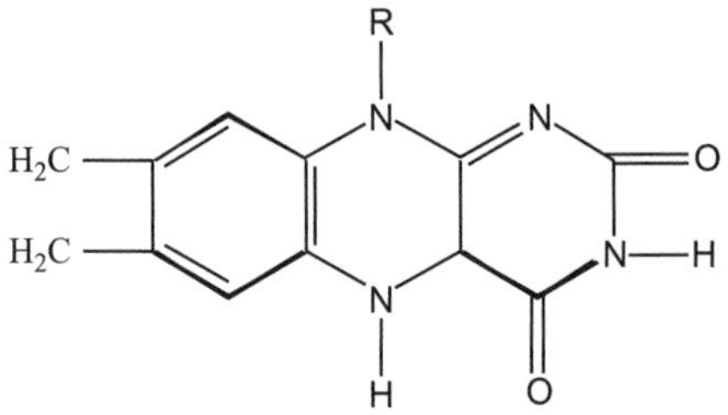

FADH

Figure 10.5. The redox reaction for FAD$^+$ to FADH [part (b) adapted from Borman (1997)].

The reaction for flavin adenine mononucleotide (FAD), given in Fig. 10.5, is analogous to that for NADH (Fig. 10.4), as presented in Chapter 9. The R in Fig. 10.4 represents the adeninine phosphate. The free-energy change is $\Delta G = +14.8$ kcal/mol, while the corresponding redox potential is $E_0' = -0.32$ V. For an equimolar mixture of NAD$^+$ and NADH (Alberts et al. 1989, p. 362; Murray et al. 1990, p. 106), the negative redox potential reflects a strong tendency of the NADH to donate electrons.

Redox potentials for selected reactions in mammalian systems are given in Table 10.2. The redox potential for H/H$_2$ is shown as -0.42 V since the standard state for biological systems is taken at pH = 7. Otherwise, the potential would be 0.0 V if the usual convention were followed of basing the redox potential on a hydrogen electrode at pH = 0. A mixture of NAD$^+$/NADH has a redox potential of -0.32 V, thus indicating that NADH has a strong potential to give up electrons. In comparison, oxygen/water has a redox potential of $+0.82$ V, and therefore O$_2$ has a strong potential to accept electrons and form water.

An example of the utility of redox potential for calculating the energy efficiency in a living cell is given by Aiba et al. (1973). If two electrons pass from NADH + H$^+$ ($E_0 = -0.32$ V) to oxygen ($E_0 = +0.82$ V), the change in potential is $\Delta E' = (0.82 - (-0.32)) = 1.14$ V, and the free energy would be

$$\Delta G' = -(2)(23.1)(1.14) = -52.7 \frac{\text{kcal}}{\text{eq}} \quad \text{at } 25\,^\circ\text{C, pH 7}$$

$$|e^-|\frac{\text{kcal}}{\text{V}\cdot\text{eq}}\bigg|\frac{\text{V}}{e^-}$$

(10.8)

Table 10.2.

Redox Potential of Selected Reaction Pairs

System	E_0 (V)
Succinate/α-ketoglutarate	−0.67
H^+/H_2	−0.42
$NAD^+/NADH$	−0.32
Lipoate; ox/red	−0.29
Acetoacetate/β-hydroxybutyrate	−0.27
Pyruvate/lactate	−0.19
Oxaloacetate/malate	−0.17
Flavoprotein–old yellow enzyme; ox/red	−0.12
Fumarate/succinate	+0.03
Cytochrome b; Fe^{3+}/Fe^{2+}	+0.08
Ubiquinone; ox/red	+0.10
Cytochrome c; Fe^{3+}/Fe^{2+}	+0.22
Cytochrome a; Fe^{3+}/Fe^{2+}	+0.29
Oxygen/water	+0.82

Source: Murray et al. (1990), Table 13.1, p. 106 (given at 25 °C, pH 7).

A very important carrier of free energy within living cells is ATP. The structures of ATP, ADP, and AMP are illustrated in Fig. 10.2a, and the process by which ATP is formed from ADP and inorganic phosphate in Fig. 10.2b. The relative rotation of three catalytic domains of a single-enzyme synthesize ATP in animal mitochondria, bacterial membranes, and plant chloroplasts. The enzyme that catalyzes the formation of ATP from ADP and P_i is ATP synthase (abbreviated ATPase). The enzyme has three binding states—loose, tight, and open—with the enzyme rotating rapidly during the process (Borman 1997). ADP and inorganic phosphate are bound to the enzyme's loose state, and then converted to ATP through an energy-driven conformational conversion from the loose state to the tight state (Fig. 10.2). The ATP is released during the enzyme's open state. The three catalytic sites of the enzyme pass through these conformational states, and produce ATP, during the course of three cycles.[3] Figure 10.3 gives a partial representation of the different catalytic sites that participate in the formation of ATP.

---

[3] The 1997 Nobel Prize in Chemistry was awarded to Paul Boyer of UCLA and John E. Walker of Cambridge for elucidating this chemical mechanism between about 1980 and 1994. Jens C. Skou of Aarhus University, Denmark, received the other half of the Chemistry Nobel Prize for discovering Na$^+$–K$^+$ ATPase, which promotes directed transport through cell membranes, and consumes about "one-third of all the ATP produced in the body in the process." This was discovered in the late 1950s (Borman 1997).

The free energy in a high-energy phosphate bond, at conditions that normally exist in a cell, is in the range from −11 to −13 kcal/mol, with an average value of −12 kcal/mol used in this calculation (Alberts et al. 1989, p. 354; Aiba et al. 1973). The redox potential would therefore suggest that the energy released from $NADH + H^+$ is sufficient to form about 4.39 high-energy phosphate bonds (= 52.7/12) (Aiba et al. 1973, p. 60). Yet, experiments have shown that only three high-energy bonds are formed. Consequently, this transfer captures about 68% of the energy (=3/4.39 × 100) in the form of high-energy phosphate bonds. The rest is dissipated as heat, which is also a product of metabolism.

The free-energy change $\Delta G^{\circ\prime}$ at *standard biochemical* conditions (pH 7.0, 20 °C, 1 M concentrations) for the hydrolysis of ATP as shown here

$$ATP + H_2O \rightarrow ADP + P_i \qquad (10.9)$$

is about −7.3 kcal/mol (Bailey and Ollis 1977). This is significantly different from the free energy for the same reaction at conditions that exist inside the cell. The standard free energy is for conditions where ATP, ADP, and P_i (inorganic phosphate) are all present at the same concentration of 1 mol/L. If the concentrations of ADP and P_i were larger than ATP, ΔG would be smaller. There is also a concentration of the three species where $\Delta G = 0$, and Eq. (10.4c) becomes Eq. (10.4d) when the concentration of ATP, ADP, and P_i in Eq. (10.9) is at equilibrium. ATP, reacting as shown in Eq. (10.9), gives a constant value of $\Delta G^{\circ\prime}$ of

$$\Delta G^{\circ} = -RT \ln K_{eq} \quad \text{where} \quad K_{eq} = \frac{[ADP]_{eq}\,[P_i]_{eq}}{[ATP]_{eq}} \qquad (10.10)$$

Therefore, ΔG in Eq. (10.4c) gives a measure of how far the reaction is from equilibrium. Alberts et al. (1989, p. 354) point out that "because the efficient conversion of ADP to ATP in mitochondria maintains such a high concentration of ATP relative to ADP and P_i, the ATP hydrolysis reaction in cells is kept very far from equilibrium and ΔG is correspondingly very negative. Without this disequilibrium, ATP hydrolysis could not be used to direct the reactions of the cell, and many biosynthetic reactions would run backward rather than forward."

HEAT: BYPRODUCT OF METABOLISM

The hydrolysis of high-energy bonds that occurs as a part of the cell's metabolism generates heat. Heat generation and dissipation are important considerations for all fermentations. For large-scale fermentations, understanding the biochemical formation of heat is important for engineering bioprocesses that remove heat from the reactor. Studies with *E. coli* K-12, grown on succinic acid, showed that the enthalpy change accompanying the growth process involves both anabolism and catabolism. The combined enthalpy change is nearly the same as catabolism alone (Battley 1991). For the purpose of these calculations, the formula weight of the cells was 24.19 Da with an atomic composition of $CH_{1.595}O_{0.374} N_{0.263} P_{0.023} S_{0.006}$ (Battley

1991). The number of available electrons in a unit carbon formula weight is 4.16. There is one carbon atom in the molecular formula from which the unit carbon formula weight is derived, and hence the degree of reduction of the biomass is given by the number of available electrons.[4] The yield coefficient for the cells (includes ions, as well as organic matter) was 41.4 g dry microbial biomass per mole of succinic acid (or about 0.35 g cells/g substrate). The stoichiometry for this conversion, as experimentally measured in an aerated fermentation, was

$$
\begin{aligned}
& C_4H_6O_4\,(aq) + 0.438\;NH\,(aq) + 0.038\;HPO_4^{2-}\,(aq) + 0.010\;HSO_4^-\,(aq) + \\
& \quad 0.086\;H^+\,(aq) + 1.747\;O_2\,(g) + 1.664\;ions \rightarrow \\
& \quad 1.664\,[CH_{1.595}O_{0.374}N_{0.263}P_{0.023}S_{0.006} + internal\;ions] + \\
& \quad 2.336\;CO_2\,(g) + 2.397\;H_2O\,(lq)
\end{aligned}
\tag{10.11}
$$

(where aq = aqueous, lq = liquid, g = gas).

These data resulted in calculation of heats in catabolism, anabolism, and metabolism for *E. coli* grown on succinic acid (Table 10.3). The calculations support the concept that a carbon substrate can be almost completely oxidized to CO_2, water, and heat through microbial transformations as represented by the following reaction (Battley 1991):

$$
\begin{aligned}
& 0.505\;C_4H_6O_4\,(aq) + 1.747\;O_2\,(aq) + 0.010\;HSO_4^-\,(aq) \rightarrow \\
& \quad 2.020\;CO_2\,(aq) + 1.515\;H_2O\,(lq) + 0.010\;HS^-\,(aq) + heat
\end{aligned}
\tag{10.12}
$$

Table 10.3.

Comparison of Heats Associated with Growth of *E. Coli*[a]

Process	$\dfrac{kcal}{mol\;Succinic\;Acid}$	$\dfrac{kcal}{g\;Succinic\;Acid}$
Anabolism	+3.5	0.03
Catabolism	−186.5	−1.58
Metabolism	−183.0	−1.55
Total	−366.0	−3.10
Oxidation	−374.4	−3.17

[a]Basis of calculation: 1 M succinic acid (at MW = 118) at pH = 7 and temperature of 25 °C.
Source: Data from Battley (1991), Table IV.

---

[4] A detailed analysis of methods, results, and impacts of different assumptions for calculating structure is discussed by Battley (1991). The formula, briefly summarized here, should be further interpreted in the context of the detailed discussion in the article by Battley (1991) if exact calculations of the relatively small heat of anabolism are to be carried out.

These results generally apply to aerated, microbial fermentations, which can give a net heat release that is close to the heat generated if the molecule (succinic acid, in this example) were completely oxidized as shown in Table 10.3.

Measurements of cell mass and heat generated for the yeast *Kluveromyces fragilis* grown aerobically on deproteinized whey permeate[5] at 30 °C with pH controlled at 5.5 gave the timecourse shown in Fig. 10.6. The decrease in substrate concentration from 45 g/L to zero occurs in less than 9 h, with a proportional increase in cell mass during the same time period (Fig. 10.6a). The rate and amount of heat generated correspond to the increase in cell mass (Fig. 10.6a), while the total heat evolved follows a linear relationship with respect to dry cell mass (Fig. 10.6b) and the amount of oxygen consumed, expressed as mol O_2/L broth.[6] The heat evolved fits a straight line with a slope of about 10 kJ heat evolved/g cell mass, or 545 kJ heat evolved/mol O_2. This corresponds to about 2.4 kcal/g and 130 kcal/mol O_2, respectively. The heat generated is higher than the 444 kJ/mol reported for *K. fragilis* grown in defined media containing lactose, glucose, and galactose (Table 10.4) by other researchers, but still falls within the expected range. Similar results

Table 10.4.

Measured Yield Coefficients for *Klebsiella Fragilis* Grown in Batch Culture on Different Carbon Sources

Carbon source	Lactose	Lactose	Glucose	Galactose
Type of medium	Whey permeate	Defined	Defined	Defined
$Y_{X/S}$ $\dfrac{\text{g cells generated}}{\text{g carbon source consumed}}$	0.511	0.517	0.450	0.590
$Y_{Q/O}$ $\dfrac{\text{kJ heat generated}}{\text{mol } O_2 \text{ consumed}}$	440.0	410.0	410.0	455.0
$Y_{Q/X}$ $\dfrac{\text{kJ heat generated}}{\text{g cells generated}}$	9.43	10.6	12.7	12.8
Closure on carbon balance (%)	90.9	97.3	90.2	112.4
Closure on energy balance (%)	85.6	90.0	89.8	102.7

Source: Marison and Von Stockar (1988), Table 1, p. 1092.

---

[5] *Whey* is the liquid that remains after casein (protein) has been coagulated for the manufacture of cheese. *Deproteinized whey permeate* refers to the liquid stream that passes through a membrane system, in which the pores of the membrane are large enough to allow sugars and other lower-molecular-weight molecules to pass while retaining the higher-molecular-weight proteins. The fluid that permeates, or passes through, the membrane is the *permeate*, while the fluid on the upstream side of the membrane that is retained is referred to as the *retentate*.

[6] Oxygen is only sparingly soluble in aqueous media, with concentrations typically found at the 6–8 ppm level. Consequently, the heat evolved on the basis of oxygen consumption is calculated by dividing the total oxygen consumed (experimentally measured) by the volume of the fermentation broth.

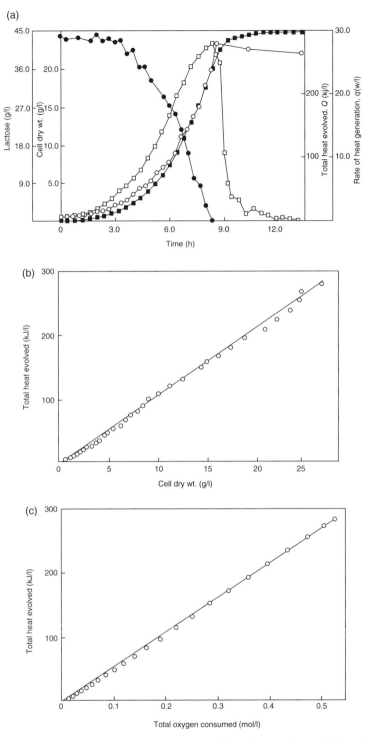

Figure 10.6. (a) Cell mass and heat generation by *Klebsiella fragilis*, at 30 °C, pH 5.5, an aeration rate of 3.5 L air/min, and agitation rate of 780 rpm. Medium contained 60 g/L whey permeate, 2.5 g/L $(NH_4)_2 PO_4$, 4 g/L yeast extract, and 0.4 g/L antifoam (polypropylene glycol P2000). Heat evolved as a function of (b) cell dry weight formed and (c) total oxygen consumed. [Symbols: o—o dry cell weight (g/L); ●—● lactose concentration (g/L); □—□ rate of heat generation (w/L); ■—■ total heat evolved kJ/L.] [Reproduced from Marison and von Stockar (1988), Ref. 50, Figs. 1 and 2, pp. 1091, 1092.]

were shown by strains of *Candida lipolytica* (grown on citrate, succinate, and hexadecane), *Candida utilis* (grown on glucose, acetate, ethanol, and glycerol), and *Candida boidinii* (grown on ethanol) (Marison and von Stockar 1988). The data indicated that some variability is likely to be encountered in the literature because of different growth conditions and substrates.

Heat is not only generated by biological activity in fermentation. Mechanical work loss through agitation causes heat generation, while heat loss occurs through evaporating water or heat transfer from the sides of the fermentor. These sources and sinks must be included in a complete energy balance for a fermentation process. A comprehensive study by Cooney et al. (1969) accounted for heat generated by agitation of the vessel and heats of evaporation, sensible-heat loss (due to heatup of fluid), and heat lost to surroundings, as well as the heat of fermentation. The total heats of agitation, evaporation, sensible-heat loss, and loss to surroundings were about 4 kcal/(L.h) for a 10-L fermentor filled with 9 L of fermentation broth. Their data were obtained for both complex and defined media over a range of oxygen consumption rates (Fig. 10.7a) and total oxygen consumption (Fig. 10.7b). A linear correlation was obtained for the Q_{heat}, the total heat produced as a function of oxygen consumption rate, q_{0x} for *E. coli* grown on succinic acid:

$$Q_{heat} = K_{ox} q_{ox} \tag{10.13}$$

The proportionality constant K_{0x} of 0.11 ± 0.01 kcal/mmol of oxygen is shown in the diagonal line in Fig. 10.7b. The value of 110 kcal/mol is similar to the 130 kcal/mol calculated for *Kluveromyces fragilis* (from Fig. 10.7c). A similar linear relation between rates and extents of CO_2 evolution (Figs. 10.7c and 10.7d) was observed with $K_{CO_2} = 0.11 \pm 0.02$ kcal/mmol CO_2.

The *rate* of heat production q_h was found to be a linear function of the rate of oxygen consumption q_{ox} as well, with

$$q_h = k q_{ox} = 0.124 q_{ox} \tag{10.14}$$

where k is in dimensions of

$$\frac{\dfrac{\text{kcal heat generated}}{\text{L fermentation broth} \cdot \text{h}}}{\dfrac{\text{mmol } O_2 \text{ consumed}}{\text{L fermentation broth} \cdot \text{h}}}$$

or kcal/mmole O_2. This important and practical series of experiments showed that heat production and oxygen consumption are directly proportional, and relatively constant, for a range of growth rates, substrates, and types of microorganism (Cooney et al. 1969).

This chapter has shown that biochemical reactions in living cells follow the same thermodynamic principles of free energy and electron transfer as do all other chemical reactions. Consideration of the thermodynamics of biochemical reactions is important for understanding how catabolic and anabolic metabolic pathways are integrated. The heat generated as a byproduct of biochemical reactions must be

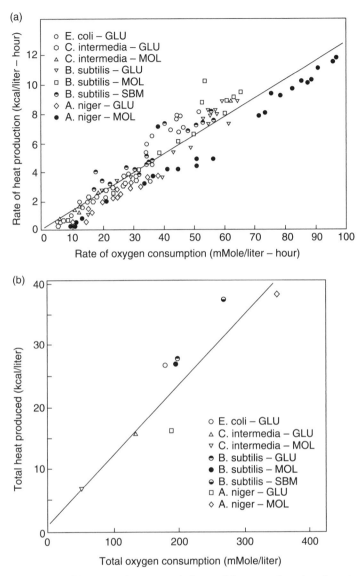

Figure 10.7. (a) Rate of heat production and (b) total heat produced as function of oxygen consumption; and (c) rate of heat production and (d) total heat produced as function of CO_2 generation. (Glu = glucose medium; MOL – molasses; SBM = soybean meal.) *Escherichia coli* and *Bacillus subtilis* were grown at pH 7 and 37 °C. *Candida intermedia* and *Aspergillus niger* were grown at pH 5.5 and 30 °C. [Reproduced from Cooney et al. (1969), Ref. 51: plots (a), (b)—Fig. 3, p. 277; plot (c)—Fig. 4, p. 279; plot (d)—Fig. 5, p. 281.]

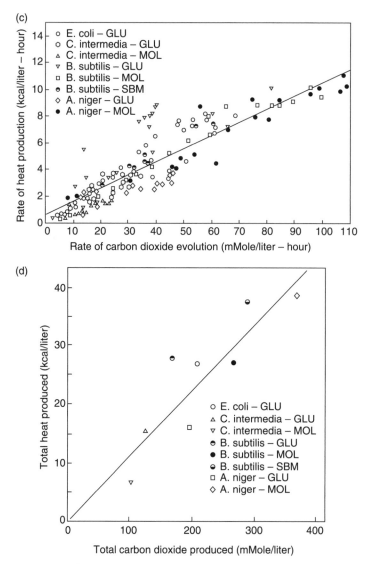

Figure 10.7. *(Continued)*

accounted for to develop an accurate energy balance for different scales of fermentations. This is particularly important for large-scale fermentations where thermal energy must be accounted for, and heat exchangers designed to remove heat from the fermentation media in order to keep the temperatures within a narrow and optimal range for cell growth.

REFERENCES

Aiba, S., A. E. Humphrey, and N. F. Millis, *Biochemical Engineering*, 2nd ed., Academic Press, New York, 1973, pp. 28–32, 56–91, 195–345.

Alberts, B., D. Bray, J. Lewis, M. Raff, K. Roberts, and J. D. Watson, *Molecular Biology of the Cell*, 2nd ed., Garland Publishing, New York, 1989, pp. 58, 65, 78, 205–216, 341, 345–348, 353–354, 359, 362–363, 373, 381–383, 601.

Bailey, J. E. and D. F. Ollis, *Biochemical Engineering Fundamentals*, McGraw-Hill, New York, 1977, pp. 3, 14–15, 221–254, 338–348, 357, 371–383.

Battley, E. H., "Calculation of Heat of Growth of Escherichia coli K-12 on Succinic Acid," *Biotechnol. Bioeng.* **37**, 334–343 (1991).

Borman, S., "Chemistry Nobel: Researchers from Three Countries Share Prize for ATP Synthase Discoveries," *Chem. Eng. News* **75**(42), 11–12 (1997).

Cooney, C. L., D. I. C. Wang, and R. I. Mateles, "Measurement of Heat Evolution and Correlation of Oxygen Consumption during Microbial Growth," *Biotechnol. Bioeng.* **11**, 269–281 (1969).

Levenspiel, O., *Chemical Reaction Engineering*, 2nd ed., Wiley, New York, 1972, pp. 212–214.

Lewin, B., *Genes VI*, Oxford Univ Press, New York, 1997, pp. 156, 1228, 1231–1232.

Marison, I. W. and U. von Stockar, "Fermentation Control and the Use of Calorimetry," *Chem. Society Trans. IBBG Meeting* **16**, 1091–1093 (1988).

Murray, R. K., P. A. Mayes, D. K. Granner, V. W. Rodwell, *Harper's Biochemistry*, 22nd ed., Appleton & Lange, Norwalk, CT, 1990, pp. 10–14, 58, 68–69, 102–106, 146, 160–161, 165, 199–200, 318–322, 336–338, 549–550, 555.

Shuler, M. L. and F. Kargi, *Bioprocess Engineering: Basic Concepts*, Prentice-Hall, Englewood Cliffs, NJ, 1992, pp. 48–56, 59, 158–159.

Smith, J. M. and H. C. van Ness, *Introduction to Chemical Engineering Thermodynamics*, 2nd ed., McGraw-Hill, New York, 1959, pp. 346–357.

CHAPTER TEN
HOMEWORK PROBLEMS

10.1. ATP hydrolysis shifts the equilibrium of a reaction to favor the products. Suppose that reaction $A \rightleftharpoons B$ has free energy $\Delta G^{\circ\prime} = +2\,\text{kJ/mol}$.

(a) Calculate K'_{eq} at $25\,°\text{C}$ for this reaction.

(b) Assuming that ATP is used by an enzyme to catalyze the reaction

$$A + ATP + H_2O \leftrightarrow B + ADP + P_i + H^+$$

calculate K'_{eq} at $25\,°\text{C}$ for this new reaction.

10.2. The enzyme fumarase in the Krebs cycle catalyzes the following reaction:

$$\text{Fumarate} + H_2O \rightleftharpoons \text{L-malate}$$

(a) Calculate the standard equilibrium constant (K') assuming that the standard Gibbs free energy for this reaction is $\Delta G^{\circ\prime} = -3.8\,\text{kJ/mol}$.

(b) At standard conditions ($25\,°\text{C}$, 1 atm), assume the starting concentrations of the reactants (fumarate and H_2O) and the product (L-malate) to be 1 M. Using the standard equilibrium constant that you calculated in part (a), calculate the final concentrations of the reactants and products after they reach equilibrium.

(c) Is the concentration of these compounds in your cells likely higher or lower than the values calculated in part (b)?

10.3. Calculate the *usable* energy efficiency of glucose metabolism for the production of ATP. Assume that oxidative phosphorylation will produce 3 ATP per NADH and 2 ATP per FADH$_2$. *Note*: Usable efficiency = η_{use} = (net $\Delta G^{o'}$ for total ATP produced/net $\Delta G^{o'}$ for 1 mol of glucose oxidation)

10.4. (a) Calculate the overall free-energy change for oxidation of pyruvate/lactate under aerobic conditions.

(b) What is the maximum number of high-energy phosphate bonds that could be formed at cellular conditions? State your assumptions.

10.5. An aerated fermentation of *Aspergillus niger* consumes oxygen at the rate of 50 mmol/(L.h), and generates CO_2 at the rate of 60 mmol/(L.h).

(a) Calculate the amount of heat generated on an hourly basis.

(b) During the course of a 24-h fermentation in a 10,000-L. tank, how much heat must be removed?

(c) Is the amount of heat generated per mole of oxygen consumed consistent with the amount of heat that would be generated in a *K. fragilis* fermentation in a defined glucose medium?

(d) Write the equations for *K. fragilis* with respect to substrate depletion, biomass (cell mass) generation, heat generation, and CO_2 produced as a function of time. Show values of constants and initial conditions, and organize your equations for solution using the Runge–Kutta integration program described in Chapters 5 and 6.

CHAPTER ELEVEN

METABOLIC PATHWAYS

INTRODUCTION

Metabolic pathways are systems of biochemical reactions that form an adaptive biochemical network used by the cell to derive energy from nutrients and to synthesize compounds needed for life functions and reproduction. Molecules that flow through some of the pathways are also a response to environmental pressures. Such molecules, referred to as *secondary metabolites*, have defensive roles, such as when a microbe secretes a β-lactam (antibiotic) that kills other bacterial species that compete with it for resources. The goal of bioprocess scientists and engineers is to capture value from the cell's metabolic pathways by identifying valuable intermediates or end products and then developing means by which these molecules will be generated at high concentrations.

Random mutation of DNA using radiation or chemical mutagens and screening survivors for desired improvements to its pathways and capability to generate targeted products has been done since the early 1950s. Molecular biology techniques developed since 1975 and sequencing of microbial genomes since 1995 have changed this approach. It is now possible to program changes into metabolic pathways by introducing new genes, modifying functions of existing ones, or blocking expression of genes in a directed manner. Microarray techniques, discussed in Chapter 14, enable screening of the altered microbes for the presence of desired traits, thus compensating for the element of uncertainty in experimental methods used to transform the microbes.

An understanding of the art and science of directing metabolic pathways to specific molecules is needed to comprehend the broad array of possibilities and products that may be derived from microbial metabolism. Chapter 9, on metabolism, introduced the reader to some of the products that are naturally generated at industrially useful concentrations. Chapter 10 discussed how the energetics of the cell are linked to glycolytic and TCA pathways, from which these products are derived. This chapter, on metabolic pathways, gives a synopsis of discussions from prior literature on auxotrophs, selected for its clarity, and on the directed focusing

Modern Biotechnology, by Nathan S. Mosier and Michael R. Ladisch
Copyright © 2009 John Wiley & Sons, Inc.

of metabolic pathways to obtain commercially useful concentrations of antibiotics, flavor enhancers, and amino acids (Aiba et al. 1973; Demain 1971; Wang et al. 1979; Bailey 1991; Stephanopoulos and Jensen 2005; Kern et al. 2007). We attempt to show how creative analysis and clever experimental techniques have made use of existing networks in wild-type organisms. These microorganisms are responsible for over half of current production of pharmaceutical, food, and feed products. This sets the stage for introducing the rationale and approaches for altering microbial metabolism by the engineering of metabolic pathways described in Chapters 12 and 13.

Living Organisms Control Metabolic Pathways at Strategic and Operational Levels

Metabolic pathways have both strategic and operational controls. Strategic control is achieved by the cell's ability to regulate genes that direct (or do not direct) the allocation of resources to the generation of enzymes and other proteins that make up the functioning components of the pathway. Operational control occurs through built-in governors of protein activity that respond to the presence and concentrations of metabolites. Metabolites regulate the activities of enzymes that have been expressed by the microbe.

Control of gene expression alters the rate of metabolism by altering the concentration of the enzymatic catalysts as well as determining the presence (or absence) of a specific metabolic pathway. This strategic genetic control affects changes relatively slowly since minutes to hours are required to transcribe the genes and synthesize the enzymes. On the other hand, operational control through feedback inhibition is rapid (seconds or less) and occurs when the product of the metabolic pathway or one of the intermediate compounds interacts with a critical enzyme. This usually occurs early in the pathway or at crucial branch points in such a way as to inhibit or slow the rate of the reaction. This slows the flux of molecules through the entire metabolic pathway from that point forward.

The focus in this chapter is the operational metabolic control of naturally occurring microbes and naturally occurring mutants of these microbes to maximize generation of a target molecule. We will examine how feedback inhibition and control of inhibiting compounds can be used to control the metabolic processes of microorganisms to maximize a desired product. In subsequent chapters, we will examine how the tools of molecular biology can be employed to specifically alter an organism's genetic code (strategic control) for purposes of metabolic engineering.

Auxotrophs Are Nutritionally Deficient Microorganisms that Enhance Product Yields in Controlled Fermentations (Relief of Feedback Inhibition and Depression)

Auxotrophs are microorganisms that require at least some reduced organic molecules, such as amino acids, which they cannot synthesize but that are essential for their growth. The missing amino acid(s) is (are) supplied as components of either synthetic or complex media in the culture medium in which the microorganisms are grown. In synthetic media, the individual amino acids essential for growth are

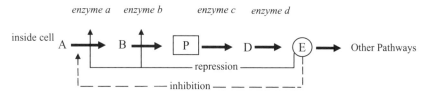

Figure 11.1. Intermediate metabolite P of an unbranched pathway is the product in controlled fermentation [modified from Wang et al. (1979), Fig. 2.1, p. 15, Wiley-Interscience, New York].

provided as individual components. In complex media, protein or undefined sources (steep water, yeast extract, peptone, etc.) provide the missing amino acids. Humans and other mammals are auxotrophs with respect to several amino acids. These amino acids (such as lysine) are called "essential" amino acids because they are essential for good health and life. For humans, these amino acids are obtained through the diet.

Examples from Wang et al. (1979) and Demain (1971) illustrate how auxotrophy is used to accumulate intermediates and end products of microbial metabolism by limiting the accumulation of inhibitory or repressive metabolites inside the cell. The metabolic pathway of Fig. 11.1 provides an example of the concepts applied to accumulate an intermediate product, denoted by the symbol P. As explained by Wang et al. (1979), the intermediates in the particular metabolic pathway are A, B, P, D, and Ⓔ. Ⓔ is required for growth, while P is the desired product that has commercial value. The reactions between each compound are assumed to be irreversible. A normal cell would produce all four metabolites (B, P, D, and Ⓔ) starting from A. However, the accumulation of the end product Ⓔ inhibits[1] the first enzyme (enzyme a) in the pathway as indicated by the dashed line in Fig. 11.1.

A second type of control may occur at a genetic level. In this case genes for enzyme a and enzyme b are downregulated so that the synthesis of these enzymes is turned off or significantly slowed. Repression by Ⓔ, indicated by solid lines in Fig. 11.1, is a form of strategic control in which a repressor molecules bind on the DNA so that transcription is blocked and production of the enzyme is repressed. This type of control is named to reflect its mechanism (e.g., the lac operon controls the genes that metabolize *lac*tose).

The consequence of the combination of repression (strategic control) and inhibition (operational control) is that the desired product, P, will not accumulate to a high concentration. If a mutant microorganism is found, or if genetic engineering techniques remove or block the gene responsible for directing synthesis of enzyme c, the product P may be synthesized in larger amounts, since Ⓔ is no longer made. While the product Ⓔ is still required for growth, it can now be added in a

---

[1] *Inhibition* in this context refers to a decrease in the rate at which an enzyme catalyzes the reaction. The product reversibly associates or binds with the protein to change its conformation or temporarily block the enzyme's active site. Either or both types of change may occur so that the access of the substrate to the enzyme's active site is inhibited or restricted, causing a decrease in the overall rate of reaction. The kinetics of inhibition are discussed in Chapter 8.

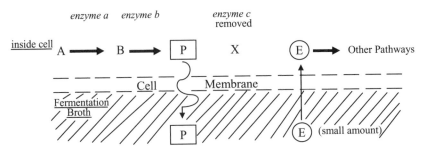

Figure 11.2. Supplementation of metabolite Ⓔ in fermentation crosses cell membrane of an auxotrophic cell [adapted from Aiba et al. (1973) and Wang et al. (1979)].

controlled amount in the fermentation broth so that it is sufficient to meet the cell's needs, but small enough to minimize or avoid repression and inhibition of enzymes a and b, as shown in Fig. 11.2. This assumes that Ⓔ is able to be taken up or transported across the cell's membrane. Since the cell is nutritionally deficient in Ⓔ, and Ⓔ is provided externally, the microorganism is said to be an auxotroph with respect to Ⓔ.

Examples of arginine auxotrophs are *Bacillus subtilis* and *Corynebacterium glutamicum* that produce large amounts of amino acids to give a fermentation broth with 16 g/L of citrulline and 26 g/L ornithine. The metabolism is based on the principles illustrated in Fig. 11.2. Ornithine or citrulline[2] represent product P in Figs. 11.1 and 11.2, while arginine is equivalent to Ⓔ.

Another challenge is to select, engineer, or control the microorganism so that destruction of the accumulated product (P) is avoided once relief of feedback inhibition and repression have been achieved. This may require that the composition of the fermentation broth be carefully controlled to maximize diffusion of the product through the cell's membrane. In some cases, it may also be possible to make the product in a form that resists intracellular degradation.

---

[2] Citrulline is an amino acid used for research purposes and has the structure

$$H_2N \quad C\,NH\,(CH_2)_3 \quad CHN - COOH$$
$$\quad\quad \| \quad\quad\quad\quad\quad\quad\quad |$$
$$\quad\quad O \quad\quad\quad\quad\quad\quad\quad H_2$$

with a molecular weight of 175.2.

Ornithine is anticholesteraemic. Its structure is

$$\quad\quad\quad NH_2$$
$$NH_2(CH_2)_3\,CHNH_2 - COOH$$

with a molecular weight of 132.2 (Atkinson and Mavituna 1983). Product concentrations of 10–30 g/L (1–3%, wt/vol basis) are typical for these types of commercial fermentations. Removal of water from fermentation products can be a major expense.

Both Branched and Unbranched Pathways Cause Feedback Inhibition and Repression (Purine Nucleotide Example)

The examples given by Figs. 11.1 and 11.2 illustrate strategies for un-branched pathways where an intermediate metabolite (P) is the desired product. Branched pathways that lead to two metabolites, where both can be inhibitory, may require a different approach. This type of pathway will exhibit multivalent, concerted, or cooperative feedback inhibition. An example from Wang et al. (1979) is illustrated in Fig. 11.3, where metabolites Ⓔ and Ⓖ affect the first enzyme in the pathway. Multivalent concerted inhibition means that Ⓔ or Ⓖ individually would have little effect on enzyme a, but when present together Ⓔ and Ⓖ completely inhibit the enzyme. Multivalent cooperative inhibition denotes an additive effect where Ⓔ alone causes some inhibition, while Ⓔ and Ⓖ together cause a greater extent of inhibition. In this case, the combined presence of Ⓔ and Ⓖ does not completely stop the enzymatic action. The end metabolites may also cause repression (strategic control), although this is not shown in Fig. 11.3.

The production of purine nucleotides is an example of where an intermediate metabolite is the desired product, and metabolites at the end of a pathway are inhibitory and/or cause repression. The nucleotides guanosine monophosphate (GMP), inosine monophosphate (IMP), and xanthine monophosphate (XMP) are commercially important since they enhance flavor when added to foods. Their production is technically challenging since their charged character inhibits cell excretion. These nucleotides are susceptible to hydrolysis by intracellular phosphatases, and are difficult to generate owing to metabolic blocks (Aiba et al. 1973).

Control of culture media components enables the permeability of the cell membrane to be increased, thereby facilitating diffusion of the product into the broth before it can be degraded by intracellular enzymes. Removal of metabolic blocks occurs through mutation and selection of special microbial strains, in which end metabolite concentrations are decreased and low levels of exogeneous nutrients, provided in the culture medium, are able to satisfy the nutrient requirement of the auxotroph. Intermediates and pathways for purine nucleotides are shown in Fig. 11.4.

The overproduction of IMP and appearance of IMP in the fermentation broth (Fig. 11.4) occur for a mutant that lacks enzyme b (S-AMP synthetase), and accumulates IMP when it is fed small amounts of exogeneous adenine. Since enzyme b is missing, IMP is converted to H (hypoxanthine), XMP, or GMP, but not AMP.

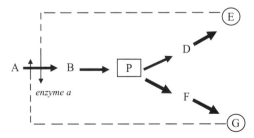

Figure 11.3. Intermediate metabolite P of a branched pathway is the product in controlled fermentation.

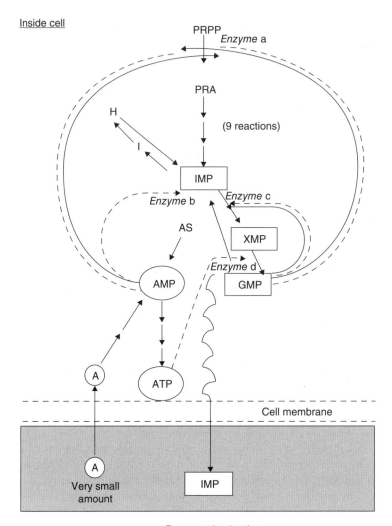

Figure 11.4. Metabolic control for the production of purine nucleotides [adapted from Aiba et al. (1973), Fig. 3.18, p. 84, and Wang et al. (1979), Fig. 2.2, p. 15].

The microorganism requires AMP for growth. Since it is deficient in this compound, the microorganism is an AMP auxotroph. The AMP requirement is met by providing controlled amounts of adenine (denoted by A in Fig. 11.4), which is a suitable precursor for AMP. AMP cannot pass through the cell membrane. Therefore, addition of exogeneous AMP is not suitable for overcoming the nutritional deficiency. Hence, adenine, which is used by the cell to synthesize AMP, is fed instead and the organism is referred to as an *adenine auxotroph* rather than an *AMP auxotroph* (Wang et al. 1979; Demain 1971).

The metabolic control in an adenine auxotroph involves several enzymes. The first enzyme in the pathway, enzyme a [phosphoribosyl–pyrophosphate (PRPP) amidotransferase] in Fig. 11.4, is not repressed or strongly inhibited since AMP is not present in a sufficient amount to act in concert with the GMP to shut down enzyme a. IMP (shown in the center of Fig. 11.4) begins to accumulate, with a small amount diverted to GMP. The GMP inhibits and represses enzyme c (IMP dehydrogenase). This shuts off formation of GMP. If ATP is present, it inhibits enzyme d, thus making the formation of GMP irreversible. *Corynebacterium glutamicum* and *Brevibacterium ammoniagenes* are examples of adenine auxotrophs that were shown by Furuya to accumulate 13 g of IMP per liter of fermentation broth (Wang et al. 1979).

An alternate pathway involving extracellular formation of IMP was reported by researchers at Kyowa Hakko Co. (Japan). An adenine auxotroph grown in a fermentation broth containing high concentrations of phosphate and magnesium but low concentrations of manganese and adenine is believed to be highly permeable to the intermediates inosine (I in Fig. 11.4), hypoxanthine (H) ribose-5-phosphate, phophoribosyl–pyrophosphate (PRPP), *and* the enzymes needed for extracellular biosynthesis of IMP. Extracellular synthesis of GMP, AMP, ADP, and ATP were also postulated (Aiba et al. 1973). An observation of this type is interesting not only from the perspective of fermentation but also in terms of conceptualizing a cell-free enzyme bioreactor that might carry out this conversion. The availability of ATP generated in this manner is of interest since it would enable highly selective, enzyme-based synthesis that requires ATP to be carried out extracellularly. Such synthesis would be useful for manufacture of high-value organic or biochemical molecules.

The Accumulation of an End Metabolite in a Branched Pathway Requires a Strategy Different from that for the Accumulation of an Intermediate Metabolite

The accumulation of end metabolites, rather than intermediates as the product of commercial interest, is likely to be difficult for an unbranched pathway of the type shown in Fig. 11.1 compared to a branched pathway. If Ⓔ in this unbranched pathway were the product of interest *and* were required for the metabolism of the cell, the strategy of supplying exogenous Ⓔ could not be used, since this approach would require high and low concentrations of Ⓔ to exist simultaneously. A low concentration would be needed to minimize repression and/or inhibition, whereas a high concentration would represent the objective of the fermentation.

A microorganism with a branched pathway is therefore needed as illustrated in Fig. 11.5. The end metabolite of pathway 1 provides the desired product of commercial interest. The second pathway would give metabolites that can be exogenously provided to meet the nutritional needs of the organism while maximizing formation of the end product of the first pathway. The strategy in this case would be to find a mutant organism or engineer an organism in which enzyme c is absent, and then provide low levels of Ⓔ as an exogenous nutrient. Product P̄ would then accumulate since inhibition of enzymes by P̄ and Ⓔ would be relieved. In effect, a decrease in the concentration of one end product (F that is formed from Ⓔ) results in the increase of another end product (P̄) (Wang et al. 1979; Demain 1971).

Multiple pathways could result in the formation of B, with each pathway regulated by a different end product. This complicates discovery and improvement of a pathway leading to high levels of P̄. Another complicating factor is that pathways that branch off from an intermediate, such as C in Fig. 11.6, often have the same characteristic as does the unbranched pathway of Fig. 11.1. The final product

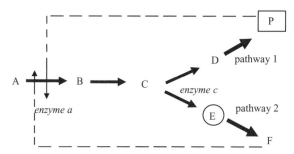

Figure 11.5. End metabolite of pathway 1 represents the desired product P in controlled fermentation; control of product production achieved by feeding Ⓔ to the fermentation [adapted from Aiba et al. (1973), Fig. 3.13, p. 78].

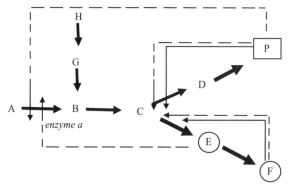

Figure 11.6. Branched metabolic pathway with complex feedback inhibition [adapted from Aiba et al. (1973), Fig. 3.13, p. 78].

may both inhibit and repress enzymes in the pathway to the right of the branch point, for example, C → D and C → Ⓔ in Fig. 11.6.

The complexity and challenges of directing microbial pathways in order to obtain products of commercial interest require the ability to rapidly screen thousands to millions of microorganisms until one is identified with the desired property (Wang et al. 1979; Demain 1971; Queener and Lively 1986; White et al. 1986). While modern biotechnology is based on directed changes in the genetic makeup of microbial cells, screening is still needed to differentiate cells in which genetic transformation was successful versus cells where it was not. Methods that have been successfully used to identify desired cells derived from *random* mutation of a basic microorganism using radiation or chemical mutagens have been based on rapid, high-throughput screening that identifies mutants that are resistant to feedback inhibition or repression (MacCabe et al. 2002; Warnecke et al. 2008). Selection of such a microorganism is achieved by plating out cells on petri dishes that contain agar[3] with toxic analogues of the desired compound (amino acid, P̲). Both the

---

[3] *Agar* is a polysaccharide that can be mixed with nutrients that support cell growth, and then poured to form a substrate that has a soft, wax-like character at room temperature. It is the material used to make up agar plates in petri dishes.

amino acid and its analog can exert inhibition or repression (as illustrated in Fig. 11.7). However, unlike the normal amino acid, its analog cannot be used for protein synthesis. Therefore the cell cannot make a protein needed for its survival (indicated by a large X in Fig. 11.7). Since P_x also inhibits/represses the cells' production of P, the cell dies from starvation. This is because the missing amino acid cannot be synthesized because of inhibition and/or repression of the pathway to P by the analog.

Conversely, mutated cells are not sensitive to the analog P_x and will grow, since the enzyme(s) in the pathway to P is (are) not repressed or inhibited by P_x and P can be produced. Since P is available to make the protein required by the cell, the cell lives. Examples of analogs used to select mutants that overproduce metabolites are given in Table 11.1.

The rapidity with which screening of the type shown in Fig. 11.7 can be done is a function of the skills and knowledge of an experienced microbiologist, as well as the ability to test for 10–100 microorganisms on a single petri dish in a 24-h time period. The colonies that might form (indicated by dots the petri shown in Fig. 11.8b) would be further screened, characterized, and propagated for eventual transfer to increasingly larger shake flasks. Large shake flasks would provide the microbial culture for a seed fermentor that, in turn, would provide the inoculums for a full-scale production system. Examples and schematic diagrams of the fermentation equipment are given in Chapter 4. Screening and selection procedures are discussed

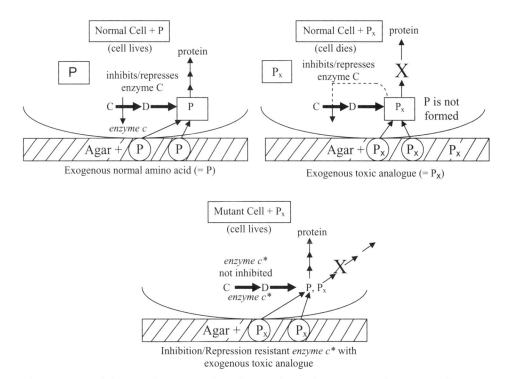

Figure 11.7. Inhibition of amino acid production by analog compound. Mutant cell accumulates desired product while wild-type cell is killed by analog compound.

Table 11.1.

Toxic Analogs (P_x) Used to Select for Microorganisms that Overproduce Metabolites

Accumulated Product	Analogues for selection
Phenylalanine	p-Fluorophenylalanine; thienylalanine
Tyrosine	p-Fluorophenylalanine; D-tyrosine
Tryptophan	5-Methyltryptophan; 6-methyltryptophan; 5-fluorotryptophan
Histidine	2-Thiazolealanine; 1,2,4-triazole-3-alanine
Proline	3,4-Dehydroproline
Valine	α-Aminobutyric acid
Isoleucine	Valine; isoleucine hydroxamate; α-aminobutyric acid; α-amino-β-hydroxyvaleric acid; O-methylthreonine
Leucine	Trifluoroleucine; 4-azaleucine
Threonine	α-Amino-β-hydroxyvaleric acid
Methionine	Ethionine; norleucine; α-methylmethionine; L-methionine-D,L-sulfoximine
Arginine	Canavanine; arginine hydroxamate; D-arginine
Adenine	2,6-Diaminopurine
Uracil	5-Fluorouracil
Hypoxanthine, inosine	5-Fluorouracil; 8-azaguanine
Guanosine	8-Azaxanthine
Nicotinic acid	3-Actetylpyridine
Pyridoxine	Isoniazid
p-Aminobenzoic acid	Sulfonamide
Thiamine	Pyrithiamine

Source: Reproduced from Wang et al. (1979), Table 2.2, p. 19.

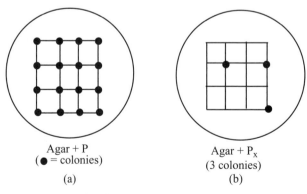

Agar + P
($\bullet$ = colonies)
(a)

Agar + P_x
(3 colonies)
(b)

Figure 11.8. Culture screening for desired auxotrophs. Colonies (represented by dots) on plate in (b) represent the desired auxotrophs.

by Queener and Lively (1986), and White et al. (1986), while inoculum development is described by Hunt and Steiber (1986) in the *Manual of Industrial Microbiology and Biotechnology*. The second edition of that *manual*, published in 1999 (edited by Demain and Davies), has been updated and expanded to include details of screening and selection procedures.

This approach has been successfully practiced for over 50 years and is the basis for the selection of strains used for industrial production of amino acids through fermentation. With advances of molecular genetic techniques, researchers can now specifically direct changes of enzymes at the genetic level that have previously been identified to control key steps leading to a given product. This second approach modifies the genetic makeup of the cell by adding or deleting genes or altering their expression (strategic control). Since the objective of these manipulations is to change the metabolism of the cell, they are collectively referred to as *metabolic engineering*, which is discussed in Chapter 12.

Metabolic engineering is more deliberate, but can also be slower if the metabolic pathway and the complex interactions between the control of gene expression and metabolism for the whole cell are not well understood. The combination of mutation and screening, followed by amplification of a desirable property identified in a mutant by genetic engineering, has the potential of combining the speed of classic microbiology and screening techniques for the discovery process, with the systematic improvement of a given pathway through genetic engineering.

Examples of microorganisms developed many years ago and in which commercially important end products are accumulated include lysine for *Corynebacterium glutamicum* and penicillin from *Penicillium chrysogenum* (Wang et al. 1979; Demain 1971). These represent mature but modern biotechnology. Lysine is an example of a primary metabolite (Type II fermentation), while penicillin is a secondary metabolite (Type III fermentation as described in Chapter 4). Primary metabolites are typically low-molecular-weight compounds that are used as building blocks for macromolecules and coenzymes on which the cell's functioning depends. The intermediates in the biosynthetic pathways that lead to the primary metabolites are sometimes also themselves classified as primary metabolites. Amino acids, purine and pyrimidine nucleotides, vitamins, and compounds in the Embden–Meyerhoff and citric acid cycle pathways are examples of primary metabolites.

Secondary metabolites are also low-molecular-weight compounds. These are formed by cells in response to stress, and can be toxic to the cells that generate them. Examples of secondary metabolites are β-lactam antibiotics (derived from two amino acids), streptomycins (derived from sugars), and tetracyclines (derived mainly from acetate or propionate units) (Evans 1965).

AMINO ACIDS

Large volumes of lysine, threonine, tryptophan, and glutamate are obtained through fermentations, while chemical syntheses are used for manufacture of methionine and its analogs, and for tryptophan. However, commercial chemical syntheses are unable to selectively produce one enantiomer (D or L) of these amino acids, resulting in a racemic mixture of both D and L forms. Except for glutamate, which is a flavor enhancer for human foods, the major high-volume uses of amino acids are as animal feed additives. They are used to formulate feed rations for poultry, pigs, sheep, and cows in order to improve amino acid content of the feed; increase egg, milk, meat, and wool production; improve animal weight gain; and possibly participate in detoxification of aflatoxin[4] in the liver (Behm et al. 1988; Tanner and Schmidtborn 1970; Degussa 1989).

The Formulation of Animal Feed Rations with Exogeneous Amino Acids Is a Major Market for Amino Acids

The D form of D/L methione and D/L tryptophan must be converted to the L form if it is to be utilized[5] by an animal. The D form of D/L methonine, with a ratio of D:L of 1:1, can be transformed into the L form by the enzyme, D-amino acid oxidase, in pigs, through the enzyme-catalyzed reaction illustrated in Fig. 11.9. Amino acid oxidase and transaminase enzymes that are specific for D/L-tryptophan carry out an analogous conversion, although cofactors are not shown (Degussa 1989).

The D form of D/L-lysine, unlike D/L-methionine, and D/L-tryptophan, cannot be converted to the L form since most animals lack the necessary amino acid isomerase that would catalyze the transformation. Hence, a chemically synthesized D/L-lysine would only be 50% and therefore not cost-effective. This provides motivation for production of lysine through microbial fermentation—100% of the lysine is in the L form. Lysine can also be produced by an enzyme-catalyzed transformation of chemically synthesized α-amino-ε-caprolactam (Degussa 1985).

---

[4] Methionine aids detoxification when it is converted to cysteine. Cysteine, in turn, is used to synthesize to glutathione; a tripeptide of glutamuylcysteinylglycine, which is needed for redox reactions involving cytochrome P450 (Degussa 1989).

[5] Except for glycine (where R = H), all α-amino acids are chiral compounds that occur in two enantiomeric or mirror-image forms:

$$\underset{\text{L-Amino acid}}{\text{R} - \text{C} \blacktriangleleft \text{COO}^-} \quad \underset{\text{D-Amino acid}}{^-\text{OOC} \blacktriangleright \text{C} - \text{R}}$$

L-Amino acid: R, C, H_3N^+, H, COO$^-$
D-Amino acid: R, C, H, NH_3^+, $^-$OOC

Proteins contain only the L form of amino acids, although some D-amino acids also exist in nature. These are involved in the synthesis of bacterial cell walls; are found in peptides such as tyrocidine and gramacidine, which have antibiotic activity; in spider venom toxins; in an opioid peptide (dermorphin) from a South American tree frog (*Phyllomedusa sauvagei*); and in antimicrobial peptides from the skin of the frog *Bombina variegata* (Kreil 1994). The presence of D-amino acids is believed to increase the biological half-life of the peptides that contain them. However, D-amino acids are not found in the metabolisms of higher living organisms, even though some animals possess D-amino acid oxidases (Degussa 1985). Racemization of the L to D forms may occur spontaneously in vitro. This property has been used as a marker to test the authenticity of ancient DNA as described in Chapter 2.

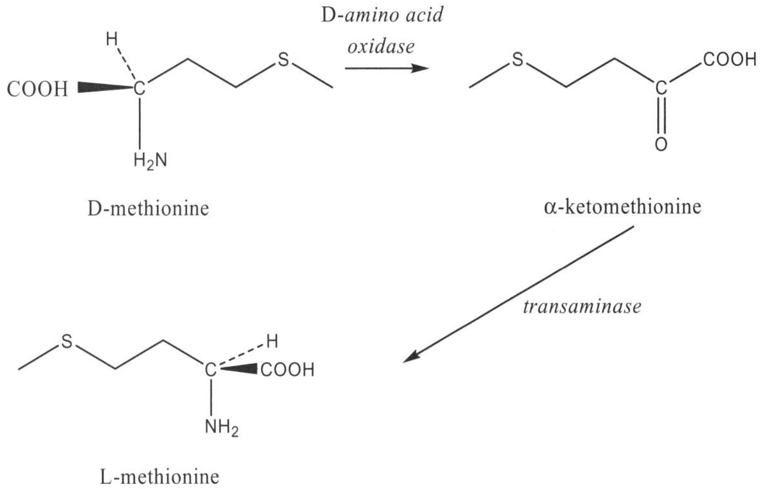

Figure 11.9. Isomerization of D-methionine to L-methionine by a two-step enzyme-catalyzed process [adapted from Tanner and Schmidtborn (1970), Fig. 27, p. 106].

Table 11.2.

Toxicity of Selected Amino Acids in Mice and Rats as Measured by Oral Administration

Amino Acid	Animal	LD_{50}^a (g Amino Acid/kg Body Weight)
L-Lysine	Rat	10
L-Tryptophan	Rat	16
D-Tryptophan	Mouse	10

[a]Based on oral dose. LD_{50} refers to the dose that is lethal to 50% of the animals.
Source: Data from Behm et al. (1988), p. 50.

A balance between amino acids already available in other feed components (such as soybeans and corn), the nutritional needs of the animal, and cost must be optimized (Tanner and Schmidtborn 1970). While amino acids could be toxic if they were to be fed in extremely large amounts (Table 11.2), introduction of amino acid imbalances in amino acid–supplemented feeds is more likely. If an amino acid is supplemented that does not constitute a limiting amino acid, a reduction in feed intake by the animal may result. This is undesirable since a goal of formulating animal feeds is to maximize weight gain as well as animal protein quality (Behm et al. 1988; Degussa 1989; Tanner and Schmidtborn 1970). The combination of biological benefits of single-component amino acids and development of microorganisms to produce them have promoted the use of some fermentation derived amino acids in agriculture.

Microbial Strain Discovery, Mutation, Screening, and Development Facilitated Introduction of Industrial, Aerated Fermentations for Amino Acid Production by *Corynbacterium glutamicum*

Corynbacterium glutamicum has catalyzed the diversification of industrial/amino acid production techniques and is presented here to provide a historical perspective on the commercialization pathway for a naturally occurring product. Prior to about 1960, food, pharmaceutical, and feed-grade amino acids were derived principally from plant and marine biomaterials or generated by chemical synthesis. After 1960, commercially viable, aerated fermentations began to be used for manufacturing amino acids as described in Chapter 1, due to the discovery and development of glutamate over producers. The common characteristic leading to glutamate overproduction was the deficiency of α-ketoglutarate dehydrogenase (indicated by the box in the lower part of Fig. 11.10), and a nutritional requirement for biotin that allows membrane permeability to be controlled so that glutamate is excreted (Aiba et al. 1973). The absence of α-ketoglutarate dehydrogenase prevents transformation of α-ketoglutarate to succinate as would normally occur in the citric acid cycle. The absence of this enzyme pushes the α-ketoglutarate to glutamate (as shown in Fig. 11.10).

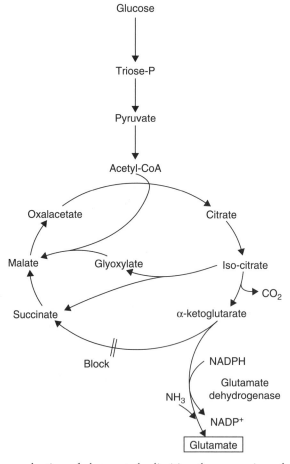

Figure 11.10. Overproduction of glutamate by limiting the expression of α-ketoglutarate dehydrogenase (block) [from Aiba et al. (1973), Fig. 3.17, p. 82].

Many genera and species have been found to be capable of producing glutamate. These include *Micrococcus, Corynebacterium, Brevibacterium*, and *Microbacterium* (Aiba et al. 1973; Wang et al. 1979). These were similar and therefore were grouped into a single genus (Wang et al. 1979).

Overproduction of Glutamate by *C. glutamicum* Depends on an Increase in Bacterial Membrane Permeability (Biotin-Deficient Mutant)

Biotin affects the permeability of bacterial cell walls because of its role in fatty acid synthesis. Changes in the lipid composition of glutamate overproducers that results in membranes that are deficient in phospholipids are caused by biotin deficiency, glycerol auxotrophy, or the addition of Tween (a detergent), fatty acid derivatives, or penicillin during logarithmic growth phase. The membranes are therefore leaky and allow glutamate to diffuse out into the fermentation broth so that it does not inhibit the enzymes that catalyze its synthesis. Hence, the absence of biotin in a microbial fermentation affects the microorganism's growth.

Biotin is a vitamin that serves as a coenzyme for acetyl-CoA carboxylase. This enzyme transforms acetyl-CoA to malonyl-CoA in the presence of ATP, as shown in Fig. 11.11 (CoA denotes coenzyme A). The production of malonyl-CoA is the initial and controlling step in fatty acid synthesis. A biotin deficiency in humans is rarely due to diet. Rather, such a deficiency reflects defects in its utilization since a large portion of the human requirement for biotin is met by synthesis from intestinal bacteria (Murray et al. 1990).

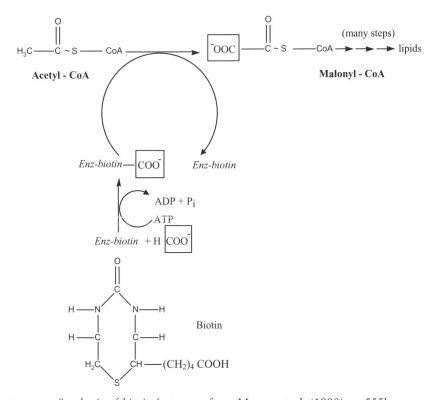

Figure 11.11. Synthesis of biotin [structure from Murray et al. (1990), p. 555].

Acetyl-CoA carboxylase (denoted as *Enz* in Fig. 11.11), is a multiprotein enzyme that contains a variable number of identical subunits. The catalyzed reaction results in free palmitate as the end product, and is the result of biochemical transformations carried out in conjunction with a complex of enzymes known as the fatty acid synthase system (Murray et al. 1990).

The role of biotin in forming fatty acids for the membrane is suggested by the effect of oleic acid on bacterial strains that require both oleic acid and biotin. Low glutamate yields result at high concentrations of oleic acid, but not at low levels. Since oleic acid is incorporated into the cell membrane, a high level was hypothesized to result in a tight membrane and low excretion of glutamate. Another observation that supports the membrane/lipid explanation is the change in cell morphology from a spherical form to a swollen rod shape and a decrease in packed cell volume in the absence of cell lysis (Wang et al. 1979).

The route to glutamate starts with glucose that passes through the Embden–Meyerhoff glycolysis pathway and enters the citric acid cycle (Fig. 11.10), which lacks α-ketoglutarate dehydrogenase. Consequently, the α-ketoglutarate cannot be converted to succinyl-CoA, and in the presence of NH_3 is transformed into glutamate by glutamate dehydrogenase instead. The supply of α-ketoglutarate is maintained when part of the isocitrate is converted to succinate and to malate through the glyoxylate cycle (Fig. 11.10). At optimal levels of biotin (about $2\,\mu g/L$ or 2 ppb, weight basis), the glutamate formed will leak out of the cell and the conversion of glucose to glutamate can approach 50% molar yields based on the sugar (glucose, MW = 180) and attain broth concentrations of 90 g glutamate/L (Aiba et al. 1973; Wang et al. 1979).

Typical concentrations of biotin in fermentations range between 1 and $5\,\mu g/L$ (1–5 ppb). An increase to $15\,\mu g/L$ increases growth rate and almost eliminates excretion of glutamate. Organic acids are accumulated instead, due to feedback inhibition from the intracellular accumulation of the glutamate, which builds up until the cells become saturated at about 50 mg glutamate/g dry weight. If the permeability barrier is altered by the addition of penicillin during the logarithmic growth phase, glutamate excretion occurs and the intracellular level decreases to 5 mg of intracellular glutamate/g cells (dry-weight basis). The cells will continue to excrete glutamate for an additional 40–50 h. Fatty acid derivatives trigger similar effects since they are not correctly incorporated into phospholipids, and therefore inhibit membrane formation, resulting in a leaky membrane as well (Wang et al. 1979).

The overproduction of amino acids *other* than glutamate, and nucleotides by auxotrophic mutants requires *high* levels of biotin. Otherwise glutamate is excreted rather than the desired products. Permeability is unidirectional, with glutamate passing to the outside only. If glutamate is lost from the cell because of permeation through the cell's membrane, the glutamate is no longer available as an intracellular precursor to other amino acids that require glutamate as a source of nitrogen derived through transamination. The control of cell permeability is an important parameter in optimizing amino acid production.

A Threonine and Methionine Auxotroph of *C. glutamicum* Avoids Concerted Feedback Inhibition and Enables Industrial Lysine Fermentations

Lysine results from a branched pathway and is an end metabolite rather than an intermediate. The pathway, with a single form of aspartokinase at the beginning, is unique to the bacteria *Corynebacterium glutamicum* and *Brevibacterium flavum*. *C. glutamicum* also differs from other bacteria since lysine does not inhibit its own pathway after the branch point. The first enzyme in the common pathway for methionine production in *E. coli* and *Salmonella typhimurium* consists of three forms of aspartokinase, with each form inhibited by a different end product. While the first step of the lysine pathway in *C. glutamicum* is catalyzed by a single enzyme, it is subject to multivalent inhibition by lysine and threonine (Fig. 11.12).

An auxotroph that requires threonine and methionine for growth was selected as a candidate for lysine production since it lacks homoserine dehydrogenase at the branch point for aspartate semialdehyde (Fig. 11.12) (Aiba et al. 1973; Wang et al. 1979). The mutant, grown on a nutrient medium containing low levels of threonine and methionine, accumulates large amounts of lysine since the concerted inhibition of the first enzyme in the pathway, aspartokinase, is bypassed, and the metabolic intermediates are shunted toward lysine. Both the first enzyme and the second enzyme (dihydrodipicolinate) of the lysine branch are neither inhibited nor repressed by lysine. L-Lysine decarboxylase is absent from the end of the pathway. These

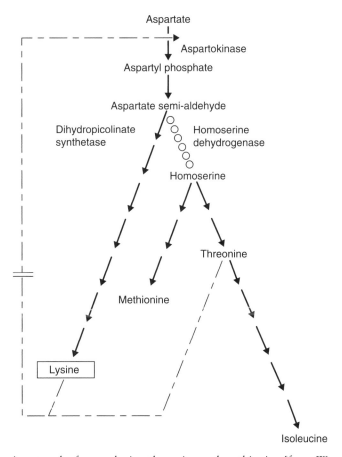

Figure 11.12. Auxotrophs for producing threonine and methionine [from Wang et al. (1979), Fig. 2.3, p. 17].

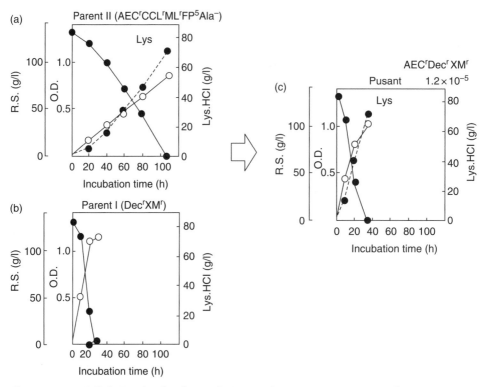

Figure 11.13. Cell fusion for developing lysine-producing microorganism. Slow-growing parent that produces lysine (A) is fused with fast-growing nonlysine producer (B). The resulting fusion product is both fast-growing and high-lysine-producing (C). [Ajinomoto, Inc. (1990)].

characteristics enabled *C. glutamicum* mutants to produce over 50 g of lysine/L fermentation broth by 1972.[6]

Cell (Protoplast) Fusion Is a Method for Breeding Amino Acid Producers that Incorporate Superior Characteristics of Each Parent (Lysine Fermentation)

Protoplast fusion is a versatile technique for inducing genetic recombination in both prokaryotes and eukaryotes. This method was significant in obtaining genetically altered organisms prior to the introduction of genetic engineering, and may continue to find use for some applications of industrial strain development (Matsushima and Blatz 1986). Hence it is summarized by an example presented here. The benefit of cell fusion is illustrated in Fig. 11.13. A slow-growing lysine producer fused with a fast-growing and vigorous carbohydrate mutant that produces little lysine gave a fast-growing lysine producer. The lysine-producing organism required 100 h to

---

[6] The 1995 global market for fermentation-derived lysine was estimated at about $600 million/year. The economic importance of the lysine market is suggested by the discovery and prosecution of a worldwide lysine cartel by the US Justice Department (Kilman 1998a,b).

achieve about 70 g/L of lysine, while only 40 h were needed for the organism based on cell fusion of the two mutants.

Protoplasts are formed by treating cells with lytic enzymes that lyse the cell wall in the presence of a hypertonic solution[7] that stabilizes the osmotic environment of the protoplasts. Protoplasts are induced to fuse and form transient diploids in the presence of an agent such as polyethylene glycol (PEG) that promotes fusion. Prokaryotic bacilli and streptomyces recombine their genomes at high frequency since the chromosomes are readily accessible in the cytoplasm. However, nuclear fusion must follow protoplast fusion in eukaryotes since the DNA is segregated from the cytoplasm by a nuclear membrane. The last step is the regeneration of viable cells from the fused protoplasts. Specific techniques are given by Matsushima and Blatz (1986).

The ability to form, fuse, and regenerate protoplasts varies greatly from organism to organism. PEG-induced fusion of cells (protoplasts) was achieved for plants in 1974, and has subsequently been applied to fuse animal, filamentous fungal, yeast, *Bacillus* spp., *Streptomyces, Brevibacterium, Streptosporangium*, and *Staphylococcus* protoplasts. Matsushima and Blatz (1986) describe and review protocols for protoplast fusion for microbial species found to be generally applicable for industrial strain development. "Easily scorable genetic markers" (e.g., nutrition, antibiotic resistance, morphology) are needed to assess the efficiency of protoplast fusion. For example, if two strains that are auxotrophic for different markers are used to prepare protoplasts, the auxotrophy of the resulting cell can be used to differentiate between regenerated cells that have genetic information from one auxotroph rather than the other.

Amino Acid Fermentations Represent Mature Technologies

In 1993, a review of the outlook of one of the leading companies in the field of amino acid production (Ajinomoto of Japan) indicated that amino acid production is a mature technology. At Ajinomoto, glutamic acid and glutamine were produced by wild-type microorganisms; lysine, by auxotrophic and regulatory mutants and microorganisms from cell fusion; threonine, by regulatory mutants and recombinant microorganisms; threonine, by regulatory mutants; and cystine, by enzyme catalysis (Ladisch and Goodman 1992). Another major amino acid producer, Kyowa Hakko Kogyo Company, Ltd., used recombinant technology to increase the production of tryptophan from 30 to 60 g/L by cloning and expressing many of the genes for the tryptophan pathway (Wang 1992). Ajinomoto's culture collection has at least 21,000 strains, of which about 5000 are bacteria, 6000 are yeast, and 10,000 are unknown. It is likely that other major manufacturers have significant culture collections. Consequently, there is a significant and diverse reserve of microorganisms for potential use in the future, with improvements that are likely to be incremental rather than revolutionary. Nonetheless, the continuous improvement of amino acid–producing

---

[7]Base components of the osmotic stabilizing solution consist of 103 g sucrose, 0.25 g K_2SO_4, and 2.03 g $MgCl_2 \cdot 6H_2O$ in 700 mL. This is then mixed with KH_2PO_4 (to give 0.5 g/L), $CaCl_2 \cdot 2H_2O$ (27.8 g/L), and TES buffer to keep the pH near 7.2 [TES = *N*-tris(hydroxymethyl) methyl-2-amino ethyl sulfonic acid]. These and other reagents are given by Matsushima and Blatz (1986).

strains and of manufacturing technologies will be an important factor in maintaining the cost-competitiveness of amino acids. Recombinant technology will play a continuing role in this process.

ANTIBIOTICS

The antibacterial property of sulfanilamide, formed in vivo in mice from Prontosil, was shown to be effective in protecting mice infected with *Streptococcus haemolyticus*. According to Evans (1965), this discovery, made in 1935, marked the beginning of a new era in the treatment of systemic microbial infections. The first commercial antibiotics were on the market within 10 years. Prior to that, other antibacterial drugs were known but were too toxic for oral or parenteral administration, or were not effectively adsorbed into the bloodstream.

Antibiotics were defined by Waksman as "chemical substances that are produced by micro-organisms and that have the capacity to inhibit and even to destroy other microorganisms." These differ from antibacterial drugs since antibacterials are obtained through chemical synthesis. A brief history by Evans (1965), paraphrased below, illustrates the role of long-term research on the discovery of practical methods for antibiotic production and applies to modern products derived through modern biotechnology.

The existence of antibiotics was recognized in 1877 by Pasteur and Jobert, who observed that anthrax were killed when contaminated by some other bacteria. Gosio found, in 1896, that myophenolic acid from *Penicillium brevicompactum* inhibited *Bacillus anthracis*, while Grieg-Smith showed *actinomycetes* produced antibacterial compounds in 1917. Fleming published his observations on the inhibition of *staphylococcus* culture by growing colonies of *P. notatum* in 1929. When Florey and Chain reexamined penicillin about 10 years later, culminating in development of methods to isolate penicillin and clinical trials in 1941, they found it to be effective and "virtually devoid of toxicity [to humans]."

Commercial manufacture was facilitated when a strain of *Penicillium chrysogenum* was discovered to be capable of growing and producing penicillin in large-scale, aerated submerged culture in 1943, as described in Chapter 1 ([see also Langlyhke (1998) and Mateles (1998)]. The metabolism of microbial production of antibiotics in the context of secondary metabolite (Type III) fermentation is summarized in the next section. A technical history of penicillin is given in a volume edited by Mateles and describes the microbiological and industrial production aspects of penicillin (Mateles 1998).

Secondary Metabolites Formed During Idiophase Are Subject to Catabolite Repression and Feedback Regulation (Penicillin and Streptomycin)

Secondary metabolites are produced by a narrow spectrum of organisms and have no general function in life processes. They are not produced during the cell's rapid growth phase (known as the *trophophase*). Rather, they are formed during a subsequent phase—the idiophase. The fermentation is Type III since the generation of product is separate from the growth phase and independent of the cell's energy

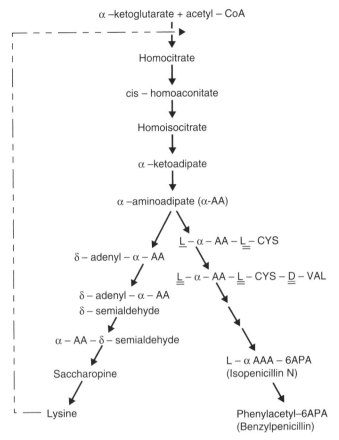

Figure 11.14. Metabolic pathway for the production of penicillin from amino acid precursors in *Penicillium chrysogenum* with feedback inhibition by lysine of homocitrate synthetase [from Wang et al. (1979), Fig. 3.3, p. 31, Wiley-Interscience].

metabolism. Antibiotics, alkaloids, and mycotoxins are secondary metabolites (Demain 1971).

The production of the antibiotic penicillin from amino acid precursors in *Penicillium chrysogenum* occurs through a branched pathway, in which lysine causes feedback inhibition of homocitrate synthetase (Fig. 11.14). The inhibition by lysine of the first enzyme in its own biosynthesis limits production of intermediates, including the α-aminodipate that is transformed to benzylpenicillin by the right branch in Fig. 11.14. Consequently, production of penicillin (structure in Fig. 11.15) is also inhibited unless α-aminodipate is added to the fermentation broth. A similar mechanism in *Streptomyces griseus* limits production of candicidin. Tyrosine, tryptophan, and phenlalanine are end products of a pathway with the candicidin intermediate, *p*-aminobenzoic acid, occurring at the branch point (Wang et al. 1979).

The production of antibiotics derived from sugars is regulated by the repression of phosphatase enzymes by phosphates at concentrations that do not inhibit growth. The biosynthesis of streptomycin by *Streptomyces griseus* involves at least three phosphatases. The intermediate is streptomycin phosphate. The

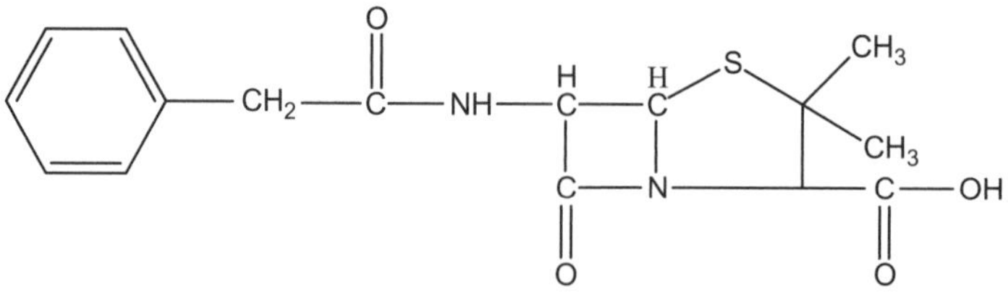

Figure 11.15. Benzyl penicillin is synthesized from two amino acids [from Evans (1965)].

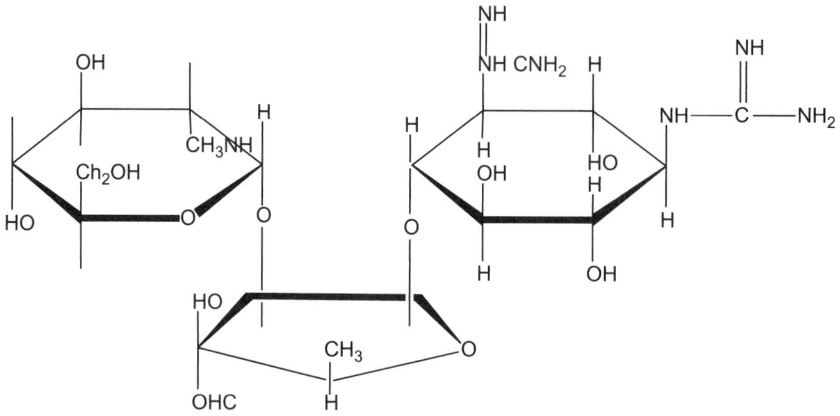

Figure 11.16. Streptomycin is synthesized from sugars [from Evans (1965)].

transformation of the biologically inactive streptomycin phosphate to the desired streptomycin[8] (structure in Fig. 11.16) is inhibited by phosphate (Wang et al. 1979; Evans 1965). Control of phosphate concentration is of obvious importance.

Catabolic repression occurs when the substrate is metabolized to intermediates that repress synthesis of enzymes that lead to formation of the secondary product. Penicillin fermentation is subject to catabolic repression when glucose is rapidly utilized during the tropophase. A mixture of glucose and lactose, which is the basis of the Jarvis and Johnson media, was found to remove repression of penicillin synthesis. On glucose exhaustion and rapid growth phase, repression of lactose utilization in *P. chrysogenum* is eliminated. The fermentation timecourse in Fig. 11.17 shows how the increase in cell mass follows the depletion of sugar up to about 40 h, during the tropophase, when enzymes for penicillin production are induced.

The idiophase and penicillin production begins at about 50 h (Fig. 11.17). Although one possible explanation was that lactose served as a precursor or inducer

---

[8] Streptomycin was discovered in 1944 by Shatz, Bugie, and Waksman after a series of disappointments in screening studies on the antimicrobial properties of metabolites from *actinomycetes*. Other antibiotics had previously been discovered that were effective but they were too toxic for human use. Streptomycin from *S. griseus* was found to be active against both Gram-negative and Gram-positive bacteria and inhibited *Mycobacterium tuberculosis*, and was soon used to treat tuberculosis. The development of resistance to this antibiotic was known, but its concomitant administration with other therapeutic agents was found to be effective (Evans 1965).

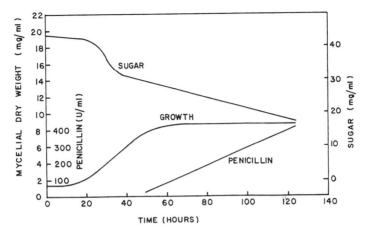

Figure 11.17. Fermentation timecourse for penicillin production. Timecourse of penicillum (aerated) fermentation in chemically defined Jarvis–Johnson medium containing glucose and lactose. Glucose is utilized during tropophase and lactose during idiophase. [From Wang et al. (1979), Fig. 3.4, p. 32.]

for penicillin, further research showed that the slow hydrolysis of lactose resulting in a slow release of glucose gave a much lower intracellular concentration of catabolites than would otherwise be present. This relieves repression of penicillin synthesis. Slow feeding of glucose has the same effect as using lactose in the medium. Hence, slow feeding of less expensive glucose replaced the use of lactose for derepression of penicillin synthesis during industrial-scale production.

From the perspective of a process scientist, lactose was initially used because its rate of consumption corresponded to maintaining the pH between 6.5 and 7.5, which is optimal for penicillin production. When a simple sugar such as glucose was used instead, it was rapidly consumed, causing the pH to increase to 8.5, where penicillin synthesis ceased. The slow feeding of glucose, such as in molasses, avoided this result (Langlyhke 1998).

The Production of Antibiotics Was Viewed as a Mature Field
Until Antibiotic-Resistant Bacteria Began to Appear

The production of antibiotics was viewed as a mature field with 50% of antibiotics manufacturers in the United States and Japan curtailing development efforts between about 1975 and 1990, although there were significant efforts to modify antibiotics in order to overcome bacterial resistance (Gross and Carey 1996). This resulted in about 160 variations of 16 basic compounds (Tanouye 1996a,b).

The unrecognized yearly cost of antibiotic resistance was estimated to be about $100 million prior to 1991, according to Walsh (1993). While this may seem small relative to the value of other types of biopharmaceuticals whose markets are in the billion-dollar range, the cost of the infections, estimated to be $4.5 billion and affecting 2 million patients in 1992 (Williams and Heymann 1998), has greatly increased since then. The perception that antibiotics represent a mature business is changing because of the increase in antibiotic-resistant microorganisms, with measures to provide incentives for new antibiotics being proposed. Drug-resistant

Staphylococcus auerus is spreading with 126,000 hospitalizations reported (Chase 2006).

Bacteria Retain Antibiotic Resistance Even When Use of the Antibiotic Has Ceased for Thousands of Generations

An example was reported for *E. coli* that carried a streptomycin resistance gene (abbreviated rpsL) (Morell 1997). The *E. coli* was kept in an antibiotic-free environment for 20,000 generations over a 10-year period. Although the gene is "known to markedly reduce the bacteria's fitness," and a single base change in the gene would revert the *E. coli* to an antibiotic-susceptible state, the *E. coli* retained its antibiotic resistance. Further research indicated that the development of antibiotic resistance was accompanied by a mutation that compensated for a loss of fitness caused by the first. This leads to the hypothesis that antibiotics that have lost their effectiveness are unlikely to become powerful medicines again.

Antibiotic Resistance Involves Many Genes (Vancomycin Example)

Vancomycin is a glycopeptide antibiotic that was isolated from *S. orientalis* and reported by Higgins et al. (1957) [cited in Evans (1965)]. It is used against Gram-positive[9] bacterial infections that are resistant to β-lactam antibiotics (such as peni-

---

[9] The Gram strain was developed by Danish physician Hans Christian Joachim Gram in 1884, and further modified by pathologist Carl Weigert in Germany. The procedure consists of placing gentian violet over a "dried smear" of cells, washing it with water, and then adding potassium triiodide to fix the dye, if it has reacted with the cells. The cells are washed with ethanol, and then stained with safranine. Cells are Gram-positive if they take on the purple color of gentian violet and Gram-negative if they retain the red color of the safranine. The dyes have the following structures (Stinson 1996):

Gentian violet   Safranine

Gram-positive cells have thick cell walls of crosslinked polysaccharide that react with gentian violet, while Gram-negative cells have thin polysaccharide walls coated with a lipid layer that reacts with safranine but resists staining with gentian violet. Antibacterial drugs that act against both Gram-positive and Gram-negative bacteria are referred to as being *wide-spectrum*.

cillin), or for which penicillin cannot be used because of patient allergies. Vancomycin binds strongly to peptidoglycan strands in the bacterial wall that terminate in the amino acids D-Ala–D-Ala. This interferes with crosslinking of newly formed peptidoglycan strands, causing a reduction in the tensile strength of the cell wall. The cells are then susceptible to lysis, due to osmotic pressure difference between the internal part of the cell and the exterior fluid. Vancomycin can be effective in fighting infections when other antibiotics fail. However, by the late 1980s vancomycin-resistant bacterial strains of *Enterococcus* began to appear, and in 1996 a strain of *Enterococcus faecium* that requires vancomycin for inducing synthesis of cell wall components was reported (News Item 1996).

Gene cloning and sequencing has delineated the mechanism by which the cell overcomes the action of this antibiotic. The metabolic pathway and the expression of resistance involves nine genes. A transmembrane protein is synthesized by the cell that detects vancomycin and sends a chemical signal to a regulatory protein that, in turn, activates transcription of genes. This leads to synthesis of peptidoglycans where D-Ala–D-Ala is replaced by D-Ala–D-lactate. The depsipeptide D-Ala–D-lactate does not strongly bind vancomycin, which itself is a heptapeptide (Walsh 1993). This neutralizes the activity of the antibiotic and makes the bacterium resistant to vancomycin.

The potential impact of vancomycin resistance is likely to be significantly greater if the genetic basis for this resistance in *Enterococci*[10] is transferred to other microorganisms (Tanouye 1996a,b). This is particularly true since vancomycin is used to treat streptococcyl or staphylococcal strains that are resistant to β-lactam antibiotics such as penicillin (Walsh 1993). Antibiotic-resistant, group A streptococcus can cause toxic shock[11] and secrete a protein, pyrogenic exotoxin A, that triggers an immune system response and causes cells that line blood vessels to be damaged, so that fluid leaks out, blood flow dwindles, and tissues die from lack of oxygen (Nowak 1994). Another organism, *Staphylococcus aureus*, is also treated with vancomycin. The first case of a partial vancomycin resistance of a staph infection was reported in Japan in 1996 (Gentry 1997).[12]

---

[10] Bacteria are found in three forms: cocci (spheres), bacilli (rods), and spirilla (spirals). The physical appearance of these 1–5-µm structures are schematically illustrated below:

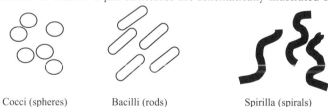

Cocci (spheres)   Bacilli (rods)   Spirilla (spirals)

Growing bacteria are in a vegetative state. Bacteria can form endospores under adverse conditions (heat, radiation, and toxins), which germinate back into vegetative form when placed in an environment that is conducive for growth. Hence, sterilization must be harsh enough to kill spores as well as vegetative cells.

[11] The sudden death of *Muppets* creator Jin Henson due to a virulent strain of streptococcus A helped to renew attention on drug-resistant bacterial infections (Wright 1990).

[12] There are 260,000 cases of staph infection per year in US hospitals. US sales of vancomycin were more than $108 million in 1996. Eli Lilly's sales of vancomycin (sold as Vancocin) were

Large quantities of antibiotics have been used in agricultural, animal-producing facilities, as well as for the treatment of human illness. About half of the antibiotics produced in the United States are used for farm animals. Animal health applications often utilize the antibiotic at subtherapeutic doses that allow resistant microbial stains to survive (Wright 1990). This form of selective pressure and the ability of resistant bacteria to use plasmids to pass on information on antibiotic resistance to nonresistant cells of different species that can also proliferate is believed by some to have helped promote antibiotic resistance.

An example of the change in viewpoint on use of antibiotics in agriculture and heightened concern about drug resistance is given by a fluoroquinolone (sacrofloxacin) approved by the US Food and Drug Administration (FDA) for use in treating bacterial infections in chickens in 1995. This antibiotic was used to fight *E. coli* infections in chickens in order to improve productivity and thereby reduce production costs for the $15 billion chicken–turkey industry (1995 dollars). Cost savings to the consumer were estimated at 1–2%. The debate surrounding the approval for this application was due to the special status of fluoroquinolones as an antibiotic of last resort for treatment of antibiotic-resistant forms of dysentery and typhoid fever. The concern was that bacteria in poultry that become resistant will pass on this trait to other bacteria that cause dysentery. Consequently, the use of the drug was tightly controlled, with only 38 veterinarians in the United States allowed to prescribe the drug for use in chickens. Anticipated use was for 150 million chickens, or about 2% of the annual US production (Hensen 1995). The use of the antibiotic is monitored closely by the FDA working together with the US Center for Disease Control and Prevention (CDC) and the US Department of Agriculture (USDA).

REFERENCES

Aiba, S., A. E. Humphrey, and N. F. Millis, *Biochemical Engineering*, 2nd ed., Academic Press, New York, 1973, pp. 28–32, 56–91, 195–345.

Ajinomoto, Inc., *Annual Report 1990*, Ajinomoto Co., Inc. 5–8 Kyobashi 1-chrome, Chuo-ku, Tokyo 104, Japan, 1990, pp. 1–15.

Atkinson, B. and F. Mavituna, *Biochemical Engineering and Biotechnology Handbook*, Nature Press, Macmillan Publishers, 1983, pp. 51–62, 320–325, 1018, 1058, 1004.

Bailey, J. E., "Toward a Science of Metabolic Engineering," *Science* 252(5013), 1668–1675 (1991).

Behm, G., D. Dressler, G. Gaus, H. Herrmann, K. Küther, and H. Tanner, "Amino Acids in Animal Nutrition," in *Arbeitsgemeinschaft für Wirkstoffe in der Tierernährung e.v. (Awt)*, Bonn, 1988, pp. 46–52.

$250 million worldwide (Gentry 1997). While guidelines for vancomycin's use restricts it to treatment of antibiotic-resistant infections, or for patients with allergy to penicillin, culture techniques that are used to identify the microbe (and therefore an alternate antibiotic) require up to 72 h. Rather than risk patient death in the meantime, physicians prescribe vancomycin. The concern was that widespread use (135,000 prescriptions per year) would accelerate development of resistance to vancomycin (Gentry 1997). This has occurred.

Chase, M., "Bacteria that Cause Ulcers May Also Boost Stomach Cancer Risk," *Wall Street J.* **B1** (March 9, 1998).

Chase, M., "Incentives for New Antibiotics Urged," *Wall Street J.* **D2** (March 1, 2006).

Degussa, A. G., DL-*Methionine—from Feed to Performance*, Frankfurt, 1989, pp. 4–5.

Degussa, A. G., *Amino Acids for Animal Nutrition*, Frankfurt, 1985, pp. 22–25.

Demain, A. L., "Overproduction of Microbial Metabolites and Enzymes Due to Alteration of Regulation," in *Adv. Biochemical Eng. 1*, T. K. Ghose and A. Fiechter, eds., Springer-Verlag, Berlin, 1971, pp. 121–142.

Demain, A. L. and J. E. Davies, eds., *Manual of Industrial Microbiology and Biotechnology*, American Society for Microbiology, 1999, pp. 32–40.

Evans, R. M., *The Chemistry of Antibiotics Used in Medicine*, Pergamon Press, Oxford, 1965, pp. 20–30, 75–81, 187.

Gentry, C., "Wide Overuse of Antibiotic Cited in Study," *Wall Street J.* **B1** (Sept. 4, 1997).

Gross, N. and J. Carey, "A Race to Squash The Superbugs: Drugmakers Final Go After the Lethal New Microbes," *Business Week* **3460**, 38–39 (1996).

Hensen, L., "Will Healthier Birds Mean Sicker People," *Business Week* **3440**, 34 (Sept. 4, 1995).

Higgins, H. M., W. H. Harrison, G. M. Wild, H. R. Bungay, and M. H. McCormick, "Vancomycin, a new antibiotic. VI. Purification and properties of vancomycin," *Antibiot. Annu.* **5**, 906–914 (1957).

Hileman, B., "Pfisteria Health Concerns Realized," *Chem. Eng. News* **75**(41), 14–15 (1997).

Hunt, G. R. and R. W. Steiber, "Inoculum Development," in *Manual of Industrial Microbiology and Biotechnology*, A. L. Demain and N. A. Solomon, eds., American Society of Microbiology, 1986, pp. 32–40.

Kern, A., E. Tilley, I. S. Hunter, M. Legisa, and A. Glieder, "Engineering Primary Metabolic Pathways of Industrial Micro-Organisms," *J. Biotechnol.* **129**(1), 6–29 (2007).

Kilman, S., "In Archer-Daniels Saga, Now Executives Face Trial: Price-Fixing Case Will Be Biggest Criminal Antitrust Showdown in Decades," *Wall Street J.* **B10** (July 9, 1998a).

Kilman, S., "Mark Whitacre Is Sentenced to 9 Years For Swindling $9.5 Million from ADM," *Wall Street J.* **B5** (March 5, 1998b).

Kreil, G., "Conversion of L to D-Amino Acids, a Post-Translational Reaction," *Science* **266**(5187), 996–997 (1994).

Ladisch, M. R. and N. Goodman, Group III Site Report: Ajinomoto Company, Inc., Kawasaki, Japan, in *JTEC Panel Report on Bioprocess Engineering in Japan*, Loyola College in Maryland, Baltimore, 1992, pp. 148–155.

Langlyhke, A. F., "The Engineer and the Biologist," in *Penicillin, a Paradigm for Biotechnology*, R. I. Matales, ed., Candida Corp., 1998, pp. 77–84.

MacCabe, A. P., M. Orejas, E. N. Tamayo, A. Villanueva, and D. Ramón, "Improving Extracellular Production of Food-Use Enzymes from *Aspergillus nidulans*," *J. Biotechnol.* **96**(1), 43–54 (2002).

Mateles, R. J., *Penicillin: A Paradigm for Biotechnology*, Candida Corp. edited volume, 1998, p. 114.

Matsushima, P. and R. H. Blatz, "Protoplast Fusion," in *Manual of Industrial Microbiology and Biotechnology*, A. L. Demain and N. A. Solomon, eds., American Society for Microbiology, Washington, DC, 1986, pp. 170–183.

Morell, V., "Antibiotic Resistance: Road of No Return," *Science* **278**(5338), 575–576 (1997).

Murray, R. K., P. A. Mayes, D. K. Granner, and V. W. Rodwell, *Harper's Biochemistry*, 22nd ed., Appleton & Lange, Norwalk, CT, 1990, pp. 10–14, 58, 68–69, 102–106, 146, 160–161, 165, 199–200, 318–322, 336–338, 549–550, 555.

News Item, "Strain of Bacteria needs Antibiotic to Survive," *Chem. Eng. News* **74**(51), 32 (1996).

Nowak, R., "Flesh-Eating Bacteria: Not New, but Still Worrisome," *Science* **264**(5166), 1665 (1994).

Queener, S. W. and D. H. Lively, "Screening and Selection for Strain Improvement," in *Manual of Industrial Microbiology and Biotechnology*, A. L. Demain and N. A. Solomon, eds., American Society of Microbiology, 1986, pp. 155–169.

Service, R. F., "*E. coli* Scare Spawns Therapy Search," *Science* **265**(5171), 475 (1994).

Stephanopoulos, G. and K. L. Jensen, "Metabolic Engineering: Developing New Products and Processes by Constructing Functioning Biosynthetic Pathways *in vivo*," *AIChE J.* **51**(12), 3091–3093 (2005).

Stinson, S. C., "Drug Firms Restock Antibacterial Arsenal: Growing Bacterial Resistance, New Disease Threats Spur Improvements to Existing Drugs and Creation of New Classes," *Chem. Eng. News* **74**(39), 75–100 (1996).

Tanner, H. and H. Schmidtborn, DL *Methionine—the Amino Acid for Animal Nutrition*, A. G. DeGussa, 1970, pp. 3, 101, 106–107.

Tanouye, E., "How Bacteria Got Smart, Created Defenses against Antibiotics," *Wall Street J.* **B1** (June 25, 1996a).

Tanouye, E., "Drug Makers Go All Out to Squash Superbugs," *Wall Street J.* **B6** (June 25, 1996b).

Walsh, C. T., "Vancomycin Resistance: Decoding the Molecular Logic," *Science* **261**(5119), 308–309 (1993).

Wang, D. I. C., Group II Site Report: Kyowa Hakko Kogyo Company, Ltd., Tokyo Research Laboratories, Machida-shi, Tokyo, 1992, pp. 106–109.

Wang, D. I. C., C. L. Cooney, A. L. Demain, P. Dunnill, A. E. Humphrey, and M. D. Lilly, *Fermentation and Enzyme Technology*, Wiley, New York, 1979, pp. 10–11, 14–36, 75–77.

Warnecke, T. E., M. D. Lynch, A. Karimpour-Fard, N. Sandoval, and R. T. Gill, "A Genomics Approach to Improve the Analysis and Design of Strain Selections," *Metab. Eng.* **10**(3–4), 154–65 (2008).

White, R. J., W. M. Maiese, and M. Greenstein, "Screening for New Products from Micro-organisms," in *Manual of Industrial Microbiology and Biotechnology*, A. L. Demain and N. A. Solomon, eds., American Society of Microbiology, 1986, pp. 24–33.

Williams, R. J. and D. L. Heymann, "Containment of Antibiotic Resistance," *Science* **279**(5354), 1153–1154 (1998).

Wright, K., "Bad News Bacteria: The Complex Genetics of Increased Virulence and Increased Resistance to Antibiotics Is Turning Certain Bacterial Threat," *Science* **249**(264), 22–24 (1990).

11.1. Some plasmids contain information so that a microorganism can produce enzymes that destroy or otherwise modify antibiotics to a less lethal form (Tanouye 1996a,b). Examples of nonpathogenic bacteria that could serve as reservoirs for genes responsible for drug resistance, and that are capable of transferring these to virulent strains (Wright 1990), have been known for some time. Resistance of *Haemophilus influenza* and *Neisseria gonorrhoeae* to ampicillin in the 1970s was traced to *E. coli*. Dysentery caused by a strain of *Shigella*, resistant to ampicillin, carbenicilin, streptomycin, sulfisoxazole, and tetracycline was traced to a plasmid that harbored resistance traits found in *E. coli* in 1983 (Wright 1990). Briefly discuss how these findings might affect transmission of disease, and international commerce. Can you cite any recent examples?

11.2. Runoff from farms has resulted in nutrient pollution that caused algal blooms. These may lead to a toxic form of *Pfisteria* through a complex series of events. These algae secrete a potent water-soluble toxin that destroys cell membranes. *Pfisteria* are normally found in a nontoxic cyst stage at the bottom of rivers but transform into an amoeboid stage when exposed to substances excreted by live fish or leached from fish tissues. Unlike other toxic dinoflagellates, of which only 2% are toxic or contain the toxin intracellularly, *Pfisteria* secretes a water-soluble neurotoxin that incapacitates the fish, and a lipid-soluble toxin that destroys parts of fish's epidermal layer, thereby wounding it (Hileman 1997). These toxins are also harmful to humans that come in contact with them through contaminated water. The control of this disease will require changes in environmental conditions leading to it. How might this affect agricultural practices? Briefly discuss and present your rationale. How would changes that you might propose to address this issue, affect consumers?

11.3. Concerted feedback inhibition of aspartokinase due to lysine and threonine occurs in the pathway shown. A homoserine dehydrogenase-deficient mutant is discovered, and makes it possible to significantly increase lysine production.

(a) Briefly explain why a homeserine dehydrogenase deficient mutant relieves the inhibition.

(b) Which amino acids in this pathway would need to be added at growth limiting concentrations? (Check the appropriate blanks).

___ Met

___ Thr

___ Ile

___ Lys

___ Gly

___ Glu

___ Gln

The pathway for answering (a) and (b) is shown below:

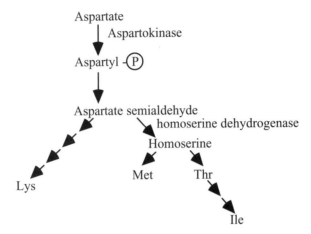

11.4. Strategies for developing auxotrophic mutants for producing product P. The microbial pathway leading to the desired product P is shown below. The dashed lines indicate inhibition caused by P_1 and P_2. The numbers indicate different enzymes. Small amounts of products P_1 and P_2 are required elsewhere in the cell for growth.

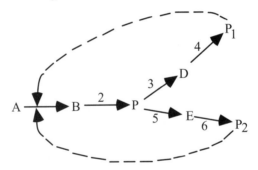

(a) The reaction pathway leading to P is subject to multivalent feedback inhibition (i.e., P_1 and P_2 alone have no effect, but together they completely inhibit the enzyme that catalyzes step 1). Both P_1 and P_2 are relatively expensive, so addition of small amounts of these required compounds is to be minimized, and, if possible, only one of the compounds (either P_1 or P_2) is to be added. If you were attempting to isolate a mutant that overproduces P, which step(s) would need to be absent? Check the appropriate number(s) numbers that correspond(s) to the steps in the reaction sequence above, and briefly explain your answer.

1. ___
2. ___
3. ___
4. ___
5. ___
6. ___

(b) The inhibition in the metabolic sequence given above is found to be cooperative with P_1 and P_2 individually inhibiting step 1 to a small degree. When both are present, the inhibition is severe. An auxotrophic mutant in P_1 and P_2 is identified which overproduces P. Which steps are likely to be absent in the mutant? Briefly explain your answer.

1. ___
2. ___
3. ___
4. ___
5. ___
6. ___

(c) Which component(s) would need to be added to the media to obtain growth of the mutant? [This answer must be consistent with (b).] Briefly explain your answer.

1. ___

2. ___

P_1. ___

P_2. ___

5. ___

6. ___

11.5. A *Corynebacterium* species that is biotin-deficient, and lacks α-ketoglutarate dehydrogenase, utilizes some part of a pathway to produce glutamate [i.e., marketed as monosodium glutamate (MSG)]. In commercial processes using this microbe, the yield is 50% of (molar) theoretical when the biotin content in the medium is maintained at 2 μg/L. Broth concentrations of glutamate reach up to 90 g/L.

(a) What key step is blocked that allows glutamate to be produced? Draw a likely pathway to glutamate formation. Explain (briefly).

(b) Biotin deficiency results in production of cell membranes deficient in phospholipids [Nakao et al., *Agric. Biol. Chem.* **37**, 2399 (1973)] and thereby allows glutamate to be excreted. When the biotin concentration is increased from 2 to 15 μg/L, the glutamate concentration in the broth levels off. In another fermentation, when penicillin is added during log growth phase, excretion of phospholipids is noted, and glutamate production appears to be triggered. Give a possible explanation of why decreased biotin levels or penicillin addition during log phase growth give high yields of glutamate.

(c) Using a different microbial strain, it is claimed that glutamate can be produced from ethanol at 66% of theoretical yield (based on alcohol consumed) to give product concentrations of 60 g/L. If ethanol costs $2.50/gal and dextrose costs $0.20/lb, which of the two processes would have lower substrate cost:

    1. Glucose → glutamate?

    2. ETOH → glutamate?

    Explain. (Show your work!) Calculate substrate costs in ¢/lb glutamate.

  (d) What is the total annual production of MSG (worldwide basis)?

  (e) Can Y_{ATP} be calculated for glutamate production? If you have insufficient data, explain what would be needed to determine this yield coefficient.

11.6. Cells are grown on a medium containing acetate as the sole carbon source. The acetate enters the TCA cycle through acetyl-CoA as shown in the diagram below:

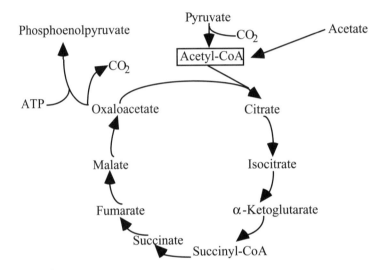

The oxaloacetate is decarboxylated to give phosphoenolpyruvate. Further careful research shows that pyruvate, glucose, and pentose-5-phosphate have accumulated in the cells. Give a possible explanation for these observations in terms of the Embden–Meyerhoff and TCA pathways. (*Note*: There is no glucose or pentose initially present in or added to the culture medium.)

11.7. *Corynebacterium in MSG Production. Corynebacterium* has both glycolytic and TCA pathways. The efficiency of conversion of glucose to glutamate is 50% of theoretical at a biotin concentration of 2 ppb (i.e., 2 µg/L) for a specially selected and mutated species of this bacterium. If the cell is "normal," it is able to produce high levels of biotin, which, in turn, promotes formation of a cell membrane with low permeability to small molecules. Glutamic acid bacteria, however, require biotin for growth. Biotin is an essential factor for the biosynthesis of fatty acids as it is the coenzyme of acetyl-CoA carboxylase. When the cells are cultivated with higher levels of biotin, production of glutamic acid occurs only when sublethal amounts of penicillin or small amounts of a detergent are added early in the exponential growth phase. When these inhibitors are added late in the exponential phase, neither inhibition of growth nor significant glutamic acid production is obtained. Use this information to analyze the possible strategies that could lead to an auxotropic or regulatory mutant of *Corynebacterium* that, under properly controlled fermentation conditions, should give high glutamic acid productivity.

(a) Which of the enzyme(s) would need to be absent or strongly inhibited to promote intracellular glutamate accumulation? Briefly explain your reasoning.

(b) A known glutamate-producing bacterium is grown with 2 ppb of biotin using a fermentation medium containing oleic acid (a fatty acid). Glutamate production is not obtained. Explain.

(c) Explain why both enzyme repression or inhibition *and* a leaky cytoplasmic membrane are needed to achieve high glutamic acid production.

(d) Assume a 30-h fermentation time, and a 20% initial glucose concentration, with other operating conditions controlled to obtain a glutamate yield that is 50% of theoretical. How many fermentors would be required to produce 250,000 tons of MSG/year (i.e., the world's supply)? Estimate the installed cost of this capacity. Assume that each fermentor has a volume of $20 \, m^3$.

11.8. More recently identified bacterial diseases include *Legionella pnuemophila* or Legionaire's disease (susceptible to erythromycin) and Lyme disease, which is transmitted by tick bites and caused by *Borrelia burgdorferi* (treated by the tetracycline doxycycline or amoxicillin). *Listeria monocytogenes* is likely to infect people with weakened immune systems, while *Helicobacter pylori* is believed to cause over 80% of stomach ulcers (Stinson 1996), and in a few cases lead to stomach cancer (Chase 1998). The bacterial pathogen *E. coli* 0157:H7 is estimated to cause 200–500 deaths/year, and is transmitted to humans through improperly cooked ground beef. While irradiation of the beef has been proposed to prevent illness, radiation is also viewed as a health threat by consumer groups (Service 1994). If these observations are proved correct, in which cases might antibiotics have a significant long-term benefit and in which cases might antibiotics *not* offer a practical solution? Briefly explain your reasoning.

11.9. The appearance of antibiotic-resistant microorganisms, as well as discovery of microbial causes of other chronic diseases, increases the importance of rapid sensing and identification of the types of drugs that effectively kill the organism. Significant opportunities exist for developing rapid testing methods. The test must be sufficiently accurate to prescribe the proper drug, while being fast enough so that a patient with a life-threatening infection doesn't die, before the result of the test is available. Once the microbe is known, charts are available that can assist the selection of the appropriate antibacterial. An example, cited by Stinson (1996), is the antibiogram (Fig. HP11.9.1) based on a system attributed to Maggs (University of Leicester, England) that uses a color-coding scheme. The darkest areas indicate the greatest chance of success for the combination of antibacterial drug(s) or antibiotic(s) against the microorganism. In some cases, a single drug will not be effective.

(a) Identify one area of the angiobiogram that indicates that a single antibiotic is not effective, but is effective if combined with another antibiotic. Briefly explain your answer.

Antibiograms give cues to prescribing antibacterials

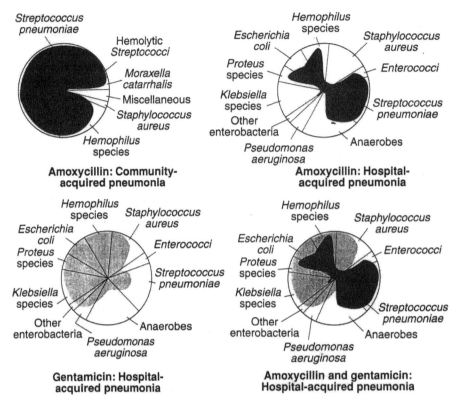

Figure HP11.9.1.   Antibiogram—graphical representation mapping susceptibility of different microorganisms to antibacterial drugs. Clinical microbiologist Tony Maggs of the University of Leicester, England, has devised as teaching aids pie-chart representations of relative prevalence of pathogens in infections and their likely susceptibility to antimicrobial drugs. For any one antibacterial drug, the pies are divided into segments according to the percent prevalence of specific pathogens in different clinical settings. Maggs colors in each segment according to the degree of susceptibility to the antibacterial. For example, in the case of amoxicillin for pneumonia acquired in the community outside of hospitals, the antibiogram is almost completely covered, indicating that physicians can expect good results by starting with that antibiotic. [Reproduced with permission from *Lancet* 345, 64 (1995).]

   (b) A vancomycin-resistant microorganism is identified where vancomycin can no longer bind to a peptide in the cell wall (see Fig. HP11.9.2). Which bond is altered by the cell to give rise to antibiotic resistance?

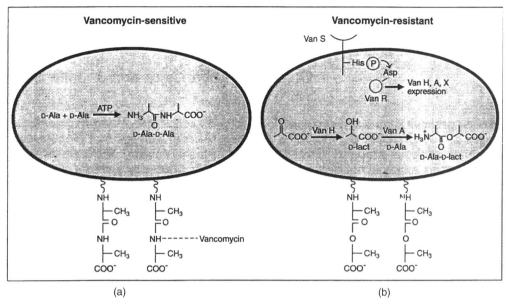

Figure HP11.9.2. Molecular logic of vancomycin resistance. Vancomycin-sensitive (a) and Vancomycin-resistant (b) bacteria differ in a critical component of their cell wall. Sensitive bacteria (a) synthesize PG strands that terminate in D-Ala–D-Ala; vancomycin binds avidly to these termini, thereby disrupting cell wall synthesis and leading to cell lysis. Resistant bacteria (b) harbor a transposable element encoding nine genes that contribute to the resistance phenotype. The gene products include a transmembrane protein (Van S) that senses the presence of the drug and transmits a signal—by transfer of a phosphoryl group—to a response regular protein (Van R) that activates transcription of the other resistance genes. The combined activities of Van H and Van A lead to synthesis of a depsipeptide, D-Ala–D-lactate, which can be incorporated into the PG strands of the cell wall. (Reprinted from Stinson, p. 98, 1996).

GENETIC ENGINEERING: DNA, RNA, AND GENES

INTRODUCTION

Metabolic engineering would not be possible without the tools of molecular biology, particularly the techniques referred to collectively as "cloning" or "genetic engineering." A bit of history on how cloning developed provides context for current developments in modern biotechnology. This chapter may be skipped, proceeding directly to Chapter 13 on metabolic engineering, if the student has taken an introductory molecular biology or biochemistry course.

The development of laboratory tools that enable foreign DNA to be introduced into microbial cells, as well as plant and animal cells, first occurred about 1969. The first commercial product derived from this technology, biosynthetic human insulin, was introduced in 1982 (Committee on Bioprocess Engineering 1992). The methods used are collectively referred to as *recombinant technology* since existing DNA from an organism is cut apart, and then recombined with foreign DNA. The enzymes that facilitated these methods are restriction endonucleases and DNA ligases. Their initial use required knowledge of partial sequences of nucleotides that make up the DNA and RNA involved in the cloning process.

Genetically engineered cells are obtained through a series of in vitro procedures that insert foreign genes into the DNA of the host cell. The cell is then propagated, resulting in additional cells that have the same foreign DNA as the originating cell. The result is clones: a colony of cells that are all descendants of a single ancestor cell (Alberts et al. 1989; Lewin 1997). The main steps in DNA cloning are obtaining a fragment of DNA that encodes the foreign gene, joining them to a vector, introducing the vector with the foreign gene into the host cell, selecting the host cells that have been genetically transformed, and confirming expression of the foreign DNA as a functional protein. There are numerous ingenious but complex approaches for cloning of foreign genes into bacteria, yeast, and mammalian and plant cells (Old and Primrose 1981; Lewin 1997). The object of cloning is to reprogram the "software" that forms the operating system of the cell—the DNA.

Modern Biotechnology, by Nathan S. Mosier and Michael R. Ladisch
Copyright © 2009 John Wiley & Sons, Inc.

DNA AND RNA

Since the information contained in the DNA is the basis of functioning organisms, the goal of gene sequencing is to define the types and locations of information in the DNA of a given organism. Once the informational (i.e., genetic) basis of metabolic processes is understood, strategies may be devised to reprogram the DNA, and generate new instructions that may treat diseases or alter biological functioning of a cell at a genetic level.

DNA Is a Double-Stranded Polymer of the Nucleotides: Thymine, Adenine, Cytosine, and Guanine

A brief synopsis of DNA structure is given here in order to relate how restriction enzymes are used to alter DNA in a directed manner. DNA, *deoxyribonucleic acid*, is a polymer formed from four basic units, called *nucleotides*. Each nucleotide is formed from deoxyribose (a sugar), phosphate, and one of four nitrogenous bases: thymine, adenine, cytosine, or guanine. When nucleotides are polymerized, the resulting DNA forms a right-handed double helix, with about 10 nucleotide pairs per turn. There are 10^3 to 10^9 base pairs in a single molecule of DNA. The nitrogenous bases form complementary hydrogen bonds such that two DNA strands of complementary sequence form a double helix. The strands match in antiparallel orientation; that is, the direction of the polymers are opposite of one another.

The pairing of nitrogenous bases occurs in a very specific and selective manner. Cytosine (C) hydrogen-bonds only with guanine (G), and thymine (T) hydrogen-bonds only with adenine (A). There are three hydrogen bonds between C and G, which makes the overall bonding stronger than A and T, which have only two hydrogen bonds. The hydrogen bonding and physical conformation of DNA gives it stability, so that it does not spontaneously dissociate unless the pH < 3 or pH > 10, or at temperatures approaching 100 °C. Even when the DNA is dissociated (i.e., denatured), it can be renatured to form a double helix again at temperatures below 65 °C as long as the sequences of the two strands are exactly complementary. Imperfect double helices can be formed at lower temperatures when the two strands are similar but not perfectly complementary.

The Information Contained in DNA Is Huge

Each DNA molecule is packaged in a separate chromosome. Each chromosome contains only a single double-stranded DNA molecule pair. The genome of the bacteria *E. coli* consists of a single DNA strand (i.e., one chromosome) containing 4.7×10^6 nucleotide pairs, while the human genome of 3×10^9 nucleotide pairs consists of 22 chromosomes and an X or Y chromosome. An individual human chromosome contains 50–250 million nucleotide pairs, and ranges in length from 1.7 to 8.5 cm when uncoiled (Alberts et al. 1989).

All of the chromosomes in an organism constitute its genome. There are four sequences in a DNA molecule containing *n* nucleotides consisting of the four fundamental monomers (A, G, T, or C). The chromosomes in a typical animal cell consist of $n = 3 \times 10^9$ nucleotides. The corresponding number of possible sequences exceeds $10^{1.8 \cdot 10^9}$. It is estimated that, in principle, 1 μmol of DNA (≈1 gram if the molecular weight is assumed to be 10^6 Da) can encode 2 gigabytes of information

(Ouyang et al. 1997) or about the same storage capacity as an iPod or flash drive. The genetic information contained in a single human cell would fill a book of more than 500,000 pages (Alberts et al. 1989).

Genes Are Nucleotide Sequences that Contain the Information Required for the Cell to Make Proteins

The information contained in the genes is a coded specification of the amino acids in a protein. Each amino acid corresponds to a sequence of three adjacent nucleotides. The proper specification of a sequence requires that the starting point of the code within a DNA molecule be properly identified by the enzymes and other cell structures that transcribe the information in the nucleotide sequence into a form and format that eventually give the proper protein structure.

Transcription Is a Process Whereby Specific Regions of the DNA (Genes) Serve as a Template to Synthesize Another Nucleotide, Ribonucleic Acid (RNA)

RNA is a single-stranded, linear polymer of nucleotides that consists of a sugar phosphate backbone that contains ribose (rather than deoxyribose as in DNA). RNA also differs from DNA since the base, uracil, replaces thymine found in the DNA. The synthesis of a messenger RNA (mRNA) molecule from a DNA template is referred to as *DNA transcription*, and is catalyzed by RNA polymerase, together with the help of other proteins known as *transcription factors*. The resulting mRNA molecules carry information for protein synthesis through the cell's cytoplasm to the ribosome, where protein synthesis occurs. Transcription also generates other types of RNA molecules that may have structural or catalytic functions.

Chromosomal DNA in a Prokaryote (Bacterium) Is Anchored to the Cell's Membrane While Plasmids Are in the Cytoplasm

The chromosomal DNA in a prokaryote forms a closed loop (or coil) that is anchored to the inner membrane of the cell. Almost all bacteria carry genetic information on this single circular piece of DNA (Kabnick and Peattie 1991). Prokaryotes contain only a single set of genetic information, and are therefore referred to as being *haploid*. In addition to chromosomal DNA, many bacteria also contain large numbers of small circular DNA molecules known as *plasmids*.

Plasmids are naturally occurring extrachromosomal DNA that carry genes and confer antibiotic resistance to the host cell. The genes in the plasmid direct production of enzymes that attack antibiotics, and thereby make the bacterium resistant against antibiotics. A cell may contain several hundred copies of a plasmid (i.e., a high copy number), so the number of genes that confer antibiotic resistance can be present in higher numbers than if the gene were located on the single chromosome. This helps a bacterial cell produce the large quantities of the enzymes needed to modify antibiotics, such as ampicillin, tetracycline, and kanamycin, rendering the antibiotics become ineffective in killing the bacterium (Alberts et al. 1989; Watson et al. 1992). The recombinant biotechnology industry was formed when the in vitro introduction of foreign genes into plasmids was demonstrated, and biologically functional plasmids containing foreign genes were shown to transform *E. coli* so that the bacterium would generate foreign proteins. In addition to the foreign gene, the plasmids contained genes that cause antibiotic resistance, and hence provided a useful marker by which microorganisms containing these plasmids could be identified—they were resistant to an antibiotic.

Chromosomal DNA in a Eukaryote (Yeast, Animal or Plant Cells) Is Contained in the Nucleus

Eukaryotes are so named because of the presence of a true (eu) nucleus (karyon) that contains the DNA in a compartment that is separate from the rest of the cell. Yeast and cells of plants and animals are eukaryotic. The DNA of a eukaryote is divided among multiple linear pieces that are attached to the nuclear membrane. While bacteria will typically have one chromosome, higher organisms have multiple chromosomes. Some eukaryotes contain plasmids, as well as chromosomal DNA. While gametes (eggs and sperm) of a eukaryotic organism are haploid, most eukaryotic cells are diploid—they contain two sets of genetic information, that is, two sets of double-stranded DNA, during stages of their life cycle associated with cell division (Kabnick and Peattie 1991).

Microorganisms Carry Genes in Plasmids Consisting of Shorter Lengths of Circular, Extrachromosomal DNA

Circular DNA in the form of plasmids has been a primary target for in vitro genetic manipulations, and has been used to introduce human genes into bacteria to obtain transformed microorganisms capable of generating human proteins or redirecting their metabolic pathways to obtain a desired product. One technique for obtaining such transformed organisms involves scission of a bacterial plasmid at a predetermined position, introduction of the foreign gene into the plasmid, and ligation (covalent bonding) of the double-stranded oligonucleotide (i.e., the gene) with the corresponding strand in the plasmid. The basic concept of this technology, illustrated in Fig. 2.3 of Chapter 2, is referred to as "genetic engineering," since a foreign gene is introduced in a directed, controlled manner, i.e., a foreign gene is engineered into the plasmid.

The new plasmid is a recombination of the original plasmid with a foreign gene. When the plasmid is successfully introduced into a host microbial cell, the cell is referred to as being *transformed*, and the microorganism is referred to as a *recombinant microorganism*. The protein generated from the gene in the recombinant plasmid is sometimes referred to as a *recombinant protein*, even though the recombination has been carried out with respect to the gene, not the protein.

Plasmids range in molecular weight from 1000 to 200,000 kDa and are widely found in prokaryotes. They can be divided into several categories. An ideal plasmid for cloning has the properties of low molecular weight, robustness, high copy number, ability to confer selectable phenotypic traits such as antibiotic resistances on host cells (enabling selection of transformed cells), and single sites for cleavage by a wide variety of restriction endonucleases (enabling selection of a restriction enzyme so that only one fragment is produced).

Plasmids are characterized as being relaxed if they are maintained in multiple copies per cell and therefore have a high copy number; or stringent, if they are found in the cell in a limited number of copies, and therefore have a low copy number. Conjugative plasmids are present as 1–3 copies/cell and have a high molecular

Table 12.1.

Properties of Natural Plasmids for Cloning DNA

Plasmid	Molecular Weight (kDa)	Single Sites for Endonucleases	Marker for Selecting Transformants	Insertional Inactivation of
pSC101	5800	*Eco*RI	Tetracycline resistance	Colicin E production
ColE1	4200	*Eco*RI	Immunity to colicin E1	Colicin E production
RSF2124	7400	*Eco*RI *Bam*HI	Ampicillin resistance	Colicin E production

Source: Reproduced from Old and Primrose (1981), Table 3.3, p. 32.

weight. Nonconjugative plasmids are usually relaxed with a high copy number and a low molecular weight. If two different types of plasmids are unable to coexist in the same cell, they are said to be "incompatible" (Old and Primrose 1981).

Naturally occurring plasmids identified during the early days of cloning were usable but not ideal. Some recombined forms of a plasmid (SC-101) was not distinguishable from reconstituted vector DNA that did not contain foreign DNA since both types of plasmids carried the gene for tetracycline resistance. The other two examples, given in Table 12.1, were selected by screening their hosts for immunity to the antibacterial peptide, colicin. However, the assay was difficult to carry out. In the case of *E. coli*, transformed with plasmid Col E1, chloramphenicol added to a late-log-phase culture of the *E. coli* caused chromosome replication to cease while the plasmid Col E1 continued to replicate. After 10–12 h a copy number of 1000–3000 c was obtained, thereby facilitating plasmid isolation carried out by either centrifugation or chromatography.

Because the original naturally occurring plasmids were less than ideal, in vitro genetic manipulations were used to generate "artificial" plasmids that contained the desirable traits. A versatile and widely used plasmid that was constructed in vitro was pBR322. This plasmid is 4362 base pairs long, contains ampicillin and tetracycline resistance genes, has the desirable replication elements, and has been completely sequenced. Eleven known enzymes cleave it at unique sites. The schematic representation of pBR322 in Fig. 12.1 shows the location of the unique cleavage sites by the indicated restriction endonucleases.

Another challenge in developing clones was achievement of the uptake of exogenous DNA by microbial species, in order to transform the original (or host) cells into cells that carried and could propagate DNA from a completely different organism. Early experiments were carried out with sectioned barley grains that were immersed in DNA from *Micrococcus lysocleikticus*. The DNA, which was radioactively labeled so that it could be detected, may have been taken up by the barley after 72 h. An experiment with the plant *Arabidopsis thaliana* was also tried. Later work showed that the DNA was not integrated into the plant cell genome. Other early experiments to introduce foreign DNA also failed because of either the inability to achieve or detect gene expression that resulted in formation of protein products or failure to achieve the replication of exogeneous DNA during culture of

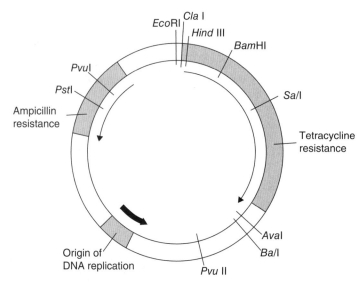

Figure 12.1. Unique cleavage sites for pBR322 [reproduced from Old and Primrose (1981), Fig. 3.5, p. 37].

transformed cells. The foreign DNA must somehow be integrated into the DNA of the host. Otherwise the DNA is not replicated and is lost during subsequent cell division. The DNA must also contain an origin of replication (i.e., a replicon) if it is to be replicated during cell division and passed on to the resulting offspring. This rationale led to the strategy of attaching foreign DNA to a suitable replicon (Old and Primrose 1981).

The first replicons (also referred to as *vectors* or "cloning vehicles") were plasmids and phages found in *E. coli*. The insertion of foreign DNA into a vector required that techniques be available for cutting and joining DNA molecules from different sources, monitoring the success of these reactions, and introducing the modified DNA back into *E. coli*. Hamilton Smith of Johns Hopkins discovered that the bacterium *Haemophilis influenzae* rapidly broke down foreign phage (virus) DNA. An enzyme in the bacterium *H. influenza* attacks DNA from *E. coli* but not from *H. influenzae*. The enzyme was found to specifically cleave DNA at sites having the sequence shown in reaction (12.1):

$$
\begin{array}{l}
\text{cleavage site} \\
\downarrow
\end{array}
$$

$$
\begin{array}{llll}
5' \text{———— G T Py} & \text{Pu A C ———— } 3' \\
3' \text{———— C A Pu} & \text{Py T G ———— } 5'
\end{array} \tag{12.1}
$$

This shorthand represents a double-stranded DNA molecule, with the 3' and 5' ends as indicated. The cleavage site occurs wherever there is the sequence

$$
\begin{array}{ll}
5' \text{———— G T Py Pu A C ———— } 3' \\
3' \text{———— C A Pu Py T G ———— } 5'
\end{array} \tag{12.2}
$$

where Py and Pu represent any of the pyrimidine or purine bases (Watson et al. 1992). This observation catalyzed development of modern recombinant techniques

(Watson et al. 1992; Old and Primrose 1981), which were first patented in 1980 by Cohen and Boyer, and form the basic techniques used for obtaining engineering, recombinant microorganisms.

Restriction Enzymes Enable Directed In Vitro Cleavage of DNA

The term *restriction enzyme* originated from experiments with a virus, phage λ, whose propagation was found to be restricted in some types of bacterial cells. Normally, this virus docks to the host cell, injects its DNA, and propagates itself within the host cell. When this virus was generated in *E. coli* strain C, and then used to infect a second strain of *E. coli* K, the virus did not propagate nearly as well in bacterial strain K as in *E. coli* C. The virus, phage λ, is therefore said to be restricted by the second host strain K. The cause of restricted propagation was found to be due to hydrolysis of foreign DNA from the invading virus by the host cell's endonuclease, specifically, a restriction endonuclease. The prefix *endo* denotes enzyme attack on nucleic acid residues that are internal to the DNA polymer. Since DNA is a polymer of nucleotides, the enzyme that hydrolyzes it is called a *nuclease*. When the endonuclease of the host cell attacks foreign DNA, its own DNA must be protected, since hydrolysis of the host cell's DNA is potentially lethal. A protection mechanism is based on differentiation of a host cell's DNA (or native DNA) from foreign DNA.

The host cell methylates nucleotides in its own DNA when these nucleotides coincide with those in foreign DNA that are attacked by the cell's endonuclease. The host cell's methylated DNA is not attacked, while the foreign viral DNA that is not methylated is hydrolyzed. However, if the phage survives one cycle of replication in the host cell, the target sequence on the viral DNA will now also be methylated. Consequently a phage recovered from *E. coli* C, and used to reinfect *E. coli* K host cells, will efficiently infect the *E. coli* K host cell since the cleavage sites of the viral DNA from *E. coli* C are now protected because of the methylation of *E. coli* C–derived DNA. Old and Primrose (1981) proposed that "host-controlled restriction acts as a mechanism by which bacteria distinguish self from non-self. It is analogous to an immunity system." Restriction may sometimes prevent infection by viruses that attack bacteria, that is, a bacteriophage such as phage λ.

The first restriction enzyme discovered had unusual properties that limited its utility for cloning. Enzymatic activity required magnesium ions, ATP,[1] and *s*-adenosylmethionine and an excess of ATP needed for cleaving the DNA. This enzyme recognizes a specific string of adjacent nucleotides in the DNA, but within this sequence, the site of cleavage varies from one molecule of DNA to the next. Hydrolysis was not specific, and defined sequences were not obtained. This type of enzyme is known as a *Type I restriction endonuclease*. A short time later, Type II restriction enzymes were discovered. These enzymes are specific with respect to the nucleotide sequences that they recognize as well as the nucleotide bonds that they cleave, within these sequences (Old and Primrose 1981).

The first Type II restriction endonuclease was reported for *Haemophilus influenzae* in 1970. This enzyme recognizes a specific target sequence in double-stranded

---

[1] Adenosine triphosphate (ATP) is a molecule that serves as a source of chemical energy in the cell. Hydrolysis of ATP to ADP and inorganic phosphate releases an energy of 11–13 kcal/mol phosphate, thus providing the energy needed for the action of the Type I restriction endonuclease.

DNA and hydrolyzed the polynucleotide chains into discrete DNA fragments of defined length and sequence. By 1981, over 150 enzymes having this activity had been identified and a systematic nomenclature was adapted (Watson et al. 1992; Old and Primrose 1981). The enzyme nomenclature was based on a three-letter abbreviation, in italics, taken from the first letter of the genus name of the host microorganism and two additional letters to specify the microorganism's strain or a distinguishing feature of the enzyme. For example, restriction enzyme from *E. coli* is abbreviated as *Eco*, and from *H. influenzae*, as *Hin*.

The enzymes are then further differentiated according to strain, and by Roman numerals if there is more than one type of enzyme in the microorganism. Thus, if a host strain is *E. coli* and it expresses two types of restriction enzymes, the two enzymes would be designated as *Eco*RI and *Eco*RII. Examples of the sites cleaved by several enzymes are presented in Table 12.2. The recognition sequence of commonly used restriction enzymes is given in Appendix 2 of Old and Primrose (1981), and the enzymes are readily available through research supply companies (Sigma 2008).

The sequence cleaved by *Eco*RI results in cohesive ends on the 5′ end of the DNA strand. This is represented in Eq. (12.3), where the dots represent bases that flank the two ends. The ends are said to be cohesive because the opposing sequences are complementary and will align and pair up when the ends are brought into close proximity under the appropriate conditions:

$$
\begin{array}{l}
5'...G\ |\ A\ A\ T\ T\ C...3' \\
3'...C\ T\ T\ A\ A\ |\ G...5'
\end{array}
$$

$$
\begin{array}{l}
5'...G \quad\quad\quad + \quad A\ A\ T\ T\ C...3' \\
3'...C\ T\ T\ A\ A \quad\quad\quad\quad G...5'
\end{array}
$$

(12.3)

The position of the DNA at which the enzyme acts is known as a *restriction site*. The cohesive ends formed [3′ … TTAA in Eq. (12.3)] are also called "sticky ends" since these sequences will associate with a complementary nucleotide sequence on a gene, or with themselves. The association of a nucleotide sequence on the end of a gene with the sequence on the DNA that has been cut by *Eco*RI, is illustrated in Eq. (12.4):

A A T T...

Gene

...T T A A

+

$$
\begin{array}{l}
5'...G \quad\quad\quad\quad\quad\quad 5'\ A\ A\ T\ T\ C...3' \\
3'...C\ T\ T\ A\ A \quad\quad\quad\quad\quad\quad G...5'
\end{array}
$$

(12.4)

↓ Anneal

$$
\begin{array}{l}
5'...G\ \textit{A A T T...} \quad\quad\quad\quad A\ A\ T\ T\ C...3' \\
3'...C\ T\ T\ A\ A \quad\quad\quad\quad\quad \textit{...T T A A}\ G...5'
\end{array}
$$

Gene

Table 12.2.

Examples of Type II Restriction Enzymes and Their Cleavage Sequences

Microbial Source of Enzyme	Abbreviated Name of Enzyme	Sequence Recognized and Cleavage Site (Denoted by Line) by the Enzyme	Type of End Formed
E. coli	*Eco*RI	5′...G|A A T T C...3′ 3′...C T T A A|G...5′	Cohesive end on 5′ strand, 6-base recognition sequence
	*Eco*RV	5′...G A T|A T C...3′ 3′...C T A|T A G...5′	Blunt ends; 6-base recognition sequence
Thermus aquaticus	*Taq*I	5′...T|C G A...3′ 3′...A G C|T...5′	Cohesive ends on 5′ strand; 4-base recognition sequence
Haemophilus haemolyticus	*Hha*I	5′...G C G|C...3′ 3′...C|G C G...5′	Cohesive ends on 3′ strand; 4-base recognition sequence
Haemophilus influenzae Rd	*Hind*II	5′...G T Py|Pu A C...3′ 3′...C A Pu|Py T G...5′	Blunt ends; 6-base recognition sequence
	*Hind*III	5′...A|A G C T T...3′ 3′...T T C G A|A...5′	Cohesive ends on 5′ strand; 6-base recognition sequence
Haemophilus aegyptius	*Hae*III	5′...G G|C C...3′ 3′...C C|G G...5′	Blunt (flush) ends, 4-base recognition sequence

Sources: Adapted from Barnum (2005), Table 3.1, p. 60; Watson et al. (1992), Table 5-1, p. 65; Old and Primrose, (1981), pp. 16–17.

The shorthand representation of schematic (12.4) identifies the sequences on the end of the gene with italic boldface font. Other symbols used in this shorthand include a dotted or solid line to denote a single strand of nucleotides, a rectangle to represent a piece of double-stranded DNA, and the number 3′ or 5′ to indicate the respective ends of a DNA molecule. The cohesive or sticky end of AATTC is said to be on the 5′ strand, since the 5′ end of the polynucleotide is exposed.

Different Type II Restriction Enzymes Give Different Patterns of Cleavage and Different Single-Stranded Terminal Sequences

Some types of restriction enzymes (*Eco*RI, *Taq*I, *Hha*I, *Hind*II, and *Hind*III in Table 12.2) cleave DNA to give fragments that have "sticky ends"—exposed single-stranded DNA that can bind with DNA having a complementary sequence. Other enzymes (*Eco*RV and *Hae*III in Table 12.1) give fragments that do not have exposed cohesive ends, and are referred to as being "blunt ends." Restriction enzymes from different sources recognize sequences that differ in length as well as nucleotide sequence. Enzymes that recognize tetranucleotide sequences would encounter a target sequence at about 1 in every 4^4 (256 nucleotide) pairs since there are four different bases, A, T, G, and C, that make up DNA. Enzymes that recognize hexanucleotide sequences would encounter a target at a frequency of one in 4^6 (1 in 4096) base pairs.

The cohesive ends formed by ECO RI may result in an intramolecular association as shown here, resulting in a closed loop of double-stranded DNA:

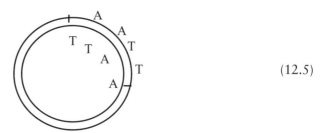

(12.5)

Alternatively, intermolecular associations between complementary nucleotide strands may form as indicated in schematic (12.6). The combination of these associations enables the introduction of a foreign gene by *inter*molecular association, and re-formation of the plasmid by *intra*molecular association. The AATT ends resulting from *Eco*RI activity will not pair with the AGCT ends formed by *Hind* III since these base sequences are not complementary. This is illustrated in schematic (12.6). However, fragments formed by the same enzyme will associate since they are complementary. Once the cohesive ends are aligned, each respective strand can be ligated or joined by DNA ligase that seal-nicks between the adjacent nucleotides in a double-stranded DNA molecule (Barnum 2005; Old and Primrose 1981).

5′ _____

A A T T _____ T T A A 5′

+

5′ _____

A A T T _____ T T A A

5′

T T A A

(12.6)

base pairing

5′ _____   5′ _____

A A T T   A A T T

                          T T A A

_____ 5′

                  5′

T T A A   T T A A

DNA Ligase Covalently Joins the Ends of DNA Fragments

Both *E. coli* and phage T4 encode the DNA ligase enzyme, although there are different cofactor requirements for the enzyme from these two different sources. DNA ligase from *E. coli* requires NAD^+, while ligase from phage T4 requires ATP as the energy source for forming a covalent bond between adjacent nucleic acids (Old and Primrose 1981). The reaction involves formation of a phosphodiester bond as illustrated in schematic (12.7a). The B symbols represent complementary bases—either G and C or A and T. The fragment is positioned by its base pairing with the complementary strand. The enzyme then transfers AMP to the phosphate on the 5′ end of the fragment to be ligated. The ligation reaction then occurs to join the 5′ and 3′ ends with the release of AMP [as shown in schemes (12.7b) and (12.7c)]. Purified DNA ligase performs this reaction in vitro. The optimum temperature for DNA ligation in vitro is 4–15 °C, although the enzyme itself has an optimum temperature of 37 °C. The higher temperature cannot be used since the base pairing is not stable. Stable bonding between the sticky ends is necessary for the ligase to properly ligate the ends of the DNA.[2]

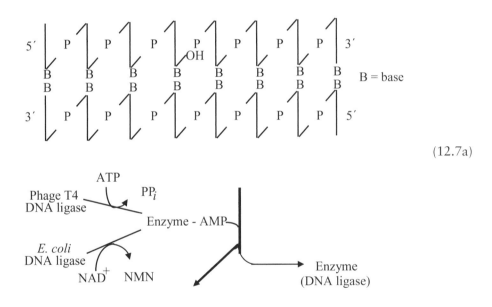

$$(12.7a)$$

---

[2] Alkaline phosphatase can be used to hydrolyze the 5′-terminal phosphate groups from the ends of the DNA that had been cleaved. This cleaves off the phosphate and prevents the ends from being ligated. When the foreign DNA is added, together with ligase, the foreign DNA is spliced into the plasmid since the foreign DNA has the necessary phosphate group. The plasmid with one nick on each strand of DNA is introduced into the host cell. Cellular repair mechanisms of the transformed host cell then complete the task of re-forming the intact plasmid by closing the nicks through ligation. There are other strategies by which recircularization of plasmids that do not contain foreign DNA is prevented (Old and Primrose 1981).

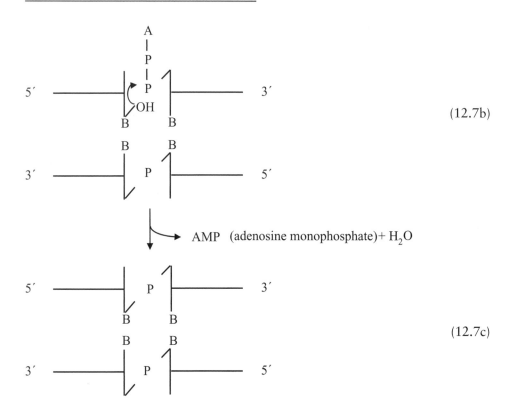

(12.7b)

AMP (adenosine monophosphate)$+ H_2O$

(12.7c)

DNA Fragments and Genes of ≤150 Nucleotides Can Be Chemically Synthesized if the Nucleotide Sequence Has Been Predetermined

An integral part of any cloning or sequencing procedure is the synthesis of oligo-nucleotide sequences to serve as primer strands, genes for constructing cloning vehicles or vectors (i.e., recombinant plasmids), or copies of DNA required for sequencing or mapping genomes. One of the early cloning experiments was the demonstration of the fusing, ligation, and expression of a chemically synthesized gene for somatostatin in 1977 by Itakura et al., as described by Old and Primrose (1981). The gene, with the corresponding amino acid sequence that results from transcription of the sequence to mRNA, and translation of the mRNA to the poly-peptide, is given in schematic (12.8):

	Met	Ala	Gly	Cys	Lys	Asn	Phe	Trp	Lys	Thr	Phe	Ser	Cys	Stop	Stop	
*Eco*RI																*Bam*HI
5′ AATTC	ATG	GCT	GGT	TGT	AAG	AAC	TTC	TGG	AAG	ACT	TTC	TCG	TGT	TGA	TAG	
G	TAC	CGA	CCA	ACA	TTC	TTG	AAG	AAA	ACC	TTC	TGA	AGC	ACA	ACT	ATC	CTAG 5′

(12.8)

The triplets in schematic (12.8) are separated by spaces to make them easier to read. The corresponding amino acids are shown underlined above each of the triplets. It was possible to chemically synthesize this gene since the amino acid sequence of somatostatin was known, thereby making it possible to specify the corresponding oligonucleotide sequence. The gene was cloned into a plasmid, obtained by modify-ing plasmid pBR322, and ligated (Old and Primrose 1981).

Programmable instruments (e.g. Applied Biosystems' ABI 3900) can synthesize 288-mer oligonucleotides in less than 10 hours at the 40- and 200-nmol scales. A limitation of the chemistry for these synthesizes is that the yields and accuracy of chain extension decrease as the chain length increases. Thus, chemical synthesis has generally been used as the source of primers for in intro propagation of DNA from mRNA through reverse transcription[3]. However, techniques such as polymerase cycling assembly (PCA) can be used to assemble longer genes from smaller, overlapping oligonucleotide fragments. PCA is similar to PCR but without the primers associated with PCR. For each cycle, DNA is melted and overlapping single strands reanneal. DNA polymerase then extends any free 3′ ends of reannealed DNA to form fully paired DNA strains. The DNA strands continue to elongate at each cycle until they either are full-length or can no longer be extended. Thus, this method does not amplify the DNA but rather produces complete complementary strands from overlapping DNA oligomers. Genes as long as 32 kilobases have successfully been generated using this method (Smith et al. 2003; Kodumal et al. 2004).

Protein Sequences Can Be Deduced and Genes Synthesized on the Bassis of Complementary DNA Obtained from Messenger RNA

Complementary DNA (cDNA) is obtained by reverse transcription of messenger RNA (mRNA) with a sequence of nucleotides corresponding to the sequence of amino acids. A partial sequence of the mRNA is sufficient to carry out reverse transcription [schematic (12.9)]. The sequence of oligonucleotides at the ends of the mRNA must be known so that a primer of complementary nucleotides can be specified, synthesized, and hybridized with the end of the mRNA. The beginning and end of the gene must also be defined for the purpose of confirming that the correct gene has been selected.

The process by which a gene is obtained requires that the appropriate primer be selected and annealed to the 5′ end of the RNA. Many mRNA molecules carry a poly(A) tail at their 3′, so that a convenient primer is a poly-T oligonucleotide as illustrated in schematic (12.9):

$$
\begin{array}{lr}
5' & 3' \\
\underline{\qquad \text{mRNA} \qquad\qquad\qquad\qquad} \text{A A A A A A A} \\
3' \longleftarrow \text{Direction of reverse} - - - \text{T T T T T T T} \\
\qquad\qquad \text{transcription} \qquad\qquad \text{Primer} \qquad 5'
\end{array}
\tag{12.9}
$$

When incubated in a solution of the deoxynucleotides and the reverse transcriptase enzyme, the DNA molecule is formed starting from the 3′ end, with the RNA serving as a template. The dashed line represents cDNA generated from the mRNA.

Reverse transcriptase will often stop before the cDNA molecule is completely generated, and hence experimental conditions must be carefully optimized to

---

[3] Reverse transcription uses an enzyme to generate DNA from RNA. The enzymes that catalyze the reverse of transcription (hence their name, *reverse transcriptase*) are found in virus particles (i.e., virions) of retroviruses whose genomes consist of two copies of single-stranded RNA. Examples of retroviruses include those that cause AIDS and some types of cancer. The enzyme is used by researchers to generate copies of DNA from mRNA in vitro.

maximize yields. At the end of the mRNA, the reverse transcriptase hooks around and forms a hairpin turn of DNA. Alkali dissolves RNA but not DNA and is used to remove the RNA. *E. coli* DNA polymerase I then generates a complementary DNA molecule starting at the hairpin turn that serves as the primer. The single-stranded cDNA thus becomes a double-stranded DNA molecule. The hairpin curve is removed by the enzyme S_1 nuclease on completion of the synthesis.

Specific probes are needed that can be assayed and that react only with the target sequence of the DNA. For this purpose labeled oligonucleotides of RNA or DNA that will specifically hybridize with the gene are used, where the hybridization is assayed by a form of autoradiography known as *Southern blotting*. While mRNA could be purified and used for this purpose, it is difficult to obtain sufficient amounts of purified mRNA. Consequently, a more practical strategy is to make multiple copies of cDNA from the RNA sequence using radioactively labeled bases and the reverse transcriptase enzyme. Once formed, the labeled DNA can be readily analyzed or used as a probe since it will hybridize with complementary nucleotide sequence.

The sequence of amino acids in a protein can be deduced by correlating triplets of nucleotides to amino acid residues. However, it does not necessarily give the entire sequence of the gene itself. Introns, if present in the gene, are excised from the pre-mRNA and are therefore absent from the mRNA used by the cell to synthesize proteins. Complementary DNA gives useful information—but not the complete sequence of the DNA in the chromosome. Sequencing of the genome requires that the chromosomal DNA be generated in sufficient amounts to be sequenced directly.

GENES AND PROTEINS

Selectable Markers Are Genes that Facilitate Identification of Transformed Cells that Contain Recombinant DNA

The recombinant plasmid is introduced into the host cell, after the host cell has been treated with $CaCl_2$ in order to make the cell membrane permeable to the DNA. In the case of *E. coli* and other Gram-negative cells, only a few percent of the cells that survive $CaCl_2$ treatment take up the recombinant plasmid. The yield of transformed cells is low. The presence of a gene on the plasmid that confers antibiotic resistance to the transformed bacteria is therefore needed to enable the researcher to identify or select those cells that have been transformed.

For example, transformed cells may be grown in a medium that contains the antibiotic tetracycline. Cells with an intact recombinant plasmid with the tetracycline resistance gene (Fig. 12.1) introduced with the recombinant plasmid survive. This plasmid, if present and functioning, will direct the synthesis of enzymes that would neutralize the action of the tetracycline. Cells that are not transformed, and do not carry genes for resistance to the antibiotic, die. The gene for Tc (tetracycline) resistance on the plasmid is referred to as a selectable genetic marker.

Detection of a reconstituted vector is enabled by plasmids with two antibiotic resistance genes. For example, *Hin*dIII, *Bam*HI, and *Sal*I restriction enzymes have target sites in the tetracycline resistance gene of pBR322. Foreign DNA inserted into an antibiotic resistance gene should inactivate that gene. Hence, if the foreign DNA is inserted and ligated into the tetracycline resistance gene, and the recombined plasmid is used to transform *E. coli*, the bacteria should be resistant to ampicillin but not tetracycline. For a reconstituted vector, the transformed bacteria would be resistant to both ampicillin and tetracycline, thus indicating that the foreign DNA had not been inserted, since the tetracycline resistance was still operational. This gives an example of how a plasmid with two genes for antibiotic resistance is useful, if one of the genes contains a restriction site. Transformed cells with this type of plasmid will not be resistant to *both* antibiotics and are thereby selectable.

Another selectable marker used in engineering of recombinant organisms is a gene that enables the cell to synthesize an essential nutrient and therefore allows the cell to grow in a culture medium that is deficient in this nutrient. Cells that are not transformed, and do not contain the gene for the essential nutrient, cannot grow in a culture medium that lacks this nutrient [see Trp LE example in Ladisch (2001)]. The transformed cells are therefore selected since they grow while the host cells do not.

The synthesis of the appropriate vector (i.e., plasmid) into which a foreign gene is introduced is obviously important. In the case of somatostatin, the starting plasmid was pBR322 into which a lac control region was cloned. The lac control region consists of a promotor to which an RNA polymerase will bind and cause transcription of the genes for β-galactosidase and constitutive[4] production of the enzyme. This structure provided a selectable marker since it facilitates identification of transformed *E. coli* by using a derivatized galactose substrate that releases a blue dye when hydrolyzed by the enzyme β-galactosidase. Since the resulting plasmid is a relaxed plasmid (many copies of it would be present in the cell), a transformed cell containing the plasmid will make many more copies of the enzyme than *E. coli* that is not transformed but may contain a low amount of the enzymes. The enzyme hydrolyzes the substrate, and causes the culture medium to turn blue when there is a lot of the β-galactosidase present. Cells that do not contain the plasmid with the lac operon would not produce enzyme, and hence the culture medium would not turn blue. In this manner, the cultures with the transformed cells may be picked out by visual inspection.

If a plasmid contains two *Eco*RI restriction sites when only one site is needed, it is undesirable. Two sites would be undesirable because these would also cause a promoter region between the two sites to be cut out and lost as shown in schematic (12.10):

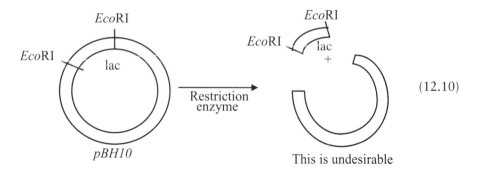

$$(12.10)$$

----

[4] A *constitutive* enzyme is produced continually, as compared to an *inducible* enzyme, which is produced in response to the appearance (or disappearance) of a signaling molecule.

Consequently, the plasmid *pBH10* was further modified to eventually give *another* plasmid with one *Eco*RI restriction site to which the synthesized gene for somatostatin was ligated. This resulted in the plasmid pSom I in schematic (12.11). The *E. coli* containing pSom I was selected by growing the bacteria in culture medium containing the antibiotic ampicillin. The transformed microorganisms were selected since they had ampicillin resistance whereas host cells that were not transformed did not contain an ampicillin resistance gene, and therefore did not grow. However, somatostatin was not found in *E. coli* that had been transformed with pSom I. The cell's protease recognized and hydrolyzed the foreign protein.

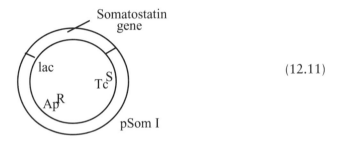

(12.11)

A Second Protein Fused to the Protein Product Is Needed to Protect the Product from Proteolysis (β-Gal-Somatostatin Fusion Protein Example)

A method of protecting the foreign protein product was developed when it was found that the somatostatin was being degraded by the cell's proteases.[5] This was achieved by creating a fusion protein consisting of β-galactosidase to which the somatostatin was attached on the –COOH terminus of the β-galactosidase. A new plasmid was created to incorporate this feature as described by Old and Primrose (1981). The resulting plasmid, pSom II-3, is given in schematic (12.12). It expressed a protein that was a combination of β-galactosidase and somatostatin, shown in schematic (12.12) (note that this drawing is not to scale—the β-galactosidase is about 30–50 times larger than the somatostatin molecule):

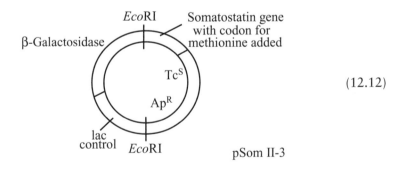

(12.12)

_____

[5] *Proteases* are enzymes that hydrolyze proteins into small pieces and/or individual amino acids, thereby destroying the functioning of the protein.

Recovery of Protein Product from Fusion Protein Requires Correct Selection of Amino Acid that Links the Two Proteins (Met Linker)

The recovery of the product protein from the fusion protein (β-galactosidase in this case) requires that the two be linked by an amino acid residue that can be hydrolyzed by a catalyst that does not attack other amino acids (i.e., selective hydrolysis) and that does not appear as an amino acid residue in the product protein itself. Otherwise, the product protein, as well as the link between it and the fusion protein, would be hydrolyzed. The fusion protein is likely to contain residues that are hydrolyzed. This is acceptable as long as the fragments that are formed can be fractionated from the product.

Methionine was chosen as the amino acid for linking the β-gal fusion protein to the somatostatin, since it is hydrolyzed by CNBr (a commonly used reagent in protein chemistry) and since it is not an amino acid found in the structure of somatostatin. Once the cell growth had been completed, and the fusion protein recovered, it was hydrolyzed with CNBr. The strategy was successful. The β-galactosidase (MW > 100 kDa) had protected the somatostatin (MW < 2000) from proteolytic attack inside the growing cells. This was confirmed when the β-galactosidase–somatostatin fusion protein was hydrolyzed and an active somatostatin molecule was obtained after further processing and purification. Since the somatostatin had no internal methionine residues, it was not destroyed during the hydrolysis step. As is the case for somatostatin, the active form of insulin does not contain methionine residues, and hence can be linked to a fusion protein via a methionine residue so that the fusion protein can be cleaved using CNBr.

Chemical Modification and Enzyme Hydrolysis Recover an Active Molecule Containing Met Residues from a Fusion Protein (β-Endorphin Example)

Another polypeptide, β-endorphin, has 31 amino acid residues and contains an internal methionine residue. The β-endorphin gene was obtained from cDNA generated by reverse transcription instead of chemical synthesis and was cloned next to the lac/β-gal region on the plasmid. The β-galactosidase protected the polypeptide, and an active hormone was obtained. However, recovery of this polypeptide could not use CNBr, since it would have attacked the methionine residue that is internal to the β-endorphin and destroyed it. An arginine linker between the β-galactosidase and the β-endorphin was used instead since there are no arginine residues that are internal to the polypeptide. The nucleotide sequence for arginine was engineered into the plasmid to be located between the β-gal and β-endorphin genes to give the corresponding fusion protein.

The proteolytic enzyme trypsin selectively attacks and hydrolyzes a protein at the arginine residue. However, it also hydrolyzes the protein at lysine residues. Lysine is one of the amino acid residues found in β-endorphin. The lysine residues were protected against hydrolysis by reacting them with citraconic anhydride after the fusion protein had been generated and recovered from the transformed bacteria. The β-endorphin/β-galactosidase fusion protein with the modified lysine residues was then treated with trypsin. The β-endorphin was released by the action of the trypsin at the arginine residue. After the β-endorphin with modified lysine was isolated, the pH was adjusted to 3.0. This caused the citraconic groups to be removed from the lysine. Further fractionation gave the active β-endorphin product (Old and Primrose 1981).

This example, as well as the production of somatostatin, illustrate that bioseparations and purification of the product from cells, other proteins, reagents, and coproducts of fusion protein processing are important. Bioseparations account for 50% or more of the manufacturing costs of many recombinant products, as mentioned in Chapter 1.

Metabolic Engineering Differs from Genetic Engineering by the Nature of the End Product

Genetic engineering alters the genetic makeup of the cell so that it will produce a molecule or protein that a wild-type organism would not otherwise make (e.g., bovine somatostatin β-endorphin or human insulin in *E. coli*). Metabolic engineer-

Table 12.3.

Snapshot of Major Fermentation Products

Classification	Product	Worldwide Production	
		Thousands of Metric Tons	Value ($ millions)
Organic acids	Citric	510	1400
	Gluconic	46	93
	Itaconic	16	68
	L-Lactic	60	150
Subtotal		632	1711
Vitamins	Vitamin C	53	425
	Riboflavin (B$_2$)	2	138
	Cyanocobalamin (B$_{12}$)	(8700 kg)	71
	Niacin	14.8	125
Subtotal		69.8	759
Amino acids	L-Aspartic	8.8	43
	L-Glutamic	640	915
	L-Lysine	150	450
	L-Phenylalanine	11	198
Subtotal		810	1606
Other	Biopesticides	3.6	126
	Nucleotides	10.1	350
	Xanthan gum	33	408
Subtotal		46.7	884
Total		1558	4960

Source: Reproduced from McCoy (1998), p. 14.

ing may address recombinant organisms or enable amplification of a preexisting gene products (protein). Metabolic engineering redirects the cell's machinery to express genes that already exist to make a product in commercially meaningful quantities. Metabolic engineering may be directed to Type I, II, or III fermentation as well as pathways involved in the secretion of these products.

Successes in this endeavor will have a significant impact on chemical and biochemical fermentation products derived from renewable resources, if efficient and cost-effective separation processes for recovery and purifying these products are developed concurrently. The recovery and purification of products derived through metabolic engineering are an important factor in the process economics of metabolically engineered organisms. The products may occur in a relatively dilute form and are coproduced with other metabolites that must usually be removed before the product can be sold. The separation methods discussed by Ladisch (2001) represent starting points of purification sequences for products from metabolically engineered organisms.

The market value of fermentation-derived products (excluding ethanol) was estimated to be on the order of $5 billion annually, approximately 10 years ago, (i.e., ca. 1998), on a worldwide basis (Table 12.3) (McCoy 1998). Significant growth is now occurring. For example, Dupont's Serona, discussed in Chapter 13, sold $100 million product, while Cargill is getting ready to ship 140,000 metric tons of Ingeo a bioplastic for food containers and textiles, in 2009. Another plastic, Mirel, by Metabolix is scheduled for introduction in 2009. These bioproducts hope to begin to address the 30 million tons of plastic discarded annually in the US (Der Hovanesian, 2008).

Fermentation ethanol for fuel and industrial purposes currently (2008) accounts for an additional $20 billion worldwide. These fermentation products are priced in the range of 10^2–10^4/metric ton. The manufacture of bioproducts derived from metabolically engineered organisms is likely to be much more sensitive to economies of scale than that of pharmaceutical recombinant products, thus requiring judicious use of bioprocess engineering principles in plant design and scaleup, and the use of agriculture to generate products ranging from plastics to enzymes.

REFERENCES

Alberts, B., D. Bray, J. Lewis, M. Raff, K. Roberts, and J. D. Watson, *Molecular Biology of the Cell*, 2nd ed., Garland Publishing, New York, 1989, pp. 6–7, 23–26, 56–57, 99–107, 162, 171–174, 178–179, 184–191, 201–210, 222–236, 341–342, 386–393, 420–421, 483–486, 514–516, 519, 557–564, 728, 762–768, 781–788, 842–848, 1031–1035.

Barnum, S. R., *Biotechnology, an Introduction*, Thomson, Brooks/Cole, Belmont, CA, 2005, pp. 18–24, 57–92.

Committee on Bioprocess Engineering, National Research Council, *Putting Biotechnology to Work: Bioprocess Engineering*, National Academy of Sciences, Washington, DC, 1992.

Der Hovanesian, M., *Business Week* **4090**, 45–47 (2008).

Kabnick, K. S. and D. A. Peattie, "*Gardia*: A Missing Link Between Prokaryotes and Eukaryotes," *Am. Scientist* **79**(1), 34–43 (1991).

Kodumal, S. J., K. G. Patel, R. Reid, H. G. Menzella, M. Welch, and D. V. Santi, "Total Synthesis of Long DNA Sequences: Synthesis of a Contiguous 32-kb Polyketide Synthase Gene Cluster," *Proc Natl Acad Sci U S A*, **101**(44), 15573–15578 (2004).

Ladisch, M. R., *Bioseparations Engineering: Principles, Practice and Economics*, Wiley, New York, 2001, pp. 520–525.

Lewin, B., *Genes*, Oxford Univ. Press, 1997, pp. 63–69, 129–133, 136–138, 153–241, 630–631, 727–774.

McCoy, M., "Chemical Makers Try Biotech Paths: Vitamins and a New DuPont Polyester Are Paving the Way for Biocatalysis Use," *Chem. Eng. News* **76**(25), 13–19 (1998).

Old, W. and S. B. Primrose, *Principles of Genetic Manipulation: An Introduction to Genetic Engineering*, 2nd ed., Univ. California Press, Berkeley, 1981, pp. 1–27, 38–43, 46, 91–92, 105, 171–179.

Ouyang, Q., P. D. Kaplan, S. Liu, and A. Libchaber, "DNA Solution of the Maximal Clique Problem," *Science* **278**(5337), 446–449 (1997).

Sigma (Sigma-aldrich.com), *2006–2007 Biochemicals Reagents and Kits for Life Science Research*, 2008, pp. 2082–2088.

Smith, H. O., C. A. Hutchison, III, C. Pfannkoch, and J. C. Venter, "Generating a Synthetic Genome by Whole Genome Assembly: [var phi]X174 Bacteriophage from Synthetic Oligo-Nucleotides," *Proc Natl Acad Sci USA*, **100**(26), 15440–15445 (2003).

Watson, J. D., M. Gilman, J. Witkowski, and M. Zoller, *Recombinant DNA*, 2nd ed., Scientific American Books, W. H. Freeman, New York, 1992, pp. 1–11, 20–48, 50–54, 63–77, 135–138, 236–237, 273–285, 460–461, 521–523, 561–563, 590–598, 603–618.

CHAPTER TWELVE
HOMEWORK PROBLEMS

12.1. The plasmid pSom I contains the gene for somatostatin. The DNA sequence for pSom I indicated that the clone carrying the plasmid should produce a peptide containing somatostatin. However, somatostatin was not produced. Somatostatin degradation due to proteolytic enzymes was found to be the cause. Another approach was tried. First pSom I was digested with *Eco*RI and *Pst*I and a large fragment containing the somatostatin gene was purified by gel electrophoresis. Then pBR322 was digested with *Eco*RI and *Pst*I to obtain a small fragment. This fragment contained the part of the ampicillin-resistant gene Ap^R lost during digestion and fractionation of pSom I. Structures of pSom I and pBR322 are shown below.

(a) *Briefly* explain what *Eco*RI denotes.

(b) Circle the large fragment obtained on *Eco*RI and *Pst* digestion of pSom I.

(c) Circle the small fragment obtained on *Eco*RI and *Pst* digestion of pBR322.

(d) Give the representation of the plasmid (pSom II) that results on ligation of the large fragment from pSom I and the small fragment from pBR322. Clearly show the *Eco*RI and *Pst* sites, and genes for ampicillin and tetracycline, as well as the gene for somatostatin.

The resulting plasmid pSom II is again digested with *Eco*RI. A lac region and β-galactosidase gene (from another source) is inserted into pSom II and ligated.

(e) Give a representation of the plasmid structure which results. Be sure to mark *Eco* RI and *Pst* sites, as well as the genes in the plasmid.

This time, the clone carrying the new plasmid gives a very large protein that, when cleaved with cyanogen bromide, d-results in an active form of somastatin.

(f) Give the name of this protein.

(g) Give an explanation of why an active form of somatostatin was obtained.

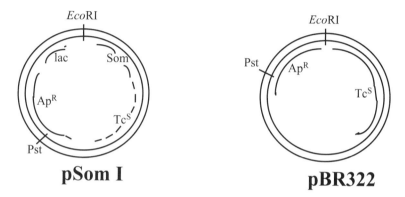

pSom I   **pBR322**

12.2. An example of labeling is given by the fragment (where Ⓟ-denotes the ^{32}P-labeled end) and is represented by

$$5' \ Ⓟ\text{-}A\ A\ T\ T\ C\ A\ T\ G\ C\ T\ 3'$$

If the restriction enzyme *Hin*dIII acts on the ^{32}P-labeled nucleotide sequence shown below

$$T\ T\ C\ A\ A\ G\ C\ T\ T\ G\ C\ A$$
$$A\ A\ G\ T\ T\ C\ G\ A\ A\ C\ G\ T$$

what are the expected fragments that result, and the location of ^{32}P labels.

12.3. *Reading Knowledge.* One of the following enzymes is a Type II restriction enzyme that gives blunt ends:

__ *Taq*I

__ *Hin*dII

__ *Hin*dIII

The following is an example of a selectable marker:

__ pBR322

__ Ampicillin resistance

__ Pvu II

Western blot refers to technique for determining size of fragments of

__ DNA

__ Protein

__ Polysaccharides

EcoRI denotes a

__ Restriction enzyme

__ Reverse transcriptase

__ Ligase

A plasmid is

__ Chromosomal DNA

__ Circular DNA

__ An inclusion body

12.4. A gene that codes for a valuable protein is engineered into a plasmid consisting of regulatory regions and the trp E gene of the trp operon. The trp operon, when turned on, initiates a sequence of events that ultimately leads to the biosynthesis of tryptophan. The "hybrid" operon consists of

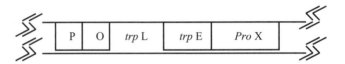

This operon, when recombined with the plasmid, is introduced into *E. coli*.

(a) The transformed *E. coli* is grown in a medium containing tryptophan, but protein corresponding to the Pro X gene is *not* formed. Explain.

(b) At the end of the fermentation, the tryptophan is depleted from the medium, and a protein containing a sequence of amino acids corresponding to trp E and protein X is detected. Explain.

12.5. The growth of *E. coli* that has been transformed with the Trp E operon coupled to proinsulin is carried out in a batch fermentation. At the end of the fermentation tryptophan is cut off in the medium and product is formed in inclusion bodies that quickly accumulate, killing the cells.

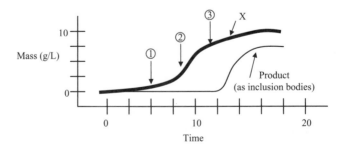

Answer the following questions:

(a) The tryptophan is cut off at point

    __ ①

    __ ②

    __ ③

(b) When the tryptophan is cut off, the cell begins to manufacture proinsulin at the point shown in the figure above.

    __ ①

    __ ②

    __ ③

(c) The cell forms an inclusion body that contains primarily

    __ DNA

    __ Proinsulin

    __ Lipids

(d) The product accumulates because

    __ The cells are near stationary phase

    __ The cells have reached a critical cell count

    __ A change in fermentation condition has induced its production

(e) The product of the fermentation is

    __ Intracellular

    __ Extracellular

CHAPTER THIRTEEN

METABOLIC ENGINEERING

INTRODUCTION

Metabolic engineering is "the purposeful modification** of intermediate metabolism using recombinant DNA techniques," (Cameron and Tong 1993). Related areas of research have resulted in terminologies of metagenomics, proteomics, and molecular evolution. *Metagenomics* is the cloning of genetic material from microorganisms that cannot be grown in the laboratory into ones that can be grown so that new forms of known genes may be identified. Proteomics use mass spectroscopy (MS) techniques to identify novel functional proteins from genes that are expressed.

Molecular evolution is the directed or random variation of selected amino acids in a protein at a genetic level, based on replacing nucleic acids in genes. These modified genes express proteins with altered functions or characteristics. A readable and well-written summary by Nedwin et al. (2005) gives an overview of current directions in enzyme discovery that has evolved from classical microbial screening, defined by Aiba et al. (1973) in the context of goals for directing activities of microorganisms, to

1. Improve production of molecules or proteins already made by the microorganism
2. Extend substrate range
3. Add new catabolic activities for degrading toxic chemicals
4. Modify cell properties
5. Add new metabolic pathways that enable the host organism to produce new chemicals.

The development of metabolic engineering has been evolutionary. Breeding and development of plants and animals has been carried out for centuries with microorganisms being selected and improved for about 100 years. Analysis of the manner in which organisms utilize carbon substrates to obtain energy and carry out biosynthesis, based on assessment of metabolic pathways, and changes to the metabolic pathways, through externally applied random methods of mutagenesis and screening, has been carried out for over 50 years. The alteration of the organism's genetic makeup by adding, deleting, or changing the characteristics of specific genes is more recent. Since about 1973, molecular biology has expanded the methods by which productivity of microorganisms can be increased for industrial-scale manufacture of value-added products as well as for expression of biotherapeutic proteins.

Directed evolution or molecular evolution are new terms that describe the application of evolutionary principles to the development of new or improved traits in organisms that have been used by microbiologists over the past 100 years and by humans, generally, since the first domestication of plants and animals more than 9000 years ago (Clutton-Brock 1999; Zohary and Hopf 2001). In the context of biotechnology, this process has previously been described as biochemical engineering (Aiba et al. 1973) and applied bacteriology (Singleton 1995). The underlying evolutionary principle of importance is based on selection that acts on variability in the traits of a population of organisms. Variability has a genetic basis; thus those traits have a probability of being passed on to offspring of the organism. Selection favors organisms that are better able to compete for limited resources in the environment, thus leaving more offspring than organisms less able to compete. Over time and multiple generations of organisms, the descendants of the organisms best able to compete will dominate.

From an engineering perspective, there are ways to "direct" the evolution of organisms: the environment (i.e., selective pressure) and genetic variability in the population. In microbiology, prior to about 1976, this was accomplished by altering the environment (i.e., growing conditions) by controlling media composition, temperature, pH, and other parameters to favor the selection of microorganisms with desirable traits that grow more quickly under the altered conditions. Genetic variability arose either from naturally occurring variation in the genomes of the microorganisms, or from alterations in the genetic sequences due to spontaneous, random mutations or random mutations induced by humans through applying mutagenic chemicals or radiation to the microbes.

Few mutations have any effect on the fitness of the organism (i.e., the mutation is neutral). Some mutations will render the organism less fit and therefore less able to compete. A very few will make the organism more fit (i.e., better able to compete). The power of selection lies in the result that neutral and deleterious mutations are weeded out and the beneficial mutations survive. Through repeated generations that are affected by selection, novel traits are "designed" through an algorithm of iterative trial and error.

The key change in directed evolution since about 1976 has been the ability to be vastly more specific in the generation of genetic variability on which selection may be applied. The tools of molecular biology have removed a great deal of the randomness and reduced the number of cycles of selective culturing required to produce organisms with the desired traits. In addition, since genes from a very different organism, such as humans, may be inserted into the target organism, this has allowed the development of microorganisms or mammalian cells that produce insulin, the "clot-busting" hydrolytic proenzyme tissue plasminogen activator (tPA), and the red blood cell production promoting hormone erythropoietin (EPO) (Ladisch and Kohlmann 1992; Ladisch 2001). These molecules would be next to impossible

products by their excretion from the cell, are increasingly being viewed at a systems level for industrially important microorganisms, although this approach was proposed back in the early 1990s (Stephanopoulos 1992). This chapter provides an introduction to the methods that are currently used for metabolic engineering. The sections that address cloning, genotic modification, and restriction enzymes are based on the excellent work of Old and Primrose (1981) and Barnum (2005), as summarized in Chapter 12.

BUILDING BLOCKS

The building blocks of a cell are nucleic, amino, fatty, and carboxylic acids, and monosaccharides. These, in turn, form macromolecules and polymers on which the functioning of the cell is based and from which the products of biotechnology are derived. Key macromolecules are the polynucleotides, deoxyribonucleic acid (DNA), and ribonucleic acid (RNA). DNA stores the information, while RNA transfers and translates it to proteins that in turn catalyze formation of other cell components and enable biochemical transformations to occur in living cells.

Enzymes carry out replication of DNA, transform the information in DNA to RNA and eventually to proteins, synthesize polysaccharides and lipids, and generate other metabolites. Polysaccharides store some of the fuel used by living cells, namely, sugars in polymers. Lipids, derived from fatty acids, form membrane structures that compartmentalize several functioning parts of the cell and separate the cell's contents from the outside environment. These macromolecules and structures enable the cell to function, and define targets at which cell-altering technologies are aimed. The goals of gene-altering technologies are to introduce genetic information to enable the cell to generate a product that it would not otherwise make, or identify structures and molecules that could serve as targets for therapeutic bioproducts in microorganisms, plants, and animals.

Methods that enable gene manipulation through recombinant techniques and cell fusion are a subset of the biotechnology based on the discovery of restriction enzymes in 1970 and development of methods for creating recombinant DNA molecules (discussed in Chapters 2 and 12). These methods set the stage for rapid growth of new industries for manufacturing biobased products that could not be made in any other way (Committee on Bioprocess Engineering 1992). The science of gene manipulation resulted in procedures and instruments for the sequencing of human, microbial, and plant genomes. Technology for the recombination of DNA also enabled the reliable propagation of pieces of human DNA so that they may be studied and characterized. This led to detailed insights into the organization of DNA, and its role in controlling both normal and aberrant cell functions. Chapter 12 provides an outline of the tools that are used. This chapter discusses examples of how these tools are applied.

L-THREONINE-OVERPRODUCING STRAINS OF *E. coli* K-12

An example of metabolic engineering is given by the construction of *E. coli* that is capable of producing significant amounts of L-threonine. A fundamental understanding of the threonine pathway facilitated directed changes in the micro-organism's plasmid DNA in order to increase the activity of a rate-limiting enzyme, homoserine dehydrogenase. This approach combined previous research on muta-tion/screening methods to select a small number of overproducing mutants from a very large starting population, with recombinant techniques that manipulated the threonine operon to enhance production of threonine by factor of 3 over the original mutant (Miwa et al. 1983). The technical basis of this improvement was moderation of feedback inhibition of a single-enzyme homoserine dehydrogenase I/aspartoki-nase, with two catalytic functions. The enzyme is subject to feedback inhibition by L-threonine and catalyzes two distinct transformations: aspartate to aspartylphos-phate, and aspartate semialdehyde to homoserine. The improvement also reflected relief of repression of genes associated with threonine production when isoleucine is provided in limited amounts to an isoleucine auxotroph whose threonine produc-tion is repressed by isoleucine.

The techniques of mutation/high-throughput screening were combined with construction of a pBR322-thr plasmid using the operon for threonine production from the mutant, threonine-producing strain (denoted as βIM4). The plasmid with the threonine operon (i.e., pBR322-thr) was reintroduced into *E. coli* (βIM4) from which the operon had originally been obtained. The best strain selected by a halo formation screening technique[1] was designated as *E. coli* 29-4. This new stain carried 11 copies of the recombined plasmid. It had 5 times higher hemoserine dehydrogenase activity, had 3 times higher threonine production, and gave a final concentration of 13.4 g/L threonine from 30 g/L glucose when grown in a 20-mL shake-flask experiment at pH 7.0 and 30 °C for 72 h[2] (Miwa et al. 1983; Sano 1987).

Genetically Altered Brevibacterium lactoferrin Has Yielded Improved Amino Acid–Producing Strains

Brevibacterium lactofermentum is a major organism used by Ajinomoto Co.[3] for commercial production of amino acids. *B. lactofermentum* is a Gram-positive,

---

[1] The halo formation test utilizes minimal medium agar plates on which is spread a threonine auxotrophic mutant ("strain 43"). Threonine-producing strains to be tested are then spotted on the plate, and incubated at 30 °C for 1–2 days. Threonine-producing strains are indicated by growth of microorganisms around the spots, indicating the production of threonine. The circular growth zones around the spotted samples resemble a halo, hence giving this method its name. It is considered to be a rapid screening method, since hundreds to thousands of spots (i.e., microbial cultures) can be tested on multiple agar plates at one time (Miwa et al. 1983).

[2] The composition of the culture medium was 3% glucose, 1% $(NH_4)_2 SO_4$, 0.1% $KH_2 PO_4$, 0.1% $Mg SO_4$-$7H_2O$, 2 ppm Fe^{2+}, 2 ppm Mn^{2+}, 1 mg/mL L-aspartate, 0.45 mg/mL L-proline, 0.1 mg/mL L-isoleucine, 0.1 mg/mL L-methionine, 0.001 mg/mL thiamine-HCl, 2 mg/mL yeast RNA, and 4% $CaCO_3$ (Miwa et al. 1983).

[3] Ajinomoto started up at the beginning of the twentieth century, with sales of the seasoning called AJI-NO-MOTO. This marine derived material contained the flavor enhancer mono-sodium glutamate (MSG), as discussed in Chapter 1. By 1990, the company held 1800 patents, had a research staff of 800, and expended on research, 3.6% of total annual sales

sporeless, nonmotile, short-rod, biotin-requiring mutant. It is of the same genus as *Corynebacterium glutamicum*. Techniques developed for improving this strain through recombinant methods are summarized by Sano (1987). Challenges included the identification and construction of plasmids carrying multiple genes associated with amino acid formation and introducing these into appropriate host organisms. Appropriate host organisms were restriction-deficient mutants characterized by reduced restriction enzymatic activity so that foreign DNA introduced into the host would not be hydrolyzed. By 1987, improved strains were obtained for threonine, phenylalanine, tyrosine, and lysine (Table 13.1).

The recombinant methods for obtaining improved transformants were a new tool in the early 1980s. Combined with conventional mutation and cell (protoplast) fusion techniques, recombinant methods offered the promise of contributing to the goal of eliminating rate-limiting steps, balancing microbial cell metabolism, and constructing improved strains for the amino acid industry. However, it is not clear

Table 13.1.

Examples of Genetically Altered *Brevibacterium Lactoferrin* with Enhanced Amino Acid Production

Amino Acid	Type	Effect	Benefit	Concentration Achieved (g/L)[a]
L-Lysine	Cell (protoplast) fusion	Slow-growing lysine producer + fast-growing wild type	Faster	70
L-Threonine	Recombinant clone— homoserine kinase/ homoserine dehydrogenase	32% increase in threonine productivity	Increased concentration	33
L-Tyrosine	Shikimate kinase gene clone	24% increase in tryptophan productivity	Increased concentration	21.6
L-Phenylalanine	Prephenate dehydrogenase clone	First enzyme in pathway increased	Increased concentration	18.2

[a]Concentration: grams of amino acid per liter of culture medium.
Source: Table constructed from paper due to Sano (1987).

of $3.4 billion. Products were in seasonings, foods and food services, oils, pharmaceuticals, amino acids, and specialty chemicals. Seasonings accounted for 22.8% of net sales, while food, pharmaceutical, and specialty chemicals were about 11.7%. This snapshot gives an insight into the commercial importance of biotechnology related products in the food/feed sector (Ajinomoto 1990; Ladisch and Goodman 1992).

which recombinant strains are currently used for commercial production of amino acids.

Metabolic Engineering May Catalyze Development of New Processes for Manufacture of Oxygenated Chemicals

The growth potential for fermentation derived chemicals is large, but the introduction of fermentation into the chemical industry, like the field of metabolic engineering itself, has been evolutionary. Metabolic engineering can result in organisms with greatly improved capability for directing their metabolism to a dominating product. However, small amounts of numerous intermediates of metabolites leading to the product are present at the end of the fermentation. The product itself is found in an aqueous solution that usually makes up at least 90% of the mass of the system. The requirements for >99% purity for most single-component chemicals, and the costs of handling the residual components of the system (i.e., cells, spent fermentation broth, fermentation offgasses, and metabolic intermediates) on a large scale are significant, if not determining, factors in the commercial application of a new organism or biotransformation.

The removal of biologically rate-limiting steps in the organism is necessary, but not sufficient, to ensure adoption of new fermentation processes by the industry. The adaptation of new processes to make oxygenated chemicals that were previously derived from chemical synthesis will require that new biomanufacturing facilities and cost-effective separation techniques be developed. Bioprocess engineering for new processes will need to be given the same emphasis as the basic life sciences that had led to the microorganisms that would generate these products. Major developments have occurred for *E. coli* and yeasts, particularly with respect to ethanol and diols.

Gene Chips Enable Examination of Glycolytic and Citric Acid Cycle Pathways in Yeast at a Genomic Level (Yeast Genome Microarray Case Study)

Yeasts are significant in the industrial production of food, fuels, enzymes, and therapeutic proteins. The yeast *Saccharomyces cerevisiae* is widely used as the preferred microorganism to produce over 6 billion gallons of fuel ethanol per year in the United States, and an equivalent volume in Brazil. Food products include yeast flavorings and for baking. The study of genes for the glycolytic pathway and citric acid cycle in *S. cerevisiae* by DeRisi et al. (1997) provides a benchmark for study of the biological role of individual genes. It offers an experimental model of the functional genomics of other, more complex organisms as well as microorganisms of industrial importance.

Saccharomyces cerevisiae was chosen since its genes are readily recognized in a genomic sequence. Furthermore, the *cis* regulatory[4] elements of yeast are compact and close to transcription units. Much is known about *S. cerevisiae*'s genetic regulatory mechanisms (DeRisi et al. 1997). Microarray technology makes it possible to track

---

[4] The term *cis* is an abbreviation for *cistron*, which refers to a genome that expresses a protein with two mutations. A second gene that has no mutations is referred to as *wild-type*. Both mutations are present on the same chromosome for *cis* type and on opposite chromosomes for the *trans* type (Lewin 1997).

gene expression for the entire yeast genome, and to "investigate the genetic circuitry that regulates and executes this program," during the shift from anaerobic to aerobic metabolism. The change from anaerobic growth to aerobic respiration on depletion of glucose is referred to as the "diauxic shift" (Aiba et al. 1973). Fundamental changes in the carbon metabolism, protein synthesis (i.e., induction of enzymes), and carbo-hydrate storage coincide with a diauxic shift (DeRisi et al. 1997).

A yeast genome microarray was prepared on a glass slide.[5] The size of the microarray was 18×18 mm (see Chapter 14 for discussion of GeneChip technol-ogy), and contained approximately 6400 distinct DNA sequences. These sequences were yeast open reading frames[6] that had been previously determined as a conse-quence of the sequencing of the yeast genomes. The open reading frames were amplified by the polymerase chain reaction using genomic DNA, from yeast strain S288C as the template. Each PCR product was verified by agarose gel electropho-resis. The overall success rate of obtaining the proper DNA sequences on the first try was about 94.5%. The distinct sequences were then printed onto glass slides over a 2-day period using a robotic printing device.

The yeast (strain DBY7286) was grown in 8 L of YPD medium in a 10-L fer-mentor inoculated with 2 mL of a fresh, overnight culture of the yeast. The aerated fermentations were carried out under constant agitation and aeration, with glucose at an initial concentration of 20 g/L. After an initial 9 h of growth, samples were harvested in successive 2-h intervals and the mRNA isolated from the cells. The mRNA was used to generate fluorescently labeled cDNA, prepared by reverse tran-scription. The resulting cDNA was hybridized to the microarrays. Seven samples from a single fermentation resulted in 43,000 expression ratio measurements. These were organized into a database available on the Internet, to facilitate analysis of the results (DeRisi et al. 1997).

---

[5] A clear description and schematic of a microarray is given by Barnum (2005). The yeast experiment described here used glass slides (Gold Seal), cleaned for 2 h in a solution of 2 N NaOH and 70% ethanol. After rinsing in distilled water, the slides were then treated with a 1:5 dilution of poly-L-lysine adhesive solution (Sigma) for 1 h, and then dried for 5 min at 40 °C in a vacuum oven. DNA samples from 100-μL PCR reactions were purified by ethanol precipitation in 96-well microtiter plates. The resulting precipitates were resuspended in 3× standard saline citrate (SSC) and transferred to new plates for arraying. A custom-built array-ing robot was used to print on a batch of 110 slides. (Details of the design of the microarray are available at cmgm.stanford.edu/pbrown.) After printing, the microarrays were rehydrated for 30 s in a humid chamber and then snap-dried for 2 s on a hotplate (100 °C). The DNA was then ultraviolet (UV)-crosslinked to the surface by subjecting the slides to 60 mJ of energy (Stratagene Stratalinker). The rest of the poly-L-lysine surface was blocked by a 15-min incubation in a solution of 70 mM succinic anhydride dissolved in a solution consisting of 315 mL of 1-methyl-2-pyrrolidinone (Aldrich) and 35 mL of 1 M boric acid (pH 8.0). Directly after the blocking reaction, the bound DNA was denatured by a 2-min incubation in distilled water at ~95 °C. The slides were then transferred into a bath of 100% ethanol at room tem-perature, rinsed, and then spun-dry in a clinical centrifuge. Slides were stored in a closed box at room temperature until used.

[6] An open reading frame contains a series of codons (i.e., triplets of bases) that exclusively code for amino acids and lacks a termination codon. Termination codons that occur fre-quently resulting in codons in the reading frame cannot be translated or read into proteins are referred to as a "blocked reading frame" (Lewin 1997).

Figure 13.1.   Metabolic reprogramming inferred from global analysis of changes in gene expression. Only key metabolic intermediates are identified. The yeast genes encoding the enzymes that catalyze each step in this metabolic circuit are identified by name in the boxes. The genes encoding succinyl-CoA synthase and glycogen-debranching enzyme have not been explicitly identified, but the open reading frames (ORFs) YGR244 and YPR184 show significant homology to known succinyl-CoA synthase and glycogen-debranching enzymes, respectively, and are therefore included in the corresponding steps in this figure. Gray boxes with white lettering identify genes whose expression diminishes in the diauxic shift. The magnitude of induction or repression is indicated for these genes. For multimeric enzyme complexes, such as succinate dehydrogenase, the indicated fold induction (indicated by number close to each box) represents an unweighted average of all the genes listed in the box. Black-and-white boxes indicate no significant differential expression (less than twofold). The direction of the arrows connecting reversible steps indicate the direction of the flow of metabolic intermediates, inferred from the gene expression pattern, after the diauxic shift. Arrows representing steps catalyzed by genes whose expression was strongly induced are shown in bold. The broad gray bands represent major increases in the flow of metabolites after the diauxic shift, inferred from the indicated changes in gene expression. [From J. L. DeRisi, V. R. Iyer, and P. O. Brown, "Exploring the Metabolic and Genetic Control of Gene Expression on a Genomic Scale," *Science* 278 (5338), 680–686 (1997). Reprinted with permission from AAAS.]

The stages of metabolism that were observed corresponded to changes in gene expression for yeast grown in a glucose-rich medium (Fig. 13.1). Only 19 genes (0.3%) showed detectable changes in expression during the first 9h, when the exponential growth phase was dominant. As the glucose was depleted, 710 genes were induced, and 1030 genes were downregulated. The shutdown of transcription of genes encoding pyruvate decarboxylase transfers pyruvate to oxaloacetate in the citric acid cycle and away from acetaldehyde. Other changes promoted entry of glucose-6-phosphate into the carbohydrate storage pathway for trehalose as indicated by the timecourse. The concentration of glycogen and trehalose increased significantly starting at about 17h, as the glucose was depleted. While the responses of many of the genes during diauxic shift were previously known, the transcriptional responses were not previously tracked. This experiment demonstrated the utility of microarray probes for tracking changes in cellular metabolism.

One consequence of this experiment was that it showed that the balanced growth assumption is reasonable during exponential growth. It also indicated that the challenge is to develop efficient methods for interpreting the volumes of data that these types of experiments can generate, and translate patterns into useful interpretations of the macroscopic behavior of living cells. This approach has evolved into global analysis of protein activities using proteome chips [see, e.g., Zhu et al. (2001)].

The Fermentation of Pentoses to Ethanol Is a Goal of Metabolic Engineering (Recombinant Bacteria and Yeast Examples)

Pentoses, particularly xylose and arabinose, make up a major fraction of monosaccharides derived from hydrolysis of cellulosic materials. These monosaccharides

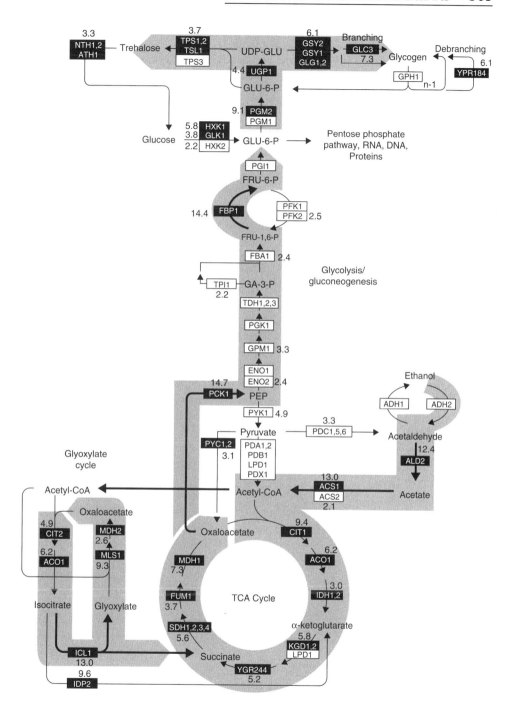

Table 13.2.

Typical Compositions of Selected Biomass Materials (Dry-Weight Basis)

Material	Hemicellulose (%) (Gives Pentoses)	Celluloses (%) (Gives Glucose)	Lignin (%)	Ash (%)	Total Weight Accounted for (%)
Spruce	26.5	42.0	28.6	0.4	97.5
Aspen	28.7	50.8	15.5	0.2	95.2
Red maple	33.0	39.0	23.0	0.2	95.2
Wheat straw	27.6	34.0	18.0	1.3	80.9
Cornstalk	32.6	33.5	11.0	1.0	78.1
Bagasse	34.0	38.0	11.0	NA	83.0
Tall fescue	24.5	30.0	3.1	NA	57.6

Source: Ladisch and Svarczkopf (1991), Table 1, p. 84.

constitute about 80% of the weight of hemicellulose. Hemicellulose accounts for about 25% of the dry weight of biomass materials that are considered as sources of sugars for fermentation processes (Ladisch and Svarczkopf 1991). The typical compositions of these materials (Table 13.2) include 30–40% cellulose, which can be hydrolyzed to its constituent monosaccharide, glucose. While glucose is readily fermentable to ethanol, pentose is not. The cost-effective utilization of biomass will require that both pentoses and hexoses be converted to ethanol during the same fermentation. The goal is often stated that both substrates are to be simultaneously fermented. However, parallel utilization is unlikely since most organisms display diauxic growth (preferential consumption of one sugar over another).

The metabolic pathways of cells that utilize pentoses require a number of extra steps in order to process the pentose into a form that can be catabolized through the glycolytic or citric acid cycle pathways. The additional metabolic steps slow the rate of pentose utilization relative to glucose. Some organisms contain pathways that efficiently utilize the pentoses, but these organisms do not process them into an end metabolite (such as ethanol) that is commercially useful. Consequently, significant efforts are being pursued to introduce capabilities of utilizing pentose into organisms that already make ethanol from glucose, or to introduce the pathway that leads to ethanol into organisms that readily utilize pentoses but do not make ethanol (Ingram et al. 1987; Sedlak and Ho 2004; Jarboe et al. 2007; Hahn-Hagerdal et al. 2007a,b).

Zymomonas mobilis is a bacterium used to ferment glucose to alcoholic beverages and has been examined for fuel ethanol production. Unlike yeast that ferments glucose to ethanol and generates 2 mol ATP/mol glucose through the Embden–Meyerhoff–Parnes glycolysis pathway, *Zymomonas* utilizes the Enter–Doudoroff pathway and produces 1 mol ATP/mol glucose under anaerobic conditions. It is the only organism known to do so. While it is a prolific glucose-fermenting organism, the wild-type strain is not capable of converting xylose to ethanol (Zhang et al. 1995). The introduction of the xylose assimilation operon and transaldolase and transketolase genes from *E. coli* into *Zymomonas* enabled the bacteria to ferment both xylose and glucose. A total of four genes were simultaneously transferred in the *Zymomonas* host, and this enabled the metabolism of *Zymomonas* to utilize xylose with the stoichiometry (Zhang et al. 1995):

$$3\,Xylose + 3\,ADP + P_i \rightarrow 5\,ethanol + 5\,CO_2 + 3\,ATP + 3H_2O \qquad (13.1)$$

The xylose is utilized through the pentose pathway that channels either C_3 or C_6 molecules into the Entner–Doudoroff pathway described in Chapter 9. A yield that was close to theoretical: 1.67 mol ethanol per mol xylose, was achieved.

Zhang et al. (1995) suggested that the lower net ATP yield results in less diversion of substrate to the formation of biomass, thus making *Zymomonas mobilis* more efficient than other organisms in producing ethanol. In this case the purposeful modification of intermediary metabolism through the use of recombinant DNA techniques (i.e., metabolic engineering) facilitated the broadening of the substrate range of a potentially useful industrial microorganism so that it could utilize mixed carbon sources that are likely to be less expensive than glucose alone. Recall that this is one of the objectives described by Aiba et al. (1973) as discussed earlier in this chapter.

Other factors impact the economics and, therefore, the use of such an organism. The final product concentration and the time required to achieve it are affected by the tolerance of the organism to high sugar concentrations encountered at the beginning of the fermentation, and high ethanol concentration toward the end. In the case of *Zymomonas mobilis*, the final product concentration of 25 g/L of ethanol was achieved in a 30-h period. This is not yet comparable to an industrial yeast-based fermentation that achieves about 100 g/L in 40 h. At least 50 g ethanol/liter broth (or 5%) is needed to attain cost-effective energy utilization in distillation processes that separate the more volatile ethanol from the less volatile water (Ladisch and Dyck 1979; Ladisch et al. 1984).

The resistance of the bacterial culture to contamination by competing organisms that are more robust than a transformed host, at the culture conditions, is also important. A competing organism that grows on the substrate more quickly than the engineered organism, while not producing the desired product, causes significant yield loses. Once a competing organism takes over the process, the process must be shut down, cleaned out, and restarted. This results in loss of the fermentation broth and the sugars that it contained, as well as a loss of time that the fermentor was in production status.

Industrial yeasts, and the conditions under which they are grown, have been selected through experience for hundreds of years. These yeasts perform well at pH = 5, 100–200 g/L sugars, and temperatures of 30–40 °C. These conditions are sometimes inhibitory or toxic to other organisms, including recombinant organisms, and therefore give yeasts a selective advantage, and decrease the chance of a non-alcohol-fermenting organism taking over. The current stains of *Zymomonas* are reportedly less robust, and therefore have yet to gain widespread acceptance in industry for fuel ethanol production, although wild-type (nonrecombinant) forms are used for beverage alcohol production.

Another factor of importance for ethanol production is the ability to process, and ultimately market, the microbial cells and other unused substrates, broth constituents, and fermentation coproducts that remain after the ethanol has been recovered. The price and markets for coproducts are a significant economic factor in the production of fuel alcohol. The dried form of the unused broth components in a corn to ethanol process, are known as "distillers' dried grain" and contain cellulose (Kim et al. 2008). *Zymomonas* are generally recognized as safe (GRAS) as an animal feed. Nonetheless, fermentation processes that utilize *Zymomonas* are still considered an early-stage technology in part because of the higher pH at which the microbe must be grown, which helps other microbes contaminate the fermentation (Mohagheghi et al. 2004).

The fermentation properties of transformed organisms are usually confirmed using pure (reagent-grade) sugars. This must be followed by demonstration that the organism is effective in sugar streams derived from hydrolysis of agricultural biomass sources. Biomass hydrolysates may contain inhibitors that result from chemical degradation of glucose to hydroxymethyl furfural or xylose to furfural. The furfurals can be toxic to some microorganisms at 100 ppm levels. Hence, tolerance to such inhibitors must also be built into the transformed organisms. Separation processes may be needed to selectively remove the inhibitors, such as demonstrated by Weil et al. 2002. The impressive feat of engineering a new pathway into an organism is only the first step required to achieve industrial use of a metabolically engineered organism. Bioprocess engineering is needed for the design of large-scale processes, definition of stable operating conditions, and generation of process designs that achieve favorable production economics while satisfying product and environmental quality standards.

Another example of metabolic engineering to alter microbes for fermenting pentose to ethanol is the introduction of genes for xylose reductase and xylitol dehydrogenase from the yeast *Pichia stipitis*, and xylulokinase from *Saccharomyces* sp. strain 1400, into a non-xylose-fermenting yeast (Ho et al. 1998; Sedlak and Ho 2004). Strain 1400 was a non-xylose-fermenting yeast. However, it is able to ferment both xylose and glucose to ethanol when it is transformed by the introduction of a plasmid containing the three genes. Previous attempts to clone bacterial xylose isomerase genes into yeast were unsuccessful, and the transformed yeast was not able to utilize xylose for growth or ferment it to ethanol. Recombinant yeasts that were transformed only with xylose reductase and xylose dehydrogenase genes give xylitol as a major byproduct. Yeast containing the three genes was able to ferment both xylose and glucose to ethanol. This result is achieved not only because of the alteration of the metabolic pathway but because of the manner in which the plasmid was constructed.

The naturally occurring xylose fermenting yeasts *Pichia stipitis* and *Candida shehatae* are able to effectively ferment xylose to ethanol, as long as glucose is absent from the fermentation media (Hahn-Hagerdal et al. 2007a,b; Slininger et al. 1990, 1991; Gong et al. 1981). If glucose and xylose are present at the same time, the glucose represses the enzymes associated with the xylose pathway, and xylose is not fermented to ethanol. The cloning of alcohol fermenting genes into the yeast was carried out so that the genes were not repressed. This was achieved by fusing the three genes to a glycolytic promoter so that it is turned on, rather than repressed, in the presence of glucose. The combined presence of the promoter with the xylulokinase, xylose reductase, and xylitol dehydrogenase genes enabled the enzymes needed for alcohol production to be produced (Table 13.3). This result enabled the

Table 13.3.

Comparison of In Vitro Enzymatic Activities in Yeasts

Yeast	Specific Activity $\left(\dfrac{\text{nmol product}}{\text{mg protein} \cdot \text{min}} \right)$		
	Xylose Reductase	Xylose Dehydrogenase	Xylulokinase
Native (wild type)	54	22	Not detected
Pichia stipitis 5773	417	410	35.2
Transformed pLNH32	411	839	100

Source: Reproduced from Ho et al. (1998), Table 1, p. 1854.

yeast to ferment xylose while minimizing formation of xylitol, which is otherwise a coproduct of ethanol fermentation (Ho et al. 1998; Sedlak and Ho 2004). The net effect was the development of a yeast with the ability to ferment both glucose and xylose, over a timecourse that begins to approach a practical range and gives an ethanol concentration that is high enough that the ethanol can be separated from water in an energy-efficient manner.

A third approach for fermentation of pentoses is based on recombinant *E. coli* into which genes for pyruvate decarboxylase (*pdc*) and alcohol dehydrogenase (*adh*) were integrated (Ingram et al. 1987; Jarboe et al. 2007). This bacterium is capable of fermenting both xylose and glucose to ethanol, although the presence of glucose in the fermentation medium inhibited xylose utilization (Beall et al. 1991). An ethanol concentration of 27 g/L was attained in 60 h by recombinant *E. coli* K011, when the fermentation was carried out at about pH 6.0 using hydrolysate obtained from corn fiber. This was an impressive result considering that previous runs with xylose-fermenting organisms were often done with reagent-grade monosaccharides rather than hydrolysates.

The corn fiber used in these experiments gave about 70% total sugars (dry-weight basis) with glucose, xylose, and arabinose at 37.2, 17.6, and 11.3, respectively, making up the major monosaccharide constituents of the corn fiber. Corn fiber is a coproduct from the wet milling of corn, derived from the corn kernel, after the starch and oil have been fractionated from the corn kernel. Corn fiber makes up about 11% of the dry weight of the kernel. This analysis and the results of Saha et al. (1998) as well as the economics of corn fiber utilization by Gulati et al. (1996) show that the fermentation of arabinose to ethanol will also be important to the economics of converting materials that contain cellulose and hemicellulose to ethanol. This has motivated the search for yeasts and bacteria that are capable of fermenting arabinose to ethanol. Screening of 116 L-arabinose-utilizing yeasts by Dien et al. (1996) identified several species that are capable of converting arabinose to low levels of ethanol (0.4% or less). These may be candidates for further development, or as a source of genetic material for engineering arabinose-fermenting capabilities into other microorganisms.

The examples of ethanol production by engineered bacteria or yeast represent works in progress, on the path to industrial application. Once microorganisms are engineered to give the productivities, product concentrations, and yields from glucose that approach figures of merit[7] indicated by the literature, (i.e., 2.5 g product/ L broth/h, 50–100 g product/L broth; and yields of ≥ 0.4 g product/g monosaccharide), these microorganisms may be incorporated into existing fermentation processes or equipment (Ladisch and Svarczkopf 1991; Gulati et al., 1996) and thereby facilitate transformation of renewable resources into value-added products.

Metabolic Engineering for a 1,3-Propanediol-Producing Organism to Obtain Monomer for Polyester Manufacture

Propanediols are industrial chemicals synthesized from nonrenewable resources. The 1,2-propanediol is a commodity chemical with a market of 1 billion lb/year. They are used in unsaturated polyester resins, liquid laundry detergents, pharmaceuticals, cosmetics, and antifreeze and deicing agents. They are synthesized by hydration of proplylene oxide as reviewed by Cameron et al. (1998). A new polyester, polytrimethylene terephthalate, will be produced from 1,3-propanediol that is reacted with terephthalic acid. The monomer, 1,3-propanediol, is a specialty chemical that has a small but potentially rapid-growth market, and is used to synthesize polymers and other organic chemicals.

The 1,3-propanediol will be manufactured via either a chemical or a fermentation route (from glucose), with market development quantities of the new polyester[8] occurring in 2002. Degussa (Germany) started up a 20 million pound/year plant to supply the 1,3-propanediol until a fermentation route is on stream. Dupont bought this process from Degussa in 2000. In 2004, DuPont formed a joint venture with Tate & Lyle, a large food bioprocessing company, to produce 1,3 propanediol for the synthesis of this new polyester, Sofona®. The final decision as to which process will be used on a commercial scale will be based on market timing, economics, and specialty applications (McCoy 1998; Mullin 2004).

The production of 4 g/L of 1,2-propanediol from glucose was observed in the strict anaerobe *Clostridium thermosaccharolyticum*. This led to attempts to transform *E. coli* with the genes for this pathway. *E. coli* was chosen since it produces methylglyoxal, which is an intermediate in the pathway to 1,2-propanediol (Fig. 13.2), so that addition of genes that result in the final two steps could yield 1,2-propanediol. This was demonstrated, although the glucose fermentation resulted in only 0.1 g/L of product, and hence was not practical.

The thorough analysis and pioneering research of Cameron et al. (1998) provides an excellent overview of the current and future challenges of metabolic engineering of a large volume chemical as well as the benefits of applying metabolic engineering to developing fermentations to obtain value-added products. This case

---

[7] *Figure of merit* is a concept that Michael Ladisch first heard from Gene Peterson [of the U.S. Department of Energy]. A figure of merit incorporates theoretically attainable levels of productivity, concentration, and selectivity. It is useful in gauging how far a fundamental research concept must develop, before it begins to approach the practical conditions needed for industrial application, or, in some cases, whether attainment of a theoretical limit will be sufficient for transfer of a new biotechnology to industry.

[8] This type of polyester was actually discovered around the late 1960s but not commercialized because of lack of a low-cost monomer.

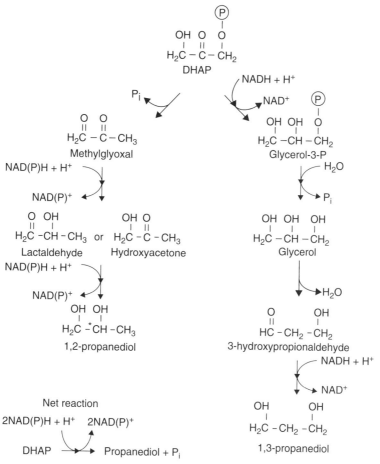

Figure 13.2. Metabolic pathways to 1,2- and 1,3-propanediol from dihydroxyacetone (DHAP), a common intermediate in sugar metabolism. The stereogenic center of 1,2-propanediol is indicated by an asterisk (*). [From Cameron and Tong 1993, Fig. 1, p. 117, Ref. 117].

study of commercialization in-progress illustrates the key elements that need to be addressed in developing a microorganism, scaling up the fermentation-based process, and obtaining a purified oxygenated chemical product from a renewable resource, namely, glucose.

The metabolic engineering of the 1,3-propanediol pathway started with the pathway for the transformation of glycerol to the product. The glycerol is transformed through the dihydroxyacetone pathway shown in Fig. 13.2. A final concentration of 70 g/L 1,3-propanediol was obtained over an 80-h time period based on *Klebsiella pneumoniae* ATCC 25995, using a fed-batch fermentation[9] to which

---

[9] A *fed-batch fermentation* is a method of microbial culture in which the substrate is fed in liquid solution to a batch vessel on an intermittent basis. Each time substrate is fed, an equal amount of the culture fluid or broth is removed. Refer to Chapter 5 for bioreactor models of this type of operation.

glycerol was fed at 20 g/L. The yield was 0.48 mol of 1,3-propanediol/mole glycerol, with other coproducts generated including 2,3-butanediol, lactate, acetate, ethanol, formate, and succinate. The volumetric productivity was 2.5 g/(L · h). These characteristics would be attractive if it were not for the high cost of glycerol ($0.50/lb or more). This led to attempts to carry out metabolic engineering in *K. pneumoniae* so that it could utilize glucose, since glucose would be a less expensive substrate.

Klebsiella pneumoniae is a bacterium that is related to *E. coli*, so it was assumed that cloning methods developed for *E. coli* should work in *K. pneumoniae*. The goal was to construct a genetic pathway that would convert glucose to 1,3-propanediol since glucose is a cost-effective substrate that costs 10–15 ¢/lb. Three strategies were considered: (1) adding the genes needed for glycerol production from glucose to an organism that already contained a pathway for glycerol to 1,3-propanediol, (2) adding the pathway for glycerol to 1,3-propanediol to an organism that transformed glucose to glycerol, or (3) adding genes for both pathways to a bacteria such as *E. coli*. Not all of the genes required to carry out the cloning were available initially, since the gene for glycerol phosphatase was not available.

The various considerations of each of these strategies is reviewed by Cameron et al. (1998), who were able to transform several microorganisms with four genes from *K. pneumoniae* needed to produce 1,3-propanediol from glucose. However, when the genes were expressed, the propanediol yield was less than 0.1 g/L, and was too small to be practically useful. While the first attempt was not successful, the methods developed and demonstrated constitute the first key steps in this type of metabolic engineering.

These results are the first steps toward developing a process for the production of the propanediols through microbial fermentation. A practical fermentation would probably require (Cameron et al. 1998): 70 g/L product, productivity of 2.5 g/(L · h), and concentrations on the order of 100 g/L. Once these productivities and concentrations are achieved, the product will still need to be recovered and purified. Distillation is not an option since the boiling points of the two propanediols are 187.4 and 214 °C at 760 mm Hg for 1,2-propanediol and 1,3-propanediol, respectively. A recovery-and-purification scheme proposed by Cameron et al. (1998) is illustrated in Fig. 13.3. There are a number of steps required, with membranes of decreasing cutoffs (i.e., pore sizes) to remove the cells and then the proteins, followed by a membrane step to concentrate the propanediol, respectively.[10] The final step of this conceptual flowsheet involves vacuum distillation that removes water and also separates other organic impurities from the final product. An alternate approach that has been proposed reacts the propanediol to form a derivative that can be extracted and further processed. This synopsis illustrates some of the challenges associated with the recovery of a bioproduct, as well as the many considerations that are involved in developing value-added products from renewable resources, based on metabolic engineering.

---

[10] *Membranes* are molecular filtration systems that are coated or synthesized to give different pore sizes through which particles or molecules, which are larger these pores, cannot pass. The principles of membrane separations are discussed in Ladisch (2001).

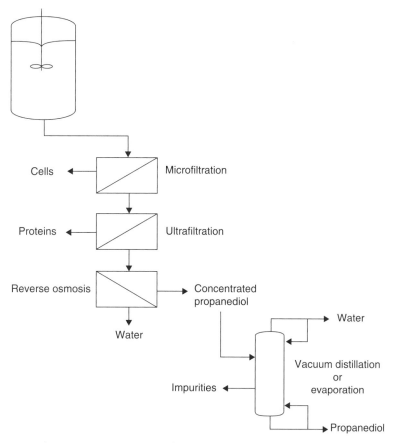

Figure 13.3.   Schematic representation of separation sequence proposed by Cameron et al. (1998) for fermentation derived 1,3-propanediol.

Redirection of Cellular Metabolism to Overproduce an Enzyme Catalyst Results in an Industrial Process for Acrylamide Production (Yamada–Nitto Process)

An example of a transformation that has made the transition from bench scale to industrial production is the generation of an enzyme catalyst from *Pseudomonas chlororaphis* B23 for production of acrylamide from acrylonitrile. The production of polymers used as flocculants, as additives, or for petroleum recovery consumes at least 200,000 tons of acrylamide/year (Nagasawa and Yamada 1989). The conventional process uses copper as a catalyst. An alternate process uses acrylonitrile derived from petrochemical sources, and converts it to acrylamide with the enzyme nitrile hydratase as the catalyst.

The enzyme-based process was scaled up to 6000 tons/year by 1989, and was reported to be producing 30,000 tons/year of acrylamide by 1992 (Nagasawa and Yamada 1989). The process is based on discovery, breeding, and selection of a bacterial microorganism, *Pseudomonas chlororaphis* B23, in which the nitrile hydratase is extremely active, while the amidase enzyme that would ordinarily

Table 13.4.

Comparison of Nitrile Hydratase and Amidase from *Pseudomonas chlororaphis* (Amidase Activity that Hydrolyzes the Desired Acrylamide Product Is Negligible)

Property	Hydratase	Amidase
Type	Inducible	
Molecular weight (kDa)		
Overall	200	106
Subunits	25	—
Special activities		
$\dfrac{\mu mol\ acrylamide}{min \cdot mg\ protein}$	1840 at 20°C	NA
$\dfrac{mol\ acrylamide}{min \cdot mol\ enzyme}$	184,000 at 20°C	NA
$\dfrac{mol\ ammonia}{min \cdot mol\ enzyme}$	NA	257 at 30°C

Source: Nagasawa and Yamada (1989).

catalyze the deamidaton of the desired product had little or no activity (Table 13.4). This technology is based on a single-enzyme conversion step, and does not entail a living microorganism. However, the methods by which the enzyme was obtained give an example of how the redirection of cellular metabolism to produce a desired product (in this case, the hydratase enzyme) resulted in an industrial process of major commercial significance.

Nagasawa and Yamada credit Galazy and his colleagues for recognizing the potential industrial use of *Brevibacterium* R312 for the production of amides and carboxylic acids (Jallageas et al. 1980; Nagasawa and Yamada 1989). Yamada and his group at Kyoto University, together with Nitto Chemical Industry Company, Ltd., proposed an enzyme-based production process for acrylamide using nitrile hydratase from *Pseudomonas chlororaphis* B23 or *Corynebacterium* N774, in a fed-batch enzyme conversion process. A conversion of 99% of acrylonitrile to acrylamide was obtained while avoiding formation of acrylic acid as a byproduct. It was possible to accumulate up to 400 g acrylamide per liter in 8 h at 10°C when resting cells of *P. chlororaphis* B23 were incubated with acrylonitrile, if the acrylonitrile was added gradually so that the nitrate hydratase activity was not inhibited. This led to development and startup of the first commercial-scale process in which cells were entrapped in a cationic acrylamide-based polymer gel. The gel immobilized the cells in particles, and the particles were readily recovered after the reaction was complete. The process was cleaner since Cu was no longer used, and a separation step to recover and recycle unreacted acrylonitrile was no longer necessary since nearly quantitative yields and selectivities were achieved.

The formation of nitrile hydratase is induced in *P. chlororaphis* in response to aliphatic nitriles and amides, while this enzyme is constitutive in *Brevibacterium* R312 and *Rhodococcus* N774 cells. For this process, nitrile hydratase is produced by *P. chlororaphis* B23 during its cultivation at 25 °C for 28 h in an optimized medium containing methacrylamide as the inducer. The nitrile hydratase activity of resting cells was assayed at 10 °C using 0.424 M acrylonitrile. Cell-free extracts were obtained by sonicating the cells, centrifugating the broth to remove the cell debris, and collecting the supernatant. The extract was assayed at 20 °C in 100 mM acrylonitrile. The nitrile hydratase enzyme catalyzes the following general reaction (Nagasawa and Yamada 1989):

$$RCN + 2H_2O \rightarrow RCOOH + NH_3 \qquad (13.2)$$

Amidase activity in the cell-free extract was assayed at 30 °C using 40 mM acrylamide by measuring the amount of ammonia and acrylic acid formed.

Figures 13.4a–13.4c summarize the strategy. A high acrylamide concentration is tolerated by nitrile hydratase from *P. chlororaphis* B23 but not by *Brevibacterium* R312 as shown in Fig. 13.4a, (where the acrynitrile level drops for *P. chlororaphis* (indicating reaction) but not for *Brevibacterium*. The acrylonitrile has the added, beneficial effect of inhibiting amidase activity (Fig. 13.4b), so that the product itself is not hydrolyzed. While, the amidase has a higher temperature stability than does the hydratase (Fig. 13.4c); its specific activity is much lower than the hydratase activity (dark circles in Fig. 13.4c). Hence, its presence in protein (enzyme) extracts from *Brevibacterium* cells has little effect on the yield of acrylamide, since deamidation of the acrylamide is minimal at 10 °C, and while the hydratase is active at 10 °C.

The production of acrylamide results from hydrolysis (i.e., addition of water) to acrylonitrile:

$$(13.3)$$

Acrylonitrile   Acrylamide

Hydrolysis of nitriles is the most common nitrile transformation, but there are relatively few cases where nitrile hydrolysis forms an amide, with the reaction stopping at the amide (Jallageas et al. 1980). Hence, the conversion of acrylonitrile to acrylamide is a special case, both biochemically and commercially. While there are a number of commercially available amides—benzamide, formamide, chloroacetamide, propionamide, salicylamide, nicotinamide, and phenylacetamide—only acrylamide is obtained by nitrile hydrolysis.

Nagasawa and Yamada (1989) attributed the high activity of the enzyme to the availability of ferric ions and PQQ as cofactors. The nitrile hydratase from *Pseudomonas* was differentiated from other bacterial sources by its tolerance to the product, acrylamide, where the enzyme maintained activity in the presence of ≤175 g/L of acrylamide at 5 °C while the enzyme from *Brevibacterium* was inactivated as shown in Fig. 13.4a. Other properties that made this system work is that large concentrations of the product, which is a substrate for the amidase enzyme, inhibited the amidase (Fig. 13.4b), while the temperature required for significant amidase activity was higher (30 °C) than for the hydratase as shown in Fig. 13.4c.

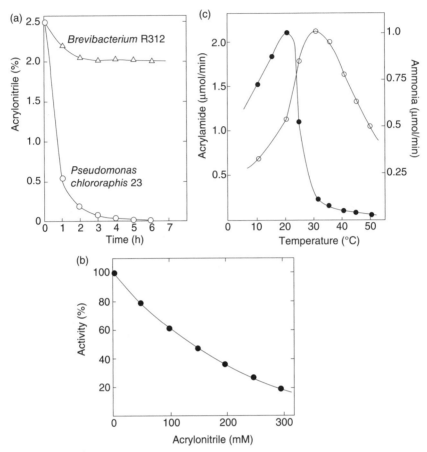

Figure 13.4. (a) Effect of acrylamide on the activity of nitrile hydratases from *Pseudomonas chlororaphis* B23 and *Brevibacterium* R312. The nitrile hydration reaction was carried out at 5 °C in the reaction mixture containing 15 units of nitrile hydratase of *P. chlororaphis* B23 (○) or *Brevibacterium* R312 (Δ), 17.5% (w/v) acrylamide, 2.5% (w/v) acrylonitrile, 0.2% (w/v) *n*-butyric acid, and 10 mM potassium phosphate buffer (pH 7.5). (b) Optimum temperature for nitrile hydratase and amidase purified from *P. chlororaphis* B23. The reactions were carried out at various temperatures for 30 min in 1.05 units of purified amidase. (c) Effect of acrylonitrile on *P. chlororaphis* B23 amidase activity. After incubation of the purified amidase (0.0286 units) in 70 mM potassium phosphate buffer (pH 7.0) containing various concentrations of acrylonitrile at 30 °C for 20 min, the hydrolysis of acrylamide was started by adding 40 mM acrylamide, followed by incubation for 20 min at 30 °C under standard conditions. Ammonia formation was determined. Relative activity is the percentage of the maximum activity attained under the experimental conditions. [Reproduced from Nagasawa and Yamada (1989): plot (a)—Fig. 4; plot (b)—Fig. 12; plot (c)—Fig. 5, all from p. 1069, Ref. 101.]

This process is an example of directed application of cellular metabolism to the benefit of an industrial process. While pathways in the microorganism were not modified through genetic engineering, screening of microorganisms for a specific characteristic and an understanding of the practical implications of the enzymes' characteristics enabled selection of a system for acrylamide production. The utility of metabolic engineering lies in its promise to amplify or modify an existing micro-organism, where thorough and wide-ranging screening of wild-type or mutated organisms has failed to result in a satisfactory organism. For ethanol and propane-diol, the pathways were introduced through directed genetic manipulation, while for acrylamide this was achieved through biocatalyst identification and isolation from a wild-type organism.

REFERENCES

Aiba, S., A. E. Humphrey, and N. F. Millis, *Biochemical Engineering*, 2nd ed., Academic Press, New York, 1973, pp. 28–32, 56–91, 195–345.

Ajinomoto, Inc., *Annual Report 1990*, Ajinomoto Co., Inc., 5-8 Kyobashi 1-chrome, Chuo-ku, Tokyo 104, Japan, 1990, pp. 1–15.

Barnum, S. R., *Biotechnology, an Introduction*, Thomson, Brooks/Cole, Belmont, CA, 2005, pp. 18–24, 57–92.

Beall, D. S., K. Ohta, and L. O. Ingram, "Parametric studies of ethanol production from xylose and other sugars by recombinant *Escherichia coli*," *Biotech. Bioeng.* **38**(3), 296–303 (1991).

Cameron, D. C. and I.-T. Tong, "Cellular and Metabolic Engineering: An Overview," *Appl. Biochem. Biotechnol.* **38**, 105–140 (1993).

Cameron, D. C., N. E. Altaras, M. L. Hoffmann, and A. J. Shaw, "Metabolic Engineering of Propanediol Pathways," *Biotechnol. Progress* **14**, 116–125 (1998).

Cirino, P. C., K. M. Mayer, and D. Umeno, "Generating Mutant Libraries Using Error-Prone PCR," *Methods Molec Biol.* **231**, 3–9 (2003).

Clutton-Brock, J., *A Natural History of Domesticated Mammals*, 2nd ed., Cambridge Univ. Press, 1999.

Committee on Bioprocess Engineering, National Research Council, *Putting Biotechnology to Work: Bioprocess Engineering*, National Academy of Sciences, Washington, DC, 1992.

DeRisi, J. L., V. R. Iyer, and P. O. Brown, "Exploring the Metabolic and Genetic Control of Gene Expression on a Genomic Scale," *Science* **278**(5338), 680–686 (1997).

Dien, B. S., C. P. Kurtzman, B. C. Saha, and R. J. Bothast, "Screening for L-arabinose Fermenting Yeasts," *Appl. Biochem. Biotechnol.* **57/58**, 233–242 (1996).

Gong, C. S., M. R. Ladisch, and G. T. Tsao, "Production of Ethanol from Wood Hemicellulose Hydrolysates by Xylose-Fermenting Yeast Mutant Candida sp. XF217, *Biotechnol. Lett.* **3**(11), 657–662 (1981).

Gulati, M., K. Kohlmann, M. R. Ladisch, R. Hespell, and R. J. Bothast, "Assessment of Ethanol Production Options for Corn Products," *Bioresource Technol.* **58**, 253–264 (1996).

Hahn-Hagerdal, B., K. Karhumaa, C. Fonseca, I. Spencer-Martins, and M. F. Gorwa-Grauslund, "Towards Industrial Pentose-Fermenting Yeast Strains," *Appl. Microbiol. Biotechnol.* **74**(5), 937–953 (2007a).

Hahn-Hagerdal, B., K. Karhumaa, M. Jeppsson, and M. F. Gorwa-Grauslund, "Metabolic Engineering for Pentose Utilization in Saccharomyces cerevisiae," *Adv. Biochem. Eng./ Biotechnol.* **108**, 147–177 (2007b).

Ho, N. W. Y., Z. Chen, and A. P. Brainard, "Genetically Engineered Saccharomyces Yeast Capable of Effective Co-fermentation of Glucose and Xylose," *Appl. Environ. Microbiol.* **64**(5), 1852–1859 (1998).

Hsu, J. T., S. Das, and S. Mahapatra, "Polymerase Chain Reaction Engineering," *Biotechnol. Bioeng.*, **55**(2), 359–366 (1997).

Ingram, L. O., T. Conway, D. P. Clark, G. W. Sewell, and J. F. Preston, "Genetic Engineering of Ethanol Production in *Escherichia coli*," *Appl. Environ. Microbiol.* **62**, 4594–4597 (1987).

Jallageas, J.-C., A. Arnaud, and P. Galzy, "Bioconversions of Nitriles and Their Applications," *Adv. Biochem. Eng.* **14**, 1–32 (1980).

Jarboe, L. R., T. B. Grabar, L. P. Yomano, K. T. Shanmugan, and L. O. Ingram, "Development of Ethanologenic Bacteria," *Adv. Biochem. Eng./Biotechnol.* **108**, 237–261 (2007).

Kim, Y., N. S. Mosier, R. Hendrickson, T. Ezeji, H. Blaschek, B. Dien, M. Cotta, B. Dale, and M. R. Ladisch, Composition of Corn Dry-Grind Ethanol By-Products: DDGS, Wet Cake and Thin Stillage, *Bioresource Technology* **99**(12), 5165–5176 (2008).

Ladisch, M. R., *Bioseparations Engineering: Principles Practice and Economics*, Wiley, New York, 2001.

Ladisch, M. R. and N. Goodman, Group III Site Report: Ajinomoto Company, Inc., Kawasaki, Japan, in *JTEC Panel Report on Bioprocess Engineering in Japan*, Loyola College in Maryland, Baltimore, 1992, pp. 148–155.

Ladisch, M. R. and J. A. Svarczkopf, "Ethanol Production and the Cost of Fermentable Sugars from Biomass," *Bioresource Technol.* **36**, 83–95 (1991).

Ladisch, M. and K. Kohlmann, "Recombinant Human Insulin," *Biotechnol. Progress* **8**(6), 469–478 (1992).

Ladisch, M., M. Voloch, J. Hong, P. Brenkowski, and G. T. Tsao, "Cornmeal Adsorber for Dehydrating Ethanol Vapors," *Industr. Eng. Chem. Proc. Design Devel.* **23**(3), 437–443 (1984).

Ladisch, M. R. and K. Dyck, "Dehydration of Ethanol: New Approach Gives Positive Energy Balance," *Science* **205**(4409), 898–900 (1979).

Lewin, B., *Genes*, Oxford Univ. Press, 1997, pp. 63–69, 129–133, 136–138, 153–241, 630–631, 727–774.

McCoy, M., "Chemical Makers Try Biotech Paths: Vitamins and a New DuPont Polyester Are Paving the Way for Biocatalysis Use," *Chem. Eng. News* **76**(25), 13–19 (1998).

Miwa, K., T. Tsuchida, O. Kurahashi, S. Nakamura, K. Sano, and H. Momose, "Construction of L-Threonine Overproducing Strains of Escherichia coli K-12 Using Recombinant DNA Techniques," *Agric. Biol. Chem.* **47**(10), 2239–2334 (1983).

Mohagheghi, A., N. Dowe, D. Schell, Y. C. Chou, C. Eddy, and M. Zhang, "Performance of a Newly Developed Integrant of Zymomonas mobilis for Ethanol Production on Corn Stover Hydrolysate," *Biotechnol. Lett.* **26**(4), 321–325 (2004).

Mullin, R., "Sustainable Specialties," *Chem. Eng. News.* **82**(45), 29–37 (2004).

Nagasawa, T. and H. Yamada, "Microbial Transformations of Nitriles," *Trends Biotechnol.* **7**(6), 153–158 (1989).

Nedwin, G. E., T. Schaefer, and P. Falholt, "Enzyme discovery – screening, cloning, evolving," *Chemical Engineering Progress*, **101**(10), 48–55 (2005).

Old, W. and S. B. Primrose, *Principles of Genetic Manipulation: An Introduction to Genetic Engineering*," 2nd ed., Univ. California Press, Berkeley, 1981, pp. 1–27, 38–43, 46, 91–92, 105, 171–179.

Petrounia, I. P. and F. H. Arnold, "Designed Evolution of Enzymatic Properties," *Current Opin. Biotechnol.* **11**, 325—330 (2000).

Saha, B. C., B. S. Dien, and R. J. Bothast, "Fuel Ethanol Production of Corn Fiber: Current Status and Technical Prospects," *Appl. Biochem. Biotechnol.* **70–72**, 115–125 (1998).

Sano, K., "Genetic Engineering of Brevibacterium lactofermentum to Breed Amino Acid Producers," *Proc. 4th European Congress on Biotechnology*, 1987, Vol. 4, pp. 735–747.

Schoemaker, H. E., D. Mink, and M. G. Wubbolts, "Dispelling the myths – biocatalysis in industrial synthesis," *Science*, **299**(5613), 1694–1697 (2003).

Sedlak, M. and N. W. Y. Ho, "Production of Ethanol from Cellulosic Biomass Hydrolysates Using Genetically Engineered *Saccharomyces* Yeast Capable of Cofermenting Glucose and Xylose," *Appl. Biochem. Biotechnol.* **113–116**, 403–405 (2004).

Singleton, P. *Bacteria in Biology, Biotechnology and Medicine*, 3rd ed., Wiley, New York, 1995.

Slininger, P. J., L. E. Branstator, R. J. Bothast, M. R. Okos, and M. R. Ladisch, "Growth, Death, and Oxygen Uptake Kinetics of Pichia stipitis on Xylose," *Biotechnol. Bioeng.* **37**, 973–980 (1991).

Slininger, P. J., L. E. Branstator, J. M. Lomont, B. S. Dien, M. R. Okos, M. R. Ladisch and B. J. Bothast, "Stoichiometry and Kinetics of Xylose Fermentation by *Pichia stipitis*," *Biochem. Eng. 6, Ann. NY Acad. Sci.* **589**, 25–40 (1990).

Stephanopoulos, G., "Fermentation Diagnosis and Control: Balancing Old Tools with New Concepts," in *Harnessing Biotechnology for the 21st Century*, M. R. Ladisch and A. Bose, eds., *Proce. 9th Intnatl. Symp. and Exposition*, American Chemical Society, Washington, DC, 1992, pp. 371–374.

Weil, J. R., B. Dien, R. Bothast, R. Hendrickson, N. S. Mosier, and M. R. Ladisch, "Removal of Fermentation Inhibitors Formed during Pretreatment of Biomass by Polymeric Adsorbents," *Industr. Eng. Chem. Res.* **41**, 6132–6138 (2002).

Wong, T. S., D. Roccatano, M. Zacharias, and U. Schwaneberg, "A Statistical Analysis of Random Mutagenesis Methods Used for Directed Protein Evolution," *J. Molec. Biol.* **355**(4), 858–871 (2006).

Zhang, M., C. Eddy, K. Deanda, M. Finkelstein, and S. Picataggio, "Metabolic Engineering of a Pentose Metabolism Pathway in Ethanologenic *Zymomonas mobilis*," *Science* **267**(5195), 240–241 (1995).

Zhu, H., M. Bilgin, R. Bangham, D. Hall, A. Casamayor, P. Bertone, N. Lan, et al., "Global Analysis of Protein Activities Using Proteome Chips," *Science* **14**, 293, 5537–5542 (2001).

Zohary, D. and M. Hopf, *Domestication of Plants in the Old World: The Origin and Spread of Cultivated Plants in West Asia, Europe, and the Nile Valley*, 3rd ed., Oxford Univ. Press, 2001.

CHAPTER THIRTEEN
HOMEWORK PROBLEMS

13.1. The fermentation of xylose to ethanol is an important step in the commercial production of ethanol from cellulose. One possible process for achieving this conversion is through isomerization of xylose to xylulose, and transformation of xylulose to hexose, and hexose to ethanol by yeast, as follows

$$A \underset{k_{-1}}{\overset{k_1}{\rightleftharpoons}} S$$

(HP13.1a)

$$S \xrightarrow[\text{yeast}]{} P$$

(HP13.1b)

where A = xylose (a pentose)

       S = xylulose (also a pentose)

       P = ethanol (the product)

Assume that the equation for cell growth is

$$\mu = \left(\frac{\mu_{max}[S]}{K_s + [S]} \right)\left(1 - \frac{P}{P_{max}} \right)^n$$

The yield coefficients for product and cell mass are

    $Y_{P/X}$ ~ product/cell mass

    $Y_{S/X}$ ~ substrate/cell mass

Initial concentration of xylose (A) is 100 g/L; xylulose (S), 1 g/L; cell mass (X), 10 g/L; and ethanol, 0.5 g/L. Set up the equations for accumulation of cell mass and product; and concentration of fermentable substrate (S, xylulose) and starting sugar (A, xylose) as a function of time:

    Initial conditions:

       A =

       S =

       X =

       P =

    Number of independent equations = _____.

$$\frac{dA}{dt} =$$

$$\frac{dS}{dt} =$$

$$\frac{dx}{dt} =$$

$$\frac{dP}{dt} =$$

13.2. The processing of biomass into ethanol involves pretreatment, hydrolysis, and fermentation, as shown in the process schematic given below:

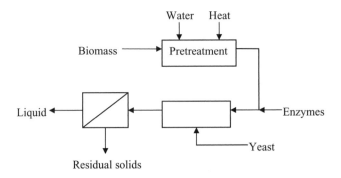

The combined action of enzymes and yeast are to convert the cellulose and hemicellulose to glucose and xylose, and subsequently ferment these to ethanol, in a single fermentation vessel. The reaction sequence is represented by

$$\text{Hemicellulose, cellulose} \xrightarrow{\text{enzymes}} \text{xylose} + \text{glucose}$$

$$\text{Xylose and glucose} \xrightarrow{\text{yeast}} \text{ethanol} + CO_2$$

These equations are simplified to the form below, for purposes of preliminary modeling:

$$B \xrightarrow{k_1} S$$

$$S \xrightarrow{\text{yeast}} P + CO_2$$

where B represents pretreated biomass, S is substrate (fermentable sugars), and P is ethanol product. The rates of fermentation for both glucose and xylose are assumed to be equivalent, due to the metabolic engineering of yeast so that xylose transport is as efficient as glucose transport. The growth equation is given by

$$\mu = \frac{\mu_{\max}S}{K_s + S}\left[1 - \frac{P}{P_{\max}}\right]^n$$

and the yield coefficients for cell mass and product are given by $Y_{X/S}$ and $Y_{P/X}$; set up the differential equations that give cell mass, substrate depletion, and product accumulation as a function of time, assuming that the glucose and xylose are utilized at the same rate at which they are formed. S is at low concentration, so that enzyme inhibition is minimal. Set up the indicated equations given below.

Biomass	$\dfrac{dB}{dt} =$
Sugars (substrate)	$\dfrac{dS}{dt} =$
Product (ethanol)	$\dfrac{dP}{dt} =$
Cell mass	$\dfrac{dX}{dt} =$

Assume that glucose is utilized more quickly than xylose (i.e., $\mu_{\max, G} > \mu_{\max, xyl}$), that the fermentation rates are represented by

$$\mu_G = \frac{\mu_{\max,G}G}{K_{S,G} + G}\left(1 - \frac{P}{P_{\max}}\right)^n$$

$$\mu_{Xy1} = \frac{\mu_{max,Xy1} Xy1}{K_{S,Xy1} + Xy1}\left(1 - \frac{P}{P_{max}}\right)^n$$

$$\mu = \mu_G + \mu_{Xy1}$$

and that hydrolysis is given by $B \xrightarrow{K_1} Xy1 + G$. Give equations that represent the change in glucose and xylose concentrations (where Xyl denotes xylose; G, glucose):

$$\frac{dG}{dt} =$$

$$\frac{dXy1}{dt} =$$

13.3. *Polymerase chain reaction engineering* (Hsu et al. 1997) describes the optimization of temperature programming and reaction time in order to maximize fidelity of DNA amplification. Taq polymerase, while thermostable by biological standards, still undergoes first-order deactivation:

$$-r_{Taq} = -\frac{d[E(t)/E_0]}{dt} = K_{Taq}\left[\frac{E(t)}{E_0}\right]^\alpha \qquad (HP13.3)$$

where for a given temperature

r_{Taq} = rate of deactivation of Taq polymerase
$E(t)$, E_0 = activity of the Taq enzyme at time t, and initially at $t = 0$
α = reaction order of thermal deactivation
K_{Taq} = deactivation constant for the Taq enzyme

(a) Assume that the deactivation reaction is first-order, and that the Arrhenius constant, $K_{d,Taq}$, $E_{d,Taq}$ are given by

$$K_{Taq} = 4.677 \times 10^{100}\ 1/s$$

$$E_{Taq} = 7.35 \times 10^5\ J/mol$$

Derive the relation for enzyme deactivation and calculate how much enzymatic activity would be lost after 10 min at 92 °C. (*Note:* R = 8.3145 J/(mol · K).

(b) How much would be lost after 10 min at 97 °C?

(c) If the Arrhenius constant ($K_{0,DNA}$) for the DNA synthesis (amplification) reaction at the same conditions, and the corresponding activation energy, $E_{0,DNA}$, are

$$K_{0,DNA} = 1.093 \times 10^{17}\ s^{-1}$$

$$E_{0,\text{DNA}} = 9.95 \times 10^4 \text{ J/mol}$$

how much DNA will be synthesized after one 10-min cycle?

(d) How much DNA will be synthesized after 10 cycles of 1 min for each cycle? For parts (c) and (d), assume that there are an excess of nucleotides (i.e., no substrate limitation).

GENOMES AND GENOMICS

INTRODUCTION

Genomics is the scientific discipline of mapping, sequencing, and analyzing genomes. Functional genomics combine bioinformatics, DNA chip technology, animal models, and other methodologies to identify and characterize genes that cause human disease, and are therefore prime targets for drug development. The genome consists of an organism's complete set of genetic material. The human genome contains about 3 billion base pairs (bp): 23 pairs of chromosomes consisting of chromosomes 1–22 and chromosome X or Y. This genome has about 400 bands that correspond to regions stained by the Giemsa[1] method. Each band consists of about 10^6–10^7 bp or 2 genome units. A genome unit is defined as the size of the genome of *E. coli* (4.6×10^6 bp) (Watson et al. 1992). About 1.5% of the 3 billion bp are genes that carry information used by the cell for synthesis of proteins (International Human Genome Sequencing Consortium 2001, 2004).

The number of chromosomes is not necessarily related to the size of the genome. The genome of *E. coli* contains about 4.6 million bp, organized into one chromosome per haploid cell. The genome of yeast consists of about 14 million bp, but it is organized into 16 chromosomes. The fruit fly, at 160 million bp, has 4 chromosomes, while *Zea mays* (corn) has about 5 billion bp in 10 chromosomes (Watson et al. 1992, p. 236). Both yeast and *E. coli* were used as tools for sequencing the human genome and are of industrial importance in fermentation processes for manufacture of fuels, chemicals, and biopharmaceuticals.

Human Genome Project

The first mapping of the human genome first started in about 1970, and it was initially thought that markers spaced at 20 cM intervals would be sufficient for mapping

---

[1] The chromosomes are first treated with trypsin, an enzyme that hydrolyzes proteins on the carboxyl side of lysine or argine amino acid residues. The trypsin attacks the proteins with which the DNA is associated. The chromosomes are then stained with Giemsa resulting in the characteristic bands (Alberts et al. 1989; Watson et al. 1992).

Modern Biotechnology, by Nathan S. Mosier and Michael R. Ladisch

genes of the human genome (1 cM is approximately equal to 10^6 base pairs).[2] Actually markers on the order of 2 cM were needed. The markers were chosen to exhibit a large extent of polymorphism, so that changes in these genes between different families or individuals could be readily detected. *Polymorphisms* are differences in amino acid sequences of proteins, or the corresponding sequence of bases in DNA.[3]

In 1980 the Human Genome Project was first proposed to sequence and define all of the human genes. However, formal work was not initiated until 1991, and the sequencing was completed in 2003. The objectives were to

1. Sequence the genomes of humans and selected model organisms.
2. Identify all of the genes.
3. Develop technologies required to achieve all of these objectives.

The National Institutes of Health (NIH) and the US Department of Energy (DOE) developed the joint research plan that led to mapping and sequencing of what was thought to be 80,000–100,000 human genes (Watson et al. 1992, Collins and Galas 1993). However, during the course of the sequencing project, scientists learned that there were only 20,000–25,000 genes (International Human Genome Sequencing Consortium 2004).

Special issues of *Science* (February 16, 2001) and *Nature* (February 15, 2001) published drafts of the sequences of the human genome. Sequences generated by the publicly sponsored Human Genome Project were published in *Nature*. The sequence from Celera Genomics was reported in *Science* (Venter et al. 2001). A press conference was held at 10 a.m., Monday, February 12, 2001, to discuss the landmark publications. By April 2003, 99% of the gene-containing part of human DNA were sequenced to 99.99% accuracy. The publication of this completed DNA sequence marked the conclusion of the publicly sponsored Human Genome Project and the transition of major research based on the knowledge gained to industry.

Markers for the genes in the chromosome are identified so that these can be located in the appropriate region of the chromosome. Physical maps of the genes

---

[2] The unit cM is an abbreviation for a *centimorgan*, named after the geneticist T. H. Morgan. The centimorgan is a genetic measure, not a physical one, that defines the frequency of recombination of two markers and occurs 1 out of 100 opportunities during meiosis (cell division during sexual reproduction). Recombination describes the process when pairs of chromosomes come together during meiosis and exchange segments (Watson et al. 1992).

[3] A chromosome with a mutant gene can be differentiated from a normal gene by the markers that it carries. Different markers result in different lengths of DNA, when the DNA is hydrolyzed by a restriction enzyme that recognizes one marker but not another. Differences as small as one base pair in the marker will result in different hydrolysis patterns. The different lengths resulting from different hydrolysis patterns are referred to as *restriction fragment length polymorphisms* (RFLP). RFLPs are inherited, just like genes, and can be used to follow chromosomes that are passed from one generation to the next. Tandem repeated DNA sequences lying between two sites where restriction enzymes hydrolyze the DNA are another form of polymorphism. Mutation of specific genes, called *alleles*, give 2 to more than 20 variants (Alberts et al. 1989). This type of polymorphism is detected by analyzing the length of fragments resulting from DNA hydrolysis by specific restriction enzymes (Watson et al. 1992).

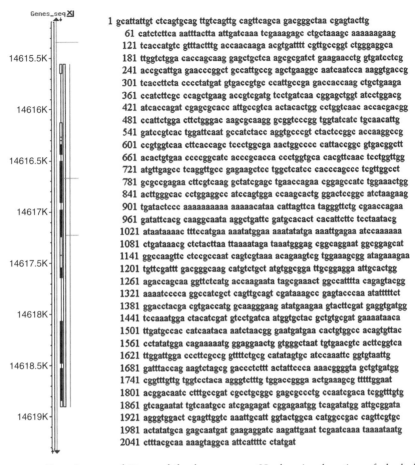

Figure 14.1. Genetic map of *Drosophila* chromosome 2L showing location of alcohol dehydrogenase with DNA sequence (from Entre Gene database http://www.ncbi.nlm. nih.gov/).

give locations of the genes to within 100 kb of their actual position (1 kb = 1000 bases) and identify the function of each gene. The sequence of the bases that make up the DNA in each gene is represented in Fig. 14.1. Once genes are located, robots carry out many repetitive operations involved in determining nucleotide sequences.

Nucleotides for some regions of the DNA may not be clear or cannot be determined because of the inability of the enzymes and biochemical methods to accurately and/or completely depolymerize the DNA into its individual nucleotides. This requires labor-intensive, manual modifications. The most time-consuming step occurs near the completion of the process that entails piecing together the sequenced fragments into the proper order. This finishing step requires intuition and computer algorithms to reconstruct sequence data back into contiguous pieces of DNA. When the sequences of DNA fragments did not overlap, the experiments were repeated to attain criteria for finished DNA sequences that must have 99.99% accuracy and may not contain more than one error in every 10,000 bases. This goal was initially not met, as was discovered when different laboratories exchanged pieces of sequenced DNA and found they could not exactly reproduce each other's sequences (Pennisi 1998a; Marshall and Pennisi 1998).

About 2–3% of the human genome was sequenced in the first 7 years. Significant progress was, however, reported for sequencing of genomes for microorganisms of industrial importance. This included *S. cerevisiae* yeast (Marshall 1997a), *E. coli* (Blattner et al. 1997), and *Bacillus subtilis*. In addition, microorganisms that cause stomach ulcers (*Helicobacter pylori*), Lyme disease (*Borrelia burgdorferi*), tuberculosis (*Myobacterium tuberculosis*), malaria (*Plasmodium falciparum*), and salmonella (*Salmonella enterica*) were well on the way to being sequenced by 1997 (Pennisi 1997a). By the end of July 2008, nearly 800 genomes for organisms representing all kingdoms (Archaea, Bacteria, Animals, Plants, Fungi, and Protists) had been completely sequenced and an additional 1500 genome projects were at various stages of completion.

The US National Institutes of Health's National Center for Biotechnology Information continually updates a list of completed and ongoing genome sequencing projects on their website (http://www.ncbi.nlm.nih.gov). The continuing challenge is to understand and assign functions to all of the identified genes. This is a formidable task when it is considered that the functions of about 40% of the genome of *E. coli*, with 4357 genes, are not known even though its entire genome of 4.6 million bp is sequenced (Durfee et al. 2008).

Deriving Commercial Potential from Information Contained in Genomes

During this period, industry sequenced parts of genomes, and made the maps publically available through large databases. For example, Perkin-Elmer, an instrument company that developed robotic machines costing $200,000–$300,000 each for gene sequencing, joined together with the nonprofit The Institute for Genomic Research (abbreviated TIGR; headed by Craig Ventor, Rockville, Maryland) and announced the goal, in May 1998, of sequencing the entire Human Genome Project by 2001 at an estimated cost of $200 million. This joint effort was based on using 230 Perkin-Elmer machines to automatically carry out many of the steps involved in sequencing human genes, and to identify biologically important DNA fragments. The sequence generated by this approach contained 2000–5000 gaps of about 58 bp each, which were closed later at an added significant cost and effort by other laboratories.

All results were made public by quarterly rather than daily release, as was the previous practice for publicly funded sequence projects. Quarterly release allowed time for companies that had this information to file for patents on sequences of potential commercial importance (Marshall and Pennisi 1998). Perkin-Elmer believed that it would make a profit by selling knowledge on what the sequences meant by finding genes in these sequences that cause diseases and discovering new medical therapies. The venture planned to try to patent "rare but pharmacologically interesting genes" with clear biomedical uses and then assemble a proprietary set of 100,000 single-nucleotide polymorphisms (SNPs) for clinical and research applications. This strategy was difficult to execute. Other companies with similar visions were Human Genome Sciences (a database of genes for creating novel drugs), HYSEQ (tools for rapid sequencing, drug discovery—collaborating with Perkin-

Elmer), INCYTE (owned a large gene and gene fragment database with plans to sell sequences and related biological information while collaborating with "gene chip" companies to do rapid gene analysis), Myriad (discovered a gene linked to breast cancer), and AXYS (searched for genes that cause diseases and drugs to treat these diseases; has studied asthma) (Carey 1998, Pennisi 1998a). Much of the promise that formed the basis of these early efforts has yet to be realized. Large companies entered the field by linking with startup companies whose entrepreneurial culture and special skills were a factor in the rapid development of new technologies (Pennisi 1998b).

Another approach is illustrated by a consortium of biotechnology companies in Germany (consisting of QIAGEN, Düsseldorf; AGOWA, Berlin; Biomax Informatics, Munich; GATC, Konstanz; and MediGenomix, Munich). These companies formed the Gene Alliance that had planned to provide combined services for large-scale genome analysis to the pharmaceutical, agricultural, and food industries. Genomes that had been sequenced by these companies included the first complete sequences of the yeast *Saccharomyces cerevisiae* in 1996 and the Gram-positive bacterium *Bacillus subtilis* in 1997. Other projects included sequencing of the yeast *Schizosaccharomyces pombe* and the model plant *Arabidopsis thaliana* (Koenig 1998). Norvatis Pharma, of Basel Switzerland, committed $250 million to create its own in-house genomics effort, The Norvatis Institute of Functional Genomics, in La Jolla, CA, by the year 2000.

The excitement of the potential impact and rapidly expanding capabilities of genome projects placed much attention on the technical aspects of defining the DNA sequences. The sequences, by themselves, are not useful for understanding disease processes, discovering new therapeutic approaches, amplifying useful traits, or introducing new traits into organisms in agriculture. Prior understanding of the functioning of the organism relative to at least one of the genes that determine a specific trait is needed.

Painstaking research, careful observations, and in some cases, decades of work in biological sciences must either precede, or follow, the identification and sequencing of genetic information. An example that illustrates this point is given by studies on genetic links to deafness that date back to the period 1928–1939, when a recessive mouse mutation on chromosome 11 arose in the progeny of an X-ray irradiated mouse. The progeny displayed deafness indicated by a lack of normal startling response and circling behavior in mice. Careful microscopy of the biological structures related to hearing showed that a defect in the gene known as shaker-2 (*sh2*) resulted in abnormal cytoskeletal morphology. Genetic mapping of the region of chromosome 11 containing the *sh2* gene showed a mutated gene that caused a single amino acid substitution in a protein (myosin). This substitution interfered with the proper functioning of tiny structures on hair cells that enable hearing (Probst et al. 1998). Another gene, on human chromosome 17, named *myo15* had similarity ("shows conserved synteny") with the *sh2* gene in mice. Mutations in this gene correlated with hereditary deafness that was previously identified in two geographically and ethnically diverse populations (Wang et al. 1998a). These results helped to expand the understanding of causes of hearing loss, and also indicated that shaker-2 mice are a useful model for studying the role of an abnormal myosin in the human auditory system, and "for the explorations of mechanisms for the delivery of functional proteins to surviving mutant hairs," to improve hearing (Probst et al. 1998).

Another example is the location of a creatine transporter gene on several chromosomes. The creatine transporter gene is mutated in the rare human hereditary disease, adrenoleukodystrophy (Pennisi 1998a). The existence of multiple copies of the same gene distributed on several chromosomes would not be anticipated solely from the sequencing of DNA.

Business models that would enable recovery of investments from genome sequences must plan for investments and resources to correlate gene sequence to protein function, and protein function to metabolic or physiological consequences. This translates to a significantly broader scope than does gene sequencing alone and adds to the time required for achieving commercialization of technology resulting from genomic discoveries. In cases where the investment horizon is too short to accommodate the added effort beyond sequencing a genome, a business based on genome sequences alone may not succeed.

The Genome for *E. coli* Consists of 4288 Genes that Code for Proteins

Markers for the different regions of the *E. coli* K-12 genome were completed in 1987. However, the sequencing of the genome of *E. coli* required an additional 10 years. Progress was initially slowed by the lack of speed and the limited accuracy of the early gene-sequencing instruments. Steadily improving technical approaches increased the rate at which the genome was sequenced. Two groups (Blaffner et al., University of Wisconsin, and Horiuchi and Mori et al., of Nagoya University) deposited the sequences of *E. coli* K-12 with GenBank in early 1997. Both groups were able to sequence 2.6 megabases in a year (1 megabase = 10^6 bases). This was the seventh genome to be made public and the largest genome for a prokaryote at that time (Pennisi 1998a).

The overall structure of the *E. coli* K-12 genome is summarized in a linear map (Blattner et al. 1997). The average distance between *E. coli* genes is 118 bp, with 70 regions larger than 600 bp. There are 4237 protein-coding genes, which account for 88% of the genome. Some of the remaining genes encode for RNAs (0.8%). There are also 0.7% of noncoding repeating sequences of bases. The remaining 11% of the DNA is for regulatory or other functions. The proteins of *E. coli* are divided into 22 functional groups, but the functions of only 60% of the genes are known.

Since *E. coli* is so easily cultured in the lab and has been studied so extensively, much was known about its characteristics. This made *E. coli* a good candidate for study to link the genes that have been sequenced to specific functions of *E. coli* proteins. For example, it is possible to determine whether small proteins are breakdown products of larger ones, or if a specific gene is associated with a single protein. As genomes of other microorganisms were sequenced, comparisons were possible that provided insights into microbial metabolism and how genes evolved. This was projected to enable development of new, appropriately targeted, antimicrobial compounds and vaccines, although major developments are yet to occur 10 years later.

Escherichia coli itself is a facultative anaerobe (i.e., grows in either the presence or absence of oxygen), colonizes the lower gut of animals, and is easily disseminated to other hosts. Pathogenic *E. coli* strains cause infections of the enteric, urinary, pulmonary, and nervous systems (Blattner et al. 1997). Understanding the genetic basis of its pathogenic traits, and perhaps being able to control them, is of obvious benefit.

DNA Sequencing Is Based on Electrophoretic Separations of Defined DNA Fragments

The basic laboratory methods for separating DNA according to size are still used today for biotechnology research. The basic principles of gel electrophoresis underlie the high-speed robotic DNA sequences that make large-scale DNA sequencing for genomics possible. Electrophoresis of DNA fragments is traditionally carried out using agarose or polyacrylamide gels cast between two glass plates. Each nucleotide in a nucleic acid polymer carries a single negative charge due to the phosphate moiety associated with each nucleotide. Therefore the DNA is strongly negatively charged and will move toward positive electrodes of an electric field. The polyacrylamide gels have a lower porosity and are used to separate DNA fragments that are less than 500 nucleotides long. The separation of larger DNA molecules requires the use of agarose gels, since the dilute solutions of agarose have sufficient porosity to allow the large DNA molecules to move through the gel, but still retain sufficient structural integrity so that the gel remains a solid at the conditions used for the separation (Alberts et al. 1989; Old and Primrose 1981). The sample that contains the DNA molecules is pipetted into small reservoirs, known as "wells," at the top of the gel as shown schematically in Fig. 14.2 (Hames and Rickwood 1990). An

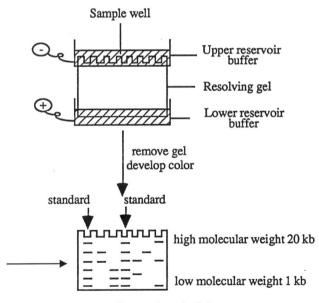

Figure 14.2. Graphical illustration of gel electrophoresis of DNA [from Hames and Rickwood (1990)].

electric potential is applied across the gel, and the charged molecules will move from the negative electrode (at the top of the gel) to the positive charge (bottom of the gel), with the lower-molecular-weight fragments moving faster than the larger fragments. The distance that the fragments migrate is inversely proportional to the log of their molecular weights. Robotic DNA sequencers separate DNA fragments using polyacrylamide gel. In these machines, the polyacrylamide gel is not encased between two glass plates. Instead, the gel fills small capillaries of glass. An electric field applied across dozens of these capillaries causes the DNA fragments to migrate through the gel. Lasers aimed at the outlet of the capillaries cause dyes associated with the DNA to fluoresce, enabling detection of the DNA for sequencing (Pang et al. 1999).

Oligonucleotides larger than 10,000 kb cannot be separated using direct current (DC). The DNA molecules stretch out to a linear form and are pulled by the electric field at a rate that is essentially independent of their length when the DNA molecule is greater than 10,000 kb. Alternating current is used in pulsed-field electrophoresis, which changes the direction of the electric field at 60 times/s. The DNA molecules reorient (or relax) each time the electric field changes direction. Since the relaxation time for the larger molecules is longer, the larger molecules move more slowly than the smaller ones and a separation is achieved (Alberts et al. 1989).

Bands of DNA are detected by soaking the entire gel in a solution of ethidium bromide, a dye that forms a complex with DNA by intercalating between the stacked bases that make up the DNA molecule. The gel is washed, and the dye that remains is bound to bands of DNA, and can be detected by photographing the orange fluorescence that results when an ultraviolet light source shines on the gel. Comparison of the resulting bands against those of a standard mixture of DNA of defined and known sizes run on the same gel enable molecular weights of the fragments to be estimated. As little as 50 ng of DNA can be detected by this method (Old and Primrose 1981).

An alternate method, known as *Southern blotting* and named after its developer, Edwin M. Southern at Oxford University, is able to detect the sequence of the DNA by hybridizing it with a probe DNA material having a known sequence that consists of bases that complement those whose presence are to be detected. The technique, developed by E. M. Southern, involves electrophoretic separation of DNA fragments obtained from the action of restriction enzymes on whole DNA, and then blotting the DNA bands onto nitrocellulose as illustrated in Fig. 14.3. The gel is laid on moist filter paper, and submersed in buffer. The nitrocellulose is laid on top of the agarose gel, and dry paper towels are stacked on top of it. The DNA diffuses or moves onto the nitrocellulose as a result of capillary action of the buffer slowly flowing through the paper as shown in Fig. 14.3. The DNA or RNA fragments stick tightly to the nitrocellulose paper. The filter paper is removed and then hybridized[4] with a radioactively labeled probe. The unbound probe is washed off,

---

[4] *DNA hybridization* or *renaturation* is the process whereby DNA will re-form a double helix from complementary single strands of DNA if kept at 65 °C for several hours. Hybridization also occurs between other single-stranded nucleic acid chains with complementary nucleotide sequences (DNA:DNA, RNA:RNA, RNA:DNA). If the sequence consisting of radioactively labeled nucleotides is used to probe for nonradioactive DNA fragments, only those strands

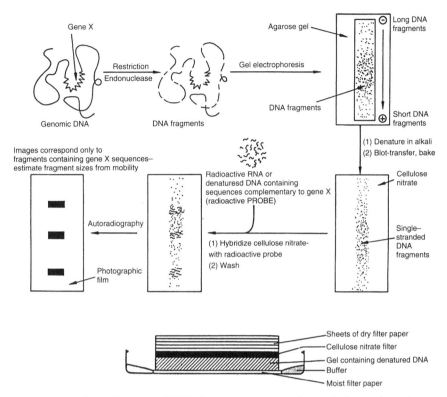

Figure 14.3. Southern blotting of DNA fragments separated by gel electrophoresis [reprinted with permission from Old and Primrose (1981), Fig. 1.4, p. 8; Figure 1.3, p. 7, University of California Press].

and the cellulose nitrate, to which the probe is hybridized to the complementary sequence, is placed on an X-ray film. The film is developed, and the bands to which the probe is bound are indicated by lines on the film (Fig. 14.3).

Two other types of blotting techniques are (Watson et al. 1992; Old and Primrose 1981)

> *Northern Blotting.* RNA is transferred to nitrocellulose paper or nylon sheet, or bound covalently to reactive paper prepared by diazotization of aminobenzylxymethyl paper. RNA analysis is carried out by electrophoretically separating the RNA on an agarose gel. The presence of different RNA bands is detected by using a radioactively labeled cDNA.

> *Western Blotting.* Protein, separated on an sodium dodecyl sulfate (SDS)–polyacrylamide gel, is blotted on nitrocellulose, and the protein is hybridized with an antibody specific to that particular protein.

that are complementary to the probe will hybridize with it. The other bands will not be detected by photoautoradiography or a similar detection technique, since they will not be associated with a radioactive probe.

Sequence-Tagged Sites (STSs) Determined from Complementary DNA (cDNA) Give Locations of Genes

A *sequence-tagged* site (STS) is a sequence of nucleotides of 200–500 bp in length from a known location on DNA. An STS is useful for mapping since it uniquely marks the section of the DNA from which the gene is obtained. Finding such sites requires knowledge, patience, and intuition (Watson et al. 1992, p. 64). The methodology for obtaining an STS is summarized by Watson et al. (1992).

Once a site is selected, mRNA transcripts that are generated when the genes expressed are used to obtain cDNA.[5] Knowledge of a small fraction of the sequence of the cDNA is sufficient to develop a unique gene markers (i.e., an STS). Comparison of cDNA from a healthy population to cDNA from people who have specific diseases has helped to identify some of the genes that are related to serious conditions.

Single-Nucleotide Polymorphisms (SNPs) Are Stable Mutations Distributed throughout the Genome that Locate Genes More Efficiently than Do STSs

Single-nucleotide polymorphisms (abbreviated SNPs and pronounced "snips") represent positions in DNA in which two alternative nucleotides may occur. SNPs are believed to be present in more than 1% in the chromosomal DNA of the human population with up to one SNP per 1000 bases of DNA (Marshall 1997a). SNPs have been used in studies that link genetic makeup of families with specific diseases or physiological characteristics of isolated populations; compare genetics of patients with a specific disease to genetics of people who are healthy; or identify changes in genes that occur in tumors (Marshall 1997a,b; Wang et al. 1998a,b). Combined with identification of SNPs located near these genes, this information allows SNPs to serve as genetic markers for disease.

SNPs are short, simple, and amenable to automated scans in digitized genetic diagnostic systems designed to analyze complete genomes of numerous individuals in search for multiple genes that contribute to diseases (Marshall 1997a; Kaiser 1997). While SNPs are not expected to be involved in disease processes, they are useful reference markers for procedures and instruments that will be able to quickly scan a genome for mutations. Figure 14.4 gives an example of a single-nucleotide polymorphism at position 15 for a fragment of DNA consisting of 30 nucleotides. The fragment and its complementary sequence (Fig. 14.4a) contain nucleotides A and T, respectively, at position 15. The shorthand representation of the DNA shows a solid line and a finely dotted line for the original and complementary strands of DNA respectively. The lines represent regions of nucleotide sequences whose nucleotide sequences are the same in both the original DNA and the DNA containing the polymorphism.

---

[5] cDNA sequences are obtained from a master catalog of human genes. While numerous clone libraries were constructed between 1987 and 1997 from human sperm and cell lines, by 1998 virtually all were to be discontinued because the National Institutes of Health (NIH) and Department of Energy (DOE) mandated that clone library donors be individuals who have given appropriate consent and are anonymous. This mandate was designed to prevent possible discrimination against DNA donors or their relatives as information from their genomes became available. These and related challenges to sequencing the human genome are discussed by Rowen et al. (1997).

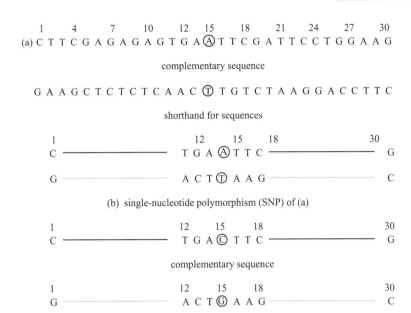

Figure 14.4. Schematic illustration of single-nucleotide polymorphisms.

The single-nucleotide polymorphism in Fig. 14.4 is indicated by the circled nucleotides at position 15, where A is replaced by C, and T by G. While the original double-stranded DNA (Fig. 14.4a) contains the sequence T G A A T T C between nucleotides 12 and 18 that would be recognized and cleaved by *Eco*RI (see Table 12.2), the single-nucleotide polymorphism (SNP) containing the sequence T G A C T T C between nucleotides 12 and 18 in Fig. 14.4b would not be recognized by this restriction enzyme. Consequently, the fragments that are formed (and can be readily detected) from the originating DNA will distinguish it from the SNP that would remain intact when treated with the same enzyme.

Individual SNPs are less informative than current genetic markers but are useful in pinpointing the exact location of genes, since these stable mutations are widely distributed throughout the human genome. Most importantly, the determination of SNPs has the potential to be automated for use in clinical application. Patterns of SNPs may be used to detect disease, or susceptibility to disease, when compared to other established genetic data (Wang et al. 1998b). A difference or change in the SNP pattern would indicate that a mutation had occurred due to an environmental rather than a hereditary factor, and thereby locate damage to the DNA that could lead to disease. Differences between SNPs in the DNA of normal genes and genes believed to cause disease can help to define relevant research and development targets. The identification of SNPs using gel sequencing and associated techniques is now enabling design and validation of oligonucleotide probes (on gene chips). These will facilitate testing of large populations of people for characteristic patterns of variation among different SNPs (Wang et al. 1998b), with certain types of patterns indicating an aberration from the norm, and the need for further testing (Table 14.1). For example, two genes linked to breast

Table 14.1.

Selected examples of Genes Identified by Sequence-Tagged Sites (STSs) for Selected Chromosomes; Summary Generated from a Human Transcript Map

Chromosome	Gene	Approximate Location Chromosome Interval	Condition Associated with a Mutated Gene
1	*gba*	21	Unable to metabolize glucocerebrosides (Gaucher's disease)
2	*msh2*	18	Familial colon cancer
3	*sclc1*	21.1 to 23	Lung cancer
4	*hd*	16	Huntington's disease
6	*iddm1*	21.3	Juvenile-onset diabetes
7	*cftr*	31	Cystic fibrosis
8	*myc*	24.1	Burkitt lymphoma
9	*cdkn2*	21	Malignant melanoma
11	*hras*	15	Cancer
	lqt1	15	Cardiac arrythmia
12	*pah*	24.1	Phenylketonuria
13	*rbi*	14	Childhood tumors of retina
	brca2	12	Breast cancer
14	*ad3*	24	Neuritic plaques found in Alzheimer's disease
16	*pkd1*	13.3	Polycystic kidney disease (causes renal failure, hypertension)
17	*tp53*	13	Cancer (p53 mutations)
	brca1	21	Early-onset breast and ovarian cancer
18	*dpc4*	21	Loss of gene causes agressive growth of pancreatic cancers
19	*apoe*	13.2	Atherosclerotic coronary artery disease
	ldlr	13.2	Extracellular accumulation of cholesterol and heart attacks
21	*sod1*	22	Amyotrophic laterial sclerosis (Lou Gehrig's disease)
X	*fmr1*	27	Mental retardation (fragile X syndrome)

Source: Hudson et al. (1996).

cancer (*brca1* and *brca2*) have hundreds of alleles,[6] some of which have been patented because they have the potential to lead to or help to direct new therapeutic approaches for treating breast cancer.

The utility of SNPs associated with alleles has increased with the development of cost-effective methods for scanning a person's genome for differences between gene fragments that may indicate an abnormal condition, and fragments that fall within a normal range of genetic variation between people. Small variations in the person's genome will be easier to detect if a library of alleles that do not cause disease, present in a broad spectrum of the population, is available to serve as a standard of reference. Automated methods using small postage-stamp-sized chips are making this feasible. A prototype chip by Affymetrix[7] was demonstrated to achieve a 90% accuracy in an initial test (Marshall 1997b).

GENSET of France, with support from Abbott Laboratories, had the goal to identify 60,000 SNPs distributed over the entire human genome, patent them, and create a map of the SNPs. The maps will be sold to researchers who would use them as a standard of comparison for genetic studies or for drug research. Abbott Laboratories planned to use specialized maps to screen genomes of patients, and determine those people who are less likely to respond to drugs in clinical trials, based on identification of variant genes in the patients (Marshall 1997b). The availability of a genetic map of SNPs and rapid analytical techniques may result in the development of predictive genetic tests. These might range from screening of newborn infants for administering prophylactic antibiotics to reduce infant mortality in healthy-appearing babies with sickle cell anemia, to low-phenylalanine diets in infants showing no symptoms but having phenylketonuria (inability to metabolize phenylalanine), thereby preventing mental retardation; or prophylactic removal of the thyroid for children at risk of thyroid cancer, due to inheritance of mutations in gene that causes this condition. However, few genetic conditions can be treated, and this has led to many issues that must be addressed. These range from possible inaccuracies in the tests, and risks due to discrimination based on results of genetic tests, as well as invasion of privacy (Kaiser 1997). Attention is now being given to the challenge of establishing appropriate policies for regulation of genetic testing and application of the results (Holtzman et al. 1997).

In April 1999, a consortium of large pharmaceutical companies and the UK Wellcome Trust philanthropy formed to find and map 300,000 common SNPs (Anonymous 2000). The SNP Consortium consists of a number of large pharmaceutical companies, including APBiotech, AstraZeneca Group PLC, Aventis, Bayer Group AG, Bristol-Myers Squibb Co., F. Hoffmann-La Roche, Glaxo Wellcome

---

[6] An *allele is* a variant of a gene. The search for alleles may expand to mapping common variations in genes that express proteins involved in activating or detoxifying drugs and chemicals that are ingested or inhaled. Genes that are exposed to environmental mutagens, such as those for cytochrome P450 and NAT, can increase cancer risk in individuals exposed to the toxins (Kaiser 1997).

[7] The company Affymetrix was formed in 1991 to research, manufacture, and sell disposable DNA probe arrays containing gene sequences on a chip, reagents for use with the chips, instruments to process probe arrays, and software to analyze and manage genetic information (Affymetrix 1998). These arrays have been given the tradename "GeneChip®" (spelled as one word), which is registered to Affymetrix, Santa Clara, CA.

PLC, IBM, Motorola, Novartis AG, Pfizer Inc., Searle, and SmithKline Beecham PLC. The SNP mapping utilized DNA from 24 individuals representing several racial groups. The project was completed in 2001, with an update in 2002. The map has been made public and can be utilized by researchers through the NIH SNP database (http://www.ncbi.nlm.nih.gov/SNP/). This database contains nearly 18 million entries for SNPs that have been identified in the human genome. However, the effort still continues to quantify the frequency of variations in these SNPs and linking these markers to risk for developing diseases.

Gene Chip Probe Array

Gene chip probe arrays consist of oligonucleotides synthesized onto the surface of a glass slide that can detect complementary sequences in DNA and RNA, with 65,536 octanucleotides ($= 4^8$) fitting into an area of $1.6\,cm^2$. These devices are fabricated through a variation of photolithography that has been used in the electronics industry to simultaneously form multiple microcircuits on a silicon computer chip by directing light through a mask. Light, focused through a mask, enables photo-directed synthesis of multiple oligomers on a silicon chip or glass surface. This technology was described by Fodor and colleagues of the Affymax Research Institute in 1991 as "light-directed, spatially addressable parallel chemical synthesis" (Fodor et al. 1991) and is the basis of gene chip probe arrays (i.e., the GeneChip®[22]). The principle of light-directed, spatially addressable, parallel chemical synthesis was initially demonstrated for peptide synthesis (Fodor et al. 1991). Proof of the success of this procedure was obtained by using a mouse monoclonal antibody[8] directed against β-endorphin. This antibody selectively binds the peptide sequence H_2N-YGGFL or H_2N-YGGFM, when these peptides display an amino-terminal Tyr (represented here by the symbol H_2N-Y). Regions to which the mouse antibody had bound were identified by the subsequent association of fluorescein-labeled goat antibody onto the mouse antibody that had bound to the tyrosine-containing peptides. Fluoresence was measured using an epifluorescence microscope with 488 nm excitation from an argon ion laser, and detection of fluoresence emission above 520 nm by a cooled photomultiplier (Affymetrix 1998).

---

[8] A monoclonal antibody is obtained from β-lymphocyte cells that secrete a single antibody so that a homogeneous preparation of antibodies can be obtained in large quantities (Alberts et al. 1989). Myelomas, which are antibody-secreting tumors, were initially used to produce antibodies, although there was no way to direct the myeloma to produce one specific type of antibody. The development of hybridomas by Köhler and Milstein, in 1975, changed this (they received a Nobel Prize for this work). They found that spleen cells (β-lymphocytes) of a mouse immunized with an antigen, when fused to a specially developed myeloma cell line that did not itself produce antibodies, resulted in hybridoma cells that retained properties of both parents. These cells would grow continuously like the myeloma cell while producing antibodies that were specific to the antigen with which the mouse had been injected. Monoclonal antibodies were used principally as diagnostic tools that detect certain types of proteins (Watson et al. 1992; Olson 1986), until about 1998, when the first successful human trials of monoclonal antibodies against breast cancer [Herceptin (trastuzumab)] and Crohn's disease—inflammation of the bowels (Avakine)—were reported (Arnst 1998). Prior to these tests, mouse monoclonal antibodies were effective in mice but not in humans since they were destroyed by the human immune system.

The yield of correct peptides per cycle was estimated to be 85–95%. For synthesis of shorter peptides, the number of synthesized molecules per $50\,\mu m^2$ would be on the order of 10^7–10^9 (assuming concentration of the peptide in the picomolar range and $0.25\,\mu L$ applied at each site). The intensity of fluorescence would still be sufficient to distinguish between adjacent regions containing a different peptide (such as the one capped with proline in our example). At a certain peptide length, however, the number of incorrectly synthesized peptides becomes too large and the concentration of correctly synthesized peptides would be so small that the detection of differences in fluorescence would become difficult. The length of the peptide that is correctly synthesized will reflect the yield or fidelity of the synthesis.

While the initial research was directed to peptides, the first large commercial applications that developed about 5 years later were for oligonucleotide arrays. In this case, relatively short oligonucleotides are synthesized on a chip where an array density of about 10^6 probes/cm^2 is estimated to be possible (Fodor et al. 1991; Ramsay 1998). A complete set of octanucleotides consisting of 65,536 probes (= 4^8 for 4 different bases) can be produced in 32 steps in 8 h. Consequently, this type of chip is referred to as a *variable detector array* (VDA) assembly.

The principles are similar to those of the peptide synthesis. The resulting array would contain multiple columns and rows. Consider, for example, part of an array consisting of 64 columns and 4 rows, where Fig. 14.5 illustrates one column of the array. The VDA assembly contains oligonucleotides that are identical within each column, except for one nucleotide, located at the center of the oligonucleotide that is varied. Figure 14.5 gives the sequences for one column where the nucleotide that has been varied is marked with a circle.

A complementary sequence of DNA in the sample that perfectly complements oligonucleotide 1 will bind (hybridize) much more strongly with oligonucleotide 1 than with oligonucleotide 2, 3, or 4. Hence, it will not be washed away during the multistep analytical procedure shown in Fig. 14.6, and will give a much stronger color than for the other immobilized oligonucleotides. The design of a chip that contains hundreds to thousands oligonucleotide sequences that contain only one variation per column, thus will give hybridization patterns that will differ for different SNPs.

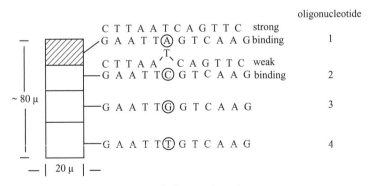

Figure 14.5. Schematic representation of oligonucleotide array.

A change in the hybridization pattern indicates the presence of a sequence variation, and with proper analysis and interpretation of the patterns, the nature of the variation, as well. In this way, a variable detector array that measures about $2\,cm^2$ has been shown to be able to detect mutations in "small, well-studied DNA targets" (such as 387 bp sequence from human immunodeficiency virus-1 genome, 3.5 kb sequence from breast cancer-associated *brca1* gene, and 16.6 kb sequence from the human mitochondrian) (Wang et al., 1998b).

This technology has proved to be an important research tool in the study of gene expression, as well as mapping of genomes. DNA chips have been used to measure expression in plant, microbial, and human systems (Ramsay 1998; Eddy and Storey 2008).

An example is shown in Fig. 14.6 that compares the images obtained from a chip designed to detect mutations in the cystic fibrosis gene. This example is

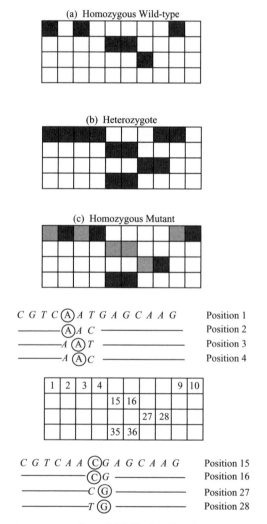

Figure 14.6. Schematic representation of DNA chip for detecting mutations [adapted from Marshall (1997b)].

transcribed from a viginette by E. Marshall (1997b). The first pattern, in Fig. 14.6a, is for the homozygous[9] wild-type (i.e., a normal) gene. The heterozygote displays additional hybridization at positions 10, 28, 35, and 36. The homozygous mutant (Fig. 14.6c) lacks the hybridization of the homozygous wild-type gene at positions 1, 3, 9, 15, 16, and 27, which indicates mutations in the gene (Marshall 1997b). The oligonucleotides on the chip that hybridized with the DNA fragments are indicated by position numbers corresponding to the array and are given by the grid at the bottom of this schematic. Straight lines denote a consistent oligonucleotide sequence, while the polymorphisms are given by the circled letters of the sequences associated with each position in the grid. Once they are identified, the single nucleotide polymorphisms must be assembled into maps that show their location on the chromosomes, if the SNPs are to be useful in human genetic studies (Wang et al. 1998b).[10] However, significant development and optimization are needed. Large-scale screening for human variation will be possible once the gene chip, PCR, clustering, and interpretation protocols become better developed and optimized.

POLYMERASE CHAIN REACTION (PCR)

The polymerase chain reaction (PCR) is an in vitro technique that enables DNA fragments to be copied in a process that is referred to as amplification. Since millions of copies can be made in a short period of time, sufficient DNA can be generated to characterize and analyze it. This technique has had immediate and beneficial impacts on tracking disease processes by characterizing genetic fingerprints of pathogenic organisms. For example, hantavirus, associated with hemorrhagic fevers and renal disease, was first isolated from striped field mice near the Hantaan River in South Korea in 1976. When a mysterious illness with these symptoms struck in New Mexico in 1993, the Centers for Disease Control (abbreviated CDC) in Atlanta quickly began to search for the cause. Within 30 days of the first death, viral genes from the victims' tissues had been propagated in sufficient amounts that the DNA could be studied, identified, and sequenced (Gomes 1997). The CDC had found the virus to be a previously unknown strain of hantavirus that destroys the lungs instead of the kidneys (Nichol et al. 1993; Gomes 1997).

---

[9] An organism or cell is *homozygous* when it has two identical alleles of a gene, where an allele is an alternative form of the gene that determines a particular characteristic (e.g., a white vs. red flower). A *heterozygote* is an organism or cell that contains alleles that are different.

[10] Error rates in both gel- and chip-based surveys determine the number of replicate samples needed to obtain meaningful results. False positives or negatives occurred for about 10% of the SNPs found. Tests with the chips showed that 98% of the loci were detectable. However, this required that 558 individual loci be separated, amplified, pooled, labeled, hybridized, and read. When the 556 loci were divided into 12 sets of 46 loci, only 90% of the samples passed the detection tests. When genotypes were determined on the basis of hybridization patterns that formed distinct (and measurable) clusters, the success rate of 99.9% was claimed for the 98% of the cases (1613/1638) to which a genotype could be assigned.

The Polymerase Chain Reaction Enables DNA to Be Copied In Vitro

The polymerase chain reaction makes copies of dissociated DNA or single-stranded copies of mRNA. The basic steps in PCR, illustrated in Fig. 14.7, are (Prowledge 1996)

1. Denaturing of the target sequence so that the polynucleotide is in a single stranded, unwound form. For DNA, this requires a temperature of 94–96 °C for 5 min (Watson et al. 1992).

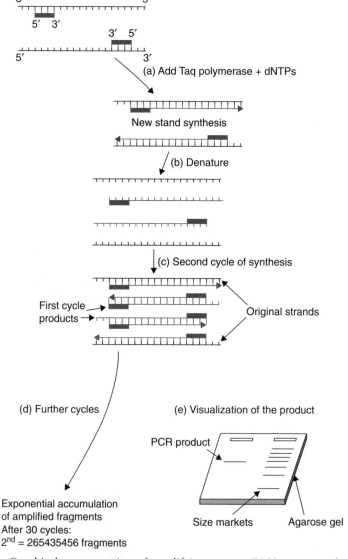

Figure 14.7. Graphical representation of amplifying a target DNA sequence through the polymerase chain reaction (PCR) [reproduced from Brown (2006), Fig. 9.2].

2. Hybridization so that the primers bind to complementary bases that flank the regions on either side of the polynucleotide of interest. This step requires about 30 s and is carried out at 30–65 °C (Watson et al. 1992).

3. Polymerization (i.e., polynucleotide or DNA synthesis) by DNA polymerase. The enzyme reads the template strand between the flanking primers, and matches the bases with complementary ribonucleotides, which are then incorporated into the primer strand. This is carried out at 72–80 °C for several minutes at pH = 8.4 in Tris buffer containing ribonucleotides representing all 4 bases (A, G, C, T).

This approach was first conceived in 1983 by Kary Mullis,[11] who was working for Cetus Corporation at the time (Cetus was purchased by Chiron in 1992 which in turn was acquired by Norvatis). See Chapter 2 for early history of this biotech company. Developmental work was done at the Cetus Corporation in 1984 and 1985, and commercialized by 1985. The method initially employed DNA polymerase from *E. coli*, which had an optimal temperature of 37 °C (Arnheim and Levinson 1990). However, the *E. coli* polymerase was inactivated by the temperatures needed to dissociate the double-stranded DNA molecule into single strands. Consequently, fresh enzyme had to be added after each cycle. Furthermore, the *E. coli* polymerase was not very specific. Errors were as high as 1%, resulting in high levels of imperfect DNA copying (Erlich 1989).

Thermally Tolerant DNA Polymerase from *Thermus aquaticus* Facilitates Automation of PCR

Polymerase from the thermophilic bacterium *Thermus aquaticus* resolved issues of specificity and speed. *T. aquaticus* (abbreviated *Taq*) is found in hotsprings and grows at 75 °C. Its enzyme has an optimum temperature of 72 °C and is sufficiently stable at 96 °C so that fresh Taq enzyme does not need to be added after each cycle of primer annealing, extension, and denaturation (Watson et al. 1992; Arnheim and Levinson 1990). Another benefit of the *Taq* polymerase is that it allows primer annealing and extension to be carried out at temperatures significantly higher than those for *E. coli* polymerase. This reduces imperfect annealing and subsequent amplification of an imperfect (nonspecific) product (Erlich 1989; MJ Research Notebook 1998). A substantially purer product is obtained. The need to add fresh enzyme after each step is eliminated. This enabled researchers to design and build automated PCR machines, and thereby accelerated the rapid application of this technology.[12]

While *Taq* polymerase is the main enzyme used for PCR, another source of thermostable DNA polymerase was isolated from an Iceland hotspring in 1990 from

---

[11] Mullis says the idea came to him during a 1983 moonlight drive in the California mountains. He received a Nobel Prize for this discovery in 1993 (Prowledge 1996). The method was developed by Mullis together with other Cetus Corporation scientists Randall Saiki, Stephen Scharf, Fred Faloona, Henry Erlich, and Norman Arnheim in 1984 and 1985 (Arnheim and Levinson 1990). DNA polymerases were discovered and first studied by Nobel Laureate Arthur Kornberg in the 1950s (National Research Council 1997).

[12] PCR is used in a variety of applications. Hence, conditions vary. The reaction conditions outlined here define a starting point.

Thermus brockianus (MJ Research Notebook 1998) and is now commercially available. PCR has enabled a broad range of investigators to work at the level of DNA, where before only large laboratories could achieve the same end. Hence, Appenzeller suggested that "the most profound effect of PCR may be to democratize the double helix" (Appenzeller, 1990).

Watson et al. (1992, p. 8) state:

> The standard PCR is typically done in a 50- or 100 μl volume and, in addition to the sample DNA, contains 50 mM KCl, 10 mM Tris/HCl (pH 8.4 at room temp.), 1.5 mM $MgCl_2$, 100 μg/ml gelatin, 0.25 μM of each primer, 200 μM of each deoxynucleotide triphosphate (dATP, dCTP, dGTP, and dTTP), and 2.5 units of Taq polymerase. The type of the DNA sample will be variable but it will usually have between 10^2 to 10^5 copies of the template (e.g., 0.1 μg DNA). A few drops of mineral oil are often added to seal the reaction and prevent condensation … The amplification can be conveniently performed in a DNA Thermal Cycler (Perkin-Elmer Instruments) using the "Step-Cycle" program set to denature at 94 °C for 20 sec, anneal at 55 °C for 20 sec, and extend at 72 °C for 30 sec for a total of 30 cycles. (The "Step-Cycle" program causes the instrument to heat and cool to the target temperatures as quickly as possible. In the current instrument, this results in a heating rate of about 0.3 °C per sec and a cooling rate of about 1 °C per sec, for an overall single cycle time of approximately 3.75 min).

Only the 5′-Terminal Primer Sequence Is Needed to Amplify the DNA by PCR

While it is not necessary to know the sequence of nucleotides in a DNA or RNA molecule to amplify it, the sequence of a small stretch of nucleotides on either side of the target must be known (Arnheim and Levinson 1990). These flanking sequences are then used to specify and synthesize single-stranded oligonucleotide fragments, or primers. These are usually about 20 nucleotides in length, with the sequence of nucleotides complementary to the bases in the respective sequence of the flanking regions (Arnheim and Levinson 1990). A mixture of the template sequence (primers), *Taq* polymerase, triphosphorylated deoxyribonucleotides, and buffer then doubles the amount of DNA each cycle, where each cycle consists of the steps illustrated in Fig. 14.7. Twenty-five cycles can be completed in about one hour with an automated instrument. The amount of DNA is doubled during each cycle, with n cycles producing 2^n as (many target nucleotides as were) present to begin with (Arnheim and Levinson 1990). Hence, two target molecules would be amplified to 2^{20} or about 10^6 million molecules in less than an hour. It is estimated that at least 95% of the product molecules will be generated with the correct sequence (Arnheim and Levinson 1990). The rate of misincorporation is estimated to be at 1 base out of 10,000 to 1 out of 200,000 per PCR cycle.

The rate of misincorporation in a naturally replicated DNA molecule is 1 in 10^9 nucleotides. This low error rate reflects the capabilities of DNA replication enzymes to remove and replace mismatched nucleotides in vivo. The *Taq* polymerase used for PCR incorporates at least one incorrect nucleotide per 200,000 since *Taq* polymerase alone does not have proofreading or repair capability. Thus the generation of genes or amplification of long sequences of genomic DNA by PCR may cause difficulty, while this error rate is acceptable for analytical procedures of shorter DNA fragments (Watson et al. 1992).

This limitation of PCR was overcome for the purpose of sequencing long fragments of genomic DNA through the use of in vivo DNA amplification using yeast or bacterial artificial chromosomes, and then fragmenting the long sequences of DNA into smaller fragments that, in turn, are amplified by PCR. After the short DNA fragments are amplified, they are sequenced. Finally, the sequences are pieced together again to give the sequence of the large fragment of genomic DNA from which they were originally derived.

The Sensitivity of PCR Can Be a Source of Significant Experimental Error

PCR amplifies targets that are present in extremely small amounts. Hence contaminants from previous experiments or the investigator's own DNA from skin, hair, or other sources can contaminate an experiment and it render useless. Great care must be taken in carrying out the experiments, including use of disposable pipette tips, and separate hoods or laboratory areas for preparing or analyzing PCR products (Arnheim and Levinson 1990).

The polymerase chain reaction has made it possible to determine sequences from DNA in remnants of extinct species and past populations. However, only a small number of ancient specimens contain DNA that is suitable for amplification. Consequently, false positives from contaminating DNA may give spurious results. One method proposed to distinguish between contaminating and actual DNA is based on the D:L-form ratio of amino acids. All of the amino acids, except for glycine, exist in D and L forms, although only the L form is used in protein biosynthesis. However, once the amino acids are isolated from an active metabolic process, they will undergo racemization in water to form equal amounts of the D and L forms. The racemization reaction occurs at conditions similar to those that cause depurination and spontaneous degradation of DNA. Therefore, the ratio of D to L forms of amino acids provides a type of internal standard against which authenticity of the DNA can be gauged (Poinar 1996).

Carefully controlled studies showed that when the samples had a D/L ratio greater than 0.08, DNA indigenous to the sample could not be recovered. Thus, the prospects for retrieving DNA sequences from dinosaur fossils are remote, with DNA sequences reported for dinosaur bones probably being due to contamination. Only insects, preserved in amber, had an extent of amino acid racemization less than 0.08, thus indicating that the DNA could have been preserved. It may be possible to isolate intact DNA from a preserved insect and amplify it using PCR (Hsu et al. 1997).

Applications of PCR Range from Obtaining Fragments of Human DNA for Sequencing to Detecting Genes Associated with Diseases

PCR has been used to detect DNA from *Helicobacter pylori* (the bacterium that causes stomach ulcers), tuberculosis, chlamydia, viral meningitis, viral hepatitis, AIDS, and cytomegalovirus. A study of the flu pandemic of 1918, which killed between 20–40 million people worldwide, including 675,000 in the United States, was made possible by using PCR to amplify DNA in tissue samples. DNA from lung tissue specimens that had been taken and preserved from a 21-year-old private who had died from the flu in 1918 indicated that the 1918 flu had probably originated from pigs and was likely a swine flu since the DNA from the victim resembled that of viruses isolated from pigs (Pennisi 1997b; Nichol et al. 1993; Gomes 1997).

The genetic variability in the population also facilitates use of PCR for forensic applications. The DNA in small samples of fresh biological materials at the scene of a crime can be amplified by PCR and then compared to the DNA of a suspect using restriction fragment separation and labeling techniques (Thornton 1989). Testing of DNA derived from the cell's nucleus has played a part in more than 30,000 criminal cases in the United States since 1988, when DNA testing was first used by the FBI. Testing is being extended to DNA derived from mitochondria since the amount of mitochondria[13] associated with each cell is much greater than a single nucleus. However, this test can be less specific than testing based on nuclear DNA, since the DNA sequences from the mitochondria are shared by siblings and maternal relatives for many generations, unlike nuclear DNA, which is different from one person to the next, except for identical twins. Mitochondrial DNA has been used to identify Czar Nicholas II's bones, to confirm the identity of the body buried in Jesse James's grave, and to identify a soldier's remains (Cohen 1997).

CONCLUSIONS

Large-scale sequencing and mapping of the genetic information in living things (genomes) since the mid-1990s has greatly increased the information (bioinformatics) available for conducting research into developmental biology, diseases, and human metabolism, as well as launching new fields of study such as comparative genomics, where insights are gained by comparing the genetic information carried by closely (or not so closely) related species.

Large-scale sequencing of genomes has been enabled by computer science (robotics and large-scale database management) as well as advances in biological sciences (PCR, DNA microchip technology, etc.). As genome sequencing becomes less costly and more routine, the cutting edge of the field is moving into mining these large bioinformatics databases for novel enzymes for industrial use, possible cures for diseases, improved understanding of the biochemistry of living organisms, and related fields of proteomics (proteins expressed in living cells), metabolomics (metabolites and their chemical transformation in living cells), and systems biology. Rapid increases in biological data are kept open to the public for general research worldwide through the maintenance of databases accessible through the Internet, along with database analysis tools to probe these enormous datasets for useful information. Our objective in this chapter was to provide a brief introduction to

---

[13] Mitochondria occupy a major portion of the cytoplasm of all eukaryotic cells, and convert energy-containing molecules, usually derived from sugars, into energy forms that can be used to drive cellular reactions. Mitochondria oxidize pyruvate (derived from glucose) and fatty acids by molecular oxygen so that 36 molecules of ATP are produced for each molecule of glucose oxidized. These organelles contain their own genome, which in humans consists of 13 protein-coding genes, 2 rRNA genes, and 22 tRNA genes, but makes up only 1% of the total cellular DNA. Mitochondria arise by the growth and division of existing mitochondria during the entire life cycle of a eukaryotic cell (Alberts et al. 1989).

this rapidly changing field. We hope that you will continue on a quest of lifelong learning to keep abreast of these advances, which are having and will continue to have enormous impacts in biotechnology, medicine, and daily life.

REFERENCES

Affymetrix 1997 Annual Report, GeneChip® Technology, *Unraveling the Mysteries of Genes*, Santa Clara, CA, 1998, p. 19.

Alberts, B., D. Bray, J. Lewis, M. Raff, K. Roberts, and J. D. Watson, *Molecular Biology of the Cell*, 2nd ed., Garland Publishing, New York, 1989, pp. 6–7, 23–26, 56–57, 99–107, 162, 171–174, 178–179, 184–191, 201–210, 222–236, 341–342, 386–393, 420–421, 483–486, 514–516, 519, 557–564, 728, 762–768, 781–788, 842–848, 1031–1035.

Anonymous, "SNP Consortium Collaborates with HGP, Publishes First Progress Reports" (Human Genome Program, US Department of Energy), *Human Genome News* **11**(1–2) (2000).

Appenzeller, T., "Democratizing the DNA Sequence," *Science* **247**(4946), 1030–1032 (1990).

Arnheim, N. and C. H. Levinson, "Polymerase Chain Reaction," *Chem. Eng. News* **68**(40), 36–47 (1990).

Arnst, C., "After 25 Years, A Big Payoff: Monoclonal Antibodies Are Generating New Excitement," *Business Week* **3580**, 147–148 (1998).

Blattner, F. R., G. Plunkett III, C. A. Bloch, N. T. Perna, V. Burland, M. Riley, J. Collado-Vides, et al., "The Complete Genome Sequence of *Escherichia coli* K-12," *Science* **277**(5331), 1453–1462 (1997).

Brown, T. A., "The Polymerase Chain Reaction," in *Gene Cloning and DNA Analysis: an Introduction*, 5th ed., Wiley-Blackwell, Oxford, 2006, pp. 181–195.

Carey, J., "The Duo Jolting the Gene Business: Craig Venter and Perkin-Elmer Target the Human Genome," *Business Week* **3579**, 70–72 (1998).

Cohen, L. P., "Innovative DNA Test Is an ID Whole Time Has Come for the FBI: Method Lets Agents Analyze a Mere Strand of Hair, but Errors Are Possible," *Wall Street J.*, A1, A6 (Dec. 19, 1997).

Collins, F. and D. Galas, "A New Five Year Plan for the US Human Genome Project," *Science* **262**(5130), 43–49 (1993).

Durfee, T., R. Nelson, S. Baldwin, G. Plunkett III, V. Burland, B. Mau, J. F. Petrosino, et al., "The Complete Genome Sequence of *Escherichia coli* DH10B: Insights into the Biology of a Laboratory Workhorse," *J. Bacteriol.*, **190**, 2597–2606 (2008).

Eddy, S. F. and K. B. Storey, "Comparative molecular physiology genomics," in *Methods in Molecular Biology vol 410*, and C. C. Martin eds, Springer, pp. 81–110 (2007).

Erlich, H. A., ed., *PCR Technology Principles and Applications for DNA Amplification*, Stockton Press, New York, 1989, pp. 1–5.

Fodor, S. P. A., J. L. Read, M. C. Pirrung, L. Stryer, A. T. Lu, and D. Solas, "Light-Directed, Spatially Addressable Parallel Synthesis," *Science* **251**, 767–773 (1991).

Gomes, Y. M., "PCR and Sero-Diagnosis of Chronic Chagas' Disease," *Appl. Biochem. Biotechnol.* **66**, 107–119 (1997).

Hames, B. D. and D. Rickwood, "One-Dimensional Polyacrylamide Gel Electrophoresis," in *Gel Electrophoresis of Proteins: A Practical Approach*, 2nd ed., B. D. Hames and D. Rickwood, eds., Oxford Univ. Press, 1990, p. 9.

Holtzman, N. A., P. D. Murphy, M. S. Watson, and P. A. Barr, "Predictive Genetic Testing: From Basic Research to Clinical Practice," *Science* **278**(5338), 602–605 (1997).

Hsu, J. T., S. Das, and S. Mahapatra, "Polymerase Chain Reaction Engineering," *Biotechnol. Bioeng.* 55(2), 359–366 (1997).

Hudson, T., L. Stein, L. Hui, J. Ma, A. Castle, J. Silva, D. Slonim, "An STS-based map of the human genome," *Am. J. Human Genet.* **59**(4): A26 (1996).

International Human Genome Sequencing Consortium, "Finishing the Euchromatic Sequence of the Human Genome," *Nature* **431**(7011), 931–945 (2004).

International Human Genome Sequencing Consortium, "Initial Sequencing and Analysis of the Human Genome," *Nature* **409**, 860–921 (2001).

Kaiser, J., "Environment Institute Lays Plan for Gene Hunt," *Science* **278**(5338), 569–570 (1997).

Koenig, R., "German Biotechs Form Gene Venture," *Science* **280**(5366), 999–1000 (1998).

Marshall, E. and Pennisi, E., "Biochip-Makers Do Battle in Court," *Science* **279**(5349), 311 (1998).

Marshall, E., "Snipping Away at Genome Patenting," *Science* **277**(5333), 1752–1753 (1997a).

Marshall, E., "Playing Chicken over Gene Markers," *Science* **278**(5346), 2046–2047 (1997b).

MJ Research Notebook, "Advertisement in Science," *Science*, **280**(5368), 1320 (1998).

National Research Council, *Forum on Intellectual Property Rights and Plant Biotechnology*, National Academy Press, Washington, DC, 1997, p. 10.

Nichol, S. T., C. F. Spiropoulo, S. Morzunou, P. E. Rollin, T. G. Ksiazek, H. Feldmann, A. Sanchez, J. Childs, S. Zaki, and C. F. Peters, "Genetic Identification of a Hantavirus Associated with an Outbreak of Acute Respiratory Illness," *Science* **262**(5135), 914–918 (1993).

Old, R. W. and S. B. Primrose, *Principles of Genetic Manipulation: An Introduction to Genetic Engineering*, 2nd ed., Univ. California Press, Berkeley, 1981, pp. 27, 38–43, 91–92, 171–179.

Olson, S., *Biotechnology: An Industry Comes of Age*, National Academy Press, Washington, DC, 1986, pp. 25–29.

Pang, H., V. Pavski, and E. S. Yeung, "DNA Sequencing Using 96-Capillary Array Electrophoresis," *J. Biochem. Biophys. Methods* **41**(2–3), 121–132 (1999).

Pennisi, E., "Sifting through and Making Sense of Genome Sequences," *Science* **280**(5370), 1692–1693 (1998a).

Pennisi, E., "Pharma Giant Creates Genomics Institute," *Science* **280**(5361), 193 (1998b).

Pennisi, E., "Laboratory Workhorse Decoded," *Science* **277**(5331), 1432–1434 (1997a).

Pennisi, E., "First Genes Isolated from the Deadly 1918 Flu Virus," *Science* **275**(5307), 1739 (1997b).

Poinar, H. N., M. Hosas, J. L. Bada, and S. Pääbo, "Amino Acid Racemization and the Preservation of Ancient DNA," *Science* **272**(5263), 864–866 (1996).

Probst, F. J., R. A. Fridell, Y. Raphael, T. L. Sauders, A. Wang, Y. Liang, R. J. Morell, et al., "Correction of Deafness in Shaker-2 Mice by an Unconventional Myosin in a BAC Transgene," *Science* **280**(5368), 1444–1447 (1998).

Prowledge, T., "The Polymerase Chain Reaction," *Breakthroughs Biosci. FASEB* 1–12 (July 11, 1996).

Ramsay, G., "DNA Chips: State of the Art," *Nature Biotechnol.* **16**(1), 40–44 (1998).

Rowen, L., G. Mahairas, and L. Hood, "Sequencing the Human Genome," *Science* 278(5338), 605–607 (1997).

Thornton, J. I., "DNA Profiling: New Tool Links Evidence to Suspects with High Certainty," *Chem. Eng. News* 67(47), 18–30 (1989).

Venter, J. C., M. D. Adams, E. W. Myers, P. W. Li, R. J. Mural, G. G. Sutton, H. O. Smith, et al., *Science*, 291(5507), 1304–1351 (2001).

Wang, A., Y. Liang, R. A. Fridell, F. J. Probst, E. R. Wilcox, J. W. Touchman, C. C. Morton, R. J. Morell, K. Noben-Trauth, S. A. Camper, and T. B. Friedman, "Association of Unconventional Myosin MYO15 Mutations with Human Nonsyndromic, Deafness, NFNB3," *Science* 280(5368), 1447–1450 (1998a).

Wang, D. G., J.-B. Fan, C.-J. Siao, A. Berno, P. Young, R. Sapolsky, G. Ghandour, et al., "Large-Scale Identification, Mapping, and Genotyping of Single-Nucleotide Polymorphisms in the Human Genome," *Science* 280(5366), 1077–1078 (1998b).

Watson, J. D., M. Gilman, J. Witkowski, and M. Zoller, *Recombinant DNA*, 2nd ed., Scientific American Books, W. H. Freeman, New York, 1992, pp. 1–11, 20–48, 50–54, 63–77, 135–138, 236–237, 273–285, 460–461, 521–523, 561–563, 590–598, 603–618.

CHAPTER FOURTEEN
HOMEWORK PROBLEMS

14.1. A gene that codes for a valuable protein is engineered into a plasmid consisting of regulatory regions and the *trp* E gene of the trp operon. The trp operon, when turned on, initiates a sequence of events that ultimately leads to the biosynthesis of tryptophan. The "hybrid" operon consists of

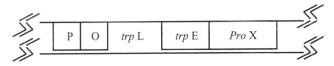

This operon, when recombined with the plasmid, is introduced into *E. coli*.

(a) The transformed *E. coli* is grown in a medium containing tryptophan, but protein corresponding to the *Pro* X gene is *not* formed. Explain.

(b) At the end of the fermentation, the tryptophan is depleted from the medium, and a protein containing a sequence of amino acids corresponding to *trp* E and protein X is detected. Explain.

14.2. The DNA from a body found, along with other evidence, was digested with restriction endonucleases, amplified using PCR, and electrophoresed to produce the map shown below:

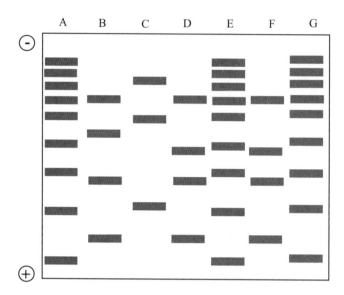

Using this map, answer the following questions:

(a) Which lanes contain reference DNA?

(b) Lane F contains DNA from the unidentified body. Which collected evidence lane(s) match(es) the DNA from the unidentified body?

(c) The DNA in the lane that matches the body is from a sample taken from the victim's hairbrush. One of the other evidence lanes is DNA from the victim's sister. Which lane is most likely the victem's sister's DNA and why?

14.3. Answer the following questions using information from the US National Institutes of Health's National Center for Biotechnology Information database (http://www.ncbi.nlm.nih.gov):

(a) *Listeria monocytogenes* is a serious foodborne pathogen that can survive and grow in refrigerated foods such as milk, cheese, and processed meats. What is the size of the *L. monocytogenes* genome? How many genes for proteins does this genome contain?

(b) The laboratory mouse (*Mus musculus*) is an important model organism for studying basic mammalian biology and human disease. How many haploid chromosomes does the mouse genome contain? What is the total size of the mouse genome (base pairs)? How many proteins (genes) have been identified in the genome?

INDEX

Modern Biotechnology, by Nathan S. Mosier and Michael R. Ladisch
Copyright © 2009 John Wiley & Sons, Inc.

411